Depleted Uranium Weapons
and International Law

A Precautionary Approach

Depleted Uranium Weapons and International Law

A Precautionary Approach

Avril McDonald / Jann K. Kleffner / Brigit Toebes

Editors

T·M·C· ASSER PRESS

Published by T·M·C·Asser press
P.O. Box 16163, 2500 BD The Hague, The Netherlands
<www.asserpress.nl>

T·M·C·Asser press' English language books are distributed exclusively by:

Cambridge University Press, The Edinburgh Building, Shaftesbury Road,
Cambridge CB2 2RU, UK,
or
for customers in the USA, Canada and Mexico:
Cambridge University Press, 100 Brook Hill Drive, West Nyack, NY 10994-2133, USA
<www.cambridge.org>

ISBN 978-90-6704-265-9

INTRODUCTION

The decision to undertake research into the legal regulation of depleted uranium (DU) weapons was motivated by a desire to offer an objective assessment of the lawfulness of the use of these weapons. Various concerns, chiefly relating to the immediate and long-term effects of the use of DU weapons on human health and the natural environment, have been raised, at least some of which remain disputed or cannot yet be verified because insufficient time has lapsed. In the light of these uncertainties, the approach adopted in the present book is a precautionary one: the main goal is to provide a thorough overview of the actual and potential legal implications of the use of DU weapons, starting from the current level of knowledge regarding their effects, but also presuming that certain adverse long-term effects of the use of these weapons, for which there is some evidence but which remain disputed, may materialise. The various contributions adopt such a precautionary approach in applying the relevant rules and principles of international law to the use of DU weapons.

An assessment of the legality of the use of DU weapons according to the applicable international legal framework depends on some basic understanding of the nature and military applications of DU and the consequences of its use in armed conflict, which are set out in Part One. All of the contributors have proceeded based on a common factual basis, provided by Dan Fahey in Chapters 1 and 2. He contextualises the following legal discussion by presenting an assessment of the known and suspected effects of DU on human health and the environment in Chapter 2. Preceding that, in Chapter 1, Dan Fahey explains what DU weapons are and why they are used by the military.

Part Two of the book presents the legal analysis of the use of DU weapons under the most relevant rules of international law in general and the law of armed conflict in particular.

First, in Chapter 3, Guido Den Dekker analyses whether DU weapons are subject to any prohibition or restriction under the law of arms control. If this were to be the case, the question whether DU use is restricted under other bodies of law, particularly the law of armed conflict, could be moot. Den Dekker also considers the prospects of a treaty being adopted which would ban or limit DU use.

In Chapter 4, Burrus Carnahan explains why the use of DU has been considered to be a matter of military necessity and explains what this means. In Chapter 5, Marten Zwanenburg analyses whether the use of DU conflicts with the principle prohibiting superfluous injury and unnecessary suffering to combatants. In Chapter 6, Jann Kleffner and Théo Boutruche address the question whether and under what conditions the use of DU weapons violates the principle of distinction between combatants and civilians. They also subject the use of DU weapons to scrutiny in the light of the principles of proportionality and precaution. In Chapter 7, Jann Kleffner examines whether DU falls foul of the prohibition of poison weapons. In Chapter 8, Erik Koppe explores whether the use of DU violates the rules of IHL concerning respect for the natural environment. In Chapter 9, Brigit Toebes examines whether the use of DU weapons may infringe international human rights law.

If DU use could be shown to breach any of the applicable primary rules, the questions of responsibility and remedies would arise. Part Three of the book thus deals with the secondary rules that would apply in cases where DU use could be shown to be unlawful. In Chapter 10, Tobias Gries and Manfred Mohr discuss the law of state responsibility as it might be applied in relation to violations of international law arising out of DU use. In Chapter 11, Avril McDonald examines the question of individual remedies for persons who have been exposed to DU and suffered ill effects.

Part Four is the concluding part of the book. In Chapter 12, Avril McDonald summarises the conclusions of the contributors regarding the legal regulation of the use of DU. She then offers a methodology for approaching the use of DU weapons in the absence of complete certainty regarding their effects. She addresses both issues of use and remediation, an approach that aims at minimising any risks for states and for individuals associated with the use of DU.

The Hague, January 2008   The Editors

SUMMARY OF CONTENTS

TABLE OF CONTENTS

LIST OF ABBREVIATIONS

ABM	Anti-Ballistic Missile
ACGIH	American Conference of Governmental Industrial Hygienists
ACHR	American Convention on Human Rights
AEPI	Army Environmental Policy Institute
AJIL	American Journal of International Law
AFDI	Annuaire Français de Droit International
AFRRI	Armed Forces Radiobiology Research Institute
AMCCOM	(US Army) Armament, Munitions, and Chemical Command
AP	Additional Protocol
API	Armour-piercing incendiary
ARDEC	Armament Research, Development and Engineering Center
ATSDR	Agency for Toxic Substances and Disease Registry
AWACS	Airborne Early Warning and Control Force
BBC	British Broadcasting Corporation
BVerfG	Bundesverfassungsgericht (Fed. Rep. of Germany)
BVerfGE	Amtliche Sammlung der Entscheidungen des Bundesverfassungsgericht
BWC	Biological Weapons Convention
BYIL	British Yearbook of International Law
CCW	Convention on Certain Conventional Weapons
CD	Conference on Disarmament
CEDAW	Convention on the Elimination of All Forms of Discrimination Against Women
CESCR	Committee on Economic, Social and Cultural Rights
CFE Treaty	Treaty on Conventional Armed Forces in Europe
CFR	Code of Federal Regulations
Chinese JIL	Chinese Journal of International Law
Chinese YIL	Chinese Yearbook of International Law
CHPPM	Center for Health Promotion and Preventative Medicine
CHR (UN)	Centre for Human Rights
CIA	Central Intelligence Agency
CIHL	Customary International Humanitarian Law
CRC	Convention on the Rights of the Child
CTBT	Comprehensive Test Ban Treaty
CTBTO	Comprehensive Test Ban Treaty Organisation
CWC	Chemical Weapons Convention
DC	Disarmament Commission
DHCC	Defense Deployment Health Clinical Center
DNA	Deoxyribonucleic acid/(US) Defense Nuclear Agency

DoD	(US) Department of Defense
DU	Depleted Uranium
ECHR	European Convention of Human Rights
ECtHR	European Court of Human Rights
EECC	Eritrea-Ethiopia Claims Commission
EJIL	European Journal of International Law
ENMOD Convention	Environmental Modification Convention
EPIL	Encyclopedia of Public International Law
ERW	Explosive remnants of war
ESC	European Social Charter
ESCOR (UN)	Economic and Social Council Official Records
EU	European Union
EURATOM	European Atomic Energy Community
Eur. Comm. HR	European Commission of Human Rights
EVRM	Europees Verdrag voor de Rechten van de Mens en Fundamentele Vrijheden
FAO	Food and Agriculture Organization
FRY	Federal Republic of Yugoslavia
GA	General Assembly (United Nations)
GAO	General Accounting Office/Government Accountability Office
GAOR (United Nations)	General Assembly Official Records
GC	Geneva Convention
GYIL	German Yearbook of International Law
Hague YIL	Hague Yearbook of International Law
HCDSC	House of Commons Defence Select Committee
HEI	High explosive incendiary
HLKO	Haager Landkriegsordnung
HRC	Human Rights Committee
HRLJ	Human Rights Law Journal
HSE	Health and Safety Executive
IACHR	Inter-American Commission on Human Rights
IAEA	International Atomic Energy Agency
ICBUW	International Coalition to Ban Uranium Weapons
ICC	International Criminal Court
ICCPR	International Covenant on Civil and Political Rights
ICESCR	International Covenant on Economic, Social and Cultural Rights
ICJ	International Court of Justice
ICJ Rep.	International Court of Justice Reports
ICLQ	International and Comparative Law Quarterly
ICRC	International Committee of the Red Cross
ICTY	International Criminal Tribunal for the former Yugoslavia
ILC	International Law Commission
IHL	International Humanitarian Law

ILM	International Legal Materials
ILO	International Labour Organization
ILR	International Law Reports
IMO	International Maritime Organization
Inter-Am. Ct. HR	Inter-American Court of Human Rights
IOM	Institute of Medicine
IRRC	International Review of the Red Cross
J Occup Environ Med	Journal of Occupational and Environmental Medicine
Leiden JIL	Leiden Journal of International Law
LTBT	Limited Test Ban Treaty
LNTS	League of Nations Treaty Series
LOAC	Law of Armed Conflict
MoD	(UK) Ministry of Defence
MTCR	Missile Technology Control Regime
NAC	North Atlantic Council
NATO	North Atlantic Treaty Organization
NGO	Non-governmental organisation
NILR	Netherlands International Law Review
NIOSH	National Institute for Occupational Safety and Health
NJCM-Bull.	Bulletin Nederlands Juristen Comité voor de Mensenrechten
NJW	Neue Juristische Wochenschrift
NMT	Nuremberg Military Tribunal
NNWS	Non-Nuclear Weapon States
NPT	Non-Proliferation Treaty
NRC	Nuclear Regulatory Commission
NSG	Nuclear Suppliers Group
NTTR	Nevada Test and Training Range
NWS	Nuclear Weapons States
NYIL	Netherlands Yearbook of International Law
OAS	Organization of American States
OSAGWI	Office of the Special Assistant to the Deputy Secretary of Defense for Gulf War Illnesses
OJ	Official Journal
OPCW	Organisation for the Prohibition of Chemical Weapons
OSCE	Organization for Security and Cooperation in Europe
OTP	Office of the Prosecutor (of the ICTR and/or ICTY)
PCIJ	Permanent Court of International Justice
POW	Prisoner of War
QBD	Queen's Bench Division
RCADI	Recueil des Cours de l'Académie de la Haye
RHDI	Revue Hellénique de Droit International

RIAA	Reports of International Arbitral Awards
SAM	Structural amorphous metals
SC	Security Council
SG	Secretary-General
SIrUS	Superfluous injury or unnecessary suffering
SIPRI	Stockholm International Peace Research Institute
SOFA	Status of Force Agreements
Stanford JIL	Stanford Journal of International Law
STEL	Short-term exposure limit
UDHR	Universal Declaration of Human Rights
UN	United Nations
UNCED	United Nations Conference on Environment and Development
UNESCO	United Nations Educational, Scientific and Cultural Organization
UNEP	United Nations Environment Programme
UNGA	United Nations General Assembly
UNSC	United Nations Security Council
UNTS	United Nations Treaty Series
USAMRIID	US Army Medical Research Institute of Infectious Diseases
VA	Veterans Administration
VCLT	Vienna Convention on the Law of Treaties
WHO	World Health Organization
Yale JIL	Yale Journal of International Law
Yb ILC	Yearbook of the International Law Commission
YIHL	Yearbook of International Humanitarian Law
ZAöRV	Zeitschrift für ausländisches öffentliches Recht und Völkerrecht

Part One

THE NATURE AND MILITARY USE OF DEPLETED URANIUM
AND THE CONSEQUENCES OF ITS USE IN ARMED CONFLICT

Chapter 1
DEPLETED URANIUM AND ITS USE IN WEAPONS

Dan Fahey

1.   INTRODUCTION

Depleted uranium (DU) munitions emerged from being a topic of relative obscurity prior to the 1991 Gulf War to become a subject of international debate by the turn of the century. The intermittent use of DU weapons in armed conflicts since 1991 only partly explains the increase in public awareness and concern.

Depleted uranium first garnered international attention in the mid 1990s when it was identified as one of the many suspected causes of illnesses among American and British veterans of the Gulf War. By the late 1990s, American veterans' groups successfully demonstrated that the US Department of Defense had concealed the extent and severity of veterans' exposures, resulting in extensive media coverage as well as increased research into the use and effects of DU weapons. In addition, humanitarian groups raised legitimate concerns about the health effects of DU in Iraq, but the government of Saddam Hussein exploited these concerns for propaganda purposes, thereby undermining their credibility.

International interest spiked again in late 2000 and early 2001 over claims that DU had caused leukaemias and other health effects among NATO troops and Balkans' civilians. Although in retrospect these concerns were rooted more in politics than in science, they did prompt Pentagon officials to lie about the existence of cancers among veterans in a US DU study as part of their efforts to quell the controversy (discussed in Chapter 2), thereby undermining the credibility of study findings and Pentagon proclamations about DU and veterans' health. By 2003 political activists had absorbed the DU issue into many different causes, including opposition to the wars in Afghanistan and Iraq. Many activists invoking the DU issue showed remarkably little regard for the accuracy or source of the information they cited, however, effectively mimicking the Pentagon practice of making claims without credible supporting evidence. Still, some activists continue to raise legitimate concerns about DU and in 2007 they succeeded in getting Belgium to become the first nation to ban the use of DU weapons in its territory.[1]

Despite more than a decade of debate and investigation, the use of DU weapons remains controversial. This chapter does not seek to resolve this controversy but rather to provide the reader with a basic understanding of DU and its use

---

[1] The politicisation of the DU question and the stances of the several interested parties are described further by D. Fahey in Chapter 2, and by A. McDonald in Chapter 12.

McDonald / Kleffner / Toebes (eds.), Depleted Uranium Weapons and International Law
© 2008, T·M·C·ASSER PRESS, *The Hague, The Netherlands and the Authors*

in munitions. It begins with a short description of the chemical properties of DU and its relation to natural uranium (section 2). The civilian and military uses of DU are discussed, concentrating on the latter (section 3), followed by a summary of the use of DU weapons in the 1991 Gulf War and the Balkans conflicts, their possible use in Afghanistan, and their use in Iraq since 2003 (section 4). Three examples of non-combat uses of DU ammunition are also provided (section 5).

2.   URANIUM AND DEPLETED URANIUM

Uranium is a naturally occurring heavy metal found in the earth's soil, rocks and oceans, but usually only in very low concentrations. Natural uranium consists of a mixture of three isotopes identified by the mass numbers U238, U235 and U234.[2] Trace amounts of natural uranium are found in drinking water and food, and the average daily human intake of uranium in food and water is about 1 microgram per day.[3]

Natural uranium is mined and processed to create highly radioactive enriched uranium[4] for use in nuclear fuel and nuclear weapons. The waste product of the uranium enrichment process is called DU because it contains less U235 and U234 (but marginally more U238) than natural uranium. DU is about 60 percent as radioactive as natural uranium, and it is chemically toxic like lead, nickel and other heavy metals. DU is 65 percent more dense than lead,[5] has a high melting point,[6] is highly pyrophoric (it ignites when it fragments), has a tensile strength comparable to most steels, and is chemically reactive.[7] DU remains radioactive throughout its decay chain, which lasts 4.5 billion years. The International Atomic Energy Agency classifies DU as a 'low specific activity' material, indicating its low-level of radioactivity.[8]

In the United States, the waste product of reprocessing spent fuel from civilian nuclear power reactors was for several decades added to DU stockpiles.[9] The DU from this source contains another uranium isotope, U236, along with small amounts of the transuranic elements[10] plutonium, americium and neptunium and

---

[2] 99.27, 0.72 and 0.0054 % by mass, respectively. World Health Organization, *Depleted Uranium: Sources, Exposure and Health Effects* (Geneva, WHO 2001) p. 3.

[3] The Royal Society, *The Health Hazards of Depleted Uranium Munitions*, Part I (London 2001) p. 2.

[4] Enriched uranium is uranium in which the U235 content has been increased from 0.7%.

[5] 18.9 g/cm³ and 11.3 g/cm³, respectively.

[6] 1132°C, 2070°F.

[7] US Institute of Medicine, *Gulf War and Health*, Vol. 1, *Depleted Uranium, Pyridostigmine Bromide, Sarin, Vaccines* (Washington, D.C., National Academy Press 2000) p. 91.

[8] International Atomic Energy Agency, *IAEA Safety Glossary, Terminology Used in Nuclear, Radiation, Radioactive Waste and Transport Safety* (Vienna, IAEA, April 2000) p. 74. International Atomic Energy Agency, 'Depleted Uranium: Questions and Answers', <http://www.iaea.org/worldatom/Press/Focus/DU/du_qaa.shtml>.

[9] US Department of Energy, *Commercial Recycling of Uranium and Plutonium from Spent Fuel*, undated, <http://www.eia.doe.gov/cneaf/nuclear/special/comrecyc.html>.

[10] That is, elements having a higher atomic number than uranium (i.e., 93 or over).

the fission product technecium-99. There is uncertainty in the United States about the extent of contamination of DU stockpiles by plutonium and other radioactive materials, but the Department of Energy is carrying out investigations and testing to clarify the ambiguities.[11]

3. USES OF DEPLETED URANIUM

3.1 The emergence of the military use of depleted uranium

Uranium has been mined since the Middle Ages but it is only in the last 100 years, and particularly in the last 50, that uranium mining has taken place on a large scale.[12] Historically, uranium has been used in the colouring of ceramics and glass, in the production of dental porcelains and as a chemical catalyst.[13] Since the Curies' discovery in 1898 of radium – which is extracted from ores containing uranium – the demand for radium for medical uses has increased and with it the mining of uranium.[14]

The discovery of nuclear fission in 1938 vastly increased the mining and processing of uranium and consequently enlarged the production and supply of DU. After World War II, increased production of enriched uranium for nuclear power plants and weapons created large stockpiles of DU in the United States.

In the late 1950s, the US Atomic Energy Commission weighed two options for the future of its growing DU stockpiles:

- 'In anticipation of its energy value as a fertile material for use in power reactors of the breeder type the depleted [uranium] could be continued to be stockpiled for the long-term future; or
- material could be exempted from restrictions with regard to nonnuclear [sic] industrial uses and placed on sale by the Commission at an established fair price.'[15]

The potential availability of large stockpiles of DU caught the attention of military planners, who used high-density alloys for ammunition, tank armour and other purposes. In addition, private industry developed commercial products made from DU, including radiation shielding and counterbalances in aircraft.[16]

---

[11] J.R. Hightower, et al., *Strategy for Characterizing Transuranics and Technicium Contamination in Depleted UF6 Cylinders*, ORNL/TM-2000/242 (Oak Ridge National Laboratory, October 2000) p. 1.

[12] WHO, *supra* n. 2, p. 23.

[13] M. Betti, 'Civil use of depleted uranium', 64 *Journal of Environmental Radioactivity* (2003) pp. 114-116.

[14] H. Nelson and R. Carmichael, *Potential Nonnuclear Uses for Depleted Uranium* (Richland, Washington, Battelle Memorial Institute, 29 January 1960) p. 3.

[15] Ibid., p. 1.

[16] Betti, *supra* n. 13, pp. 116-117.

The development of DU weapons began around 1959 in the United States[17] and in the early 1960s in the United Kingdom.[18] Weapons manufacturers developed high-density DU alloys for use in armour-piercing ammunition known as kinetic energy penetrators. This ammunition is simply a solid rod of dense metal stabilised by tail fins; there is no explosive charge. The large energy of motion (kinetic energy) of the rod, travelling at speeds between 1 and 1.8 kilometers per second, enables it to penetrate armoured vehicles such as tanks.[19]

During the 1960s and 1970s, the US military used armour-piercing ammunition made from tungsten alloy.[20] For use in munitions, tungsten is alloyed with an iron-copper, iron-nickel or nickel-copper binder.[21] In the early 1970s concerns about the high cost of tungsten alloy,[22] combined with improved performance of DU weapons,[23] prompted the US Department of Defense (DoD) to consider replacing tungsten alloy with DU in kinetic energy penetrators.

In 1974 the DoD announced its intention to switch to DU based on five considerations:

- Depleted uranium was readily available in large quantities from the stockpiles of the Atomic Energy Commission, and (at that time) was more cost effective and more readily available than tungsten;
- The use of DU in munitions did not compete with other uses for DU;
- The metallurgical properties of DU allow it to be heat-treated to varying degrees of hardness and strength;
- The pyrophoricity of DU produces burning fragments upon impact with a target, which can ignite flammable materials and cause secondary damage;
- The US Navy stated that DU was the best material available for ammunition for its (then) new Phalanx missile-defence gun.[24]

---

[17] See J.D. Edmands, et al., 'Uptake and mobility of uranium in black oaks: implications for biomonitoring depleted uranium-contaminated groundwater', 44 *Chemosphere* (2001) pp. 90-91.

[18] T. Carter, *Comparison of Kirkcudbright and Eskmeals Environmental Monitoring Data for Generalized Derived Limits for Uranium* (London, Ministry of Defence June 2002) p. vii.

[19] The Royal Society, *supra* n. 3, p. 2. R. Pengelley, 'The DU debate: what are the risks', *Jane's Defence Weekly* (15 January 2001).

[20] The density of available tungsten alloys ranges from 17 g/cm³ to 19 g/cm³, roughly equal to that of DU.

[21] International Tungsten Industry Association, *Tungsten* (London 1997), <http://www.itia.org.uk/tunstext1.htm>; Anonymous, 'RO defence 120 mm tank gun ammunition', *Jane's Defence Weekly* (8 January 2001). Tungsten alloy ammunition contains tungsten (91-93%), nickel (3-5%), and either cobalt (2-4%) or iron (2-4%).

[22] J. Middleton, 'Elimination of toxic/hazardous materials from small caliber ammunition – An overview', International Tungsten Industry Association, December 2000 Newsletter, p. 5, <http://www.itia.org.uk>.

[23] P. Bolté, 'The tank killers – tungsten v. depleted uranium', *National Defense* (May-June 1983) p. 44.

[24] Joint Technical Coordinating Group for Munitions Effectiveness (JTCG/ME), Ad Hoc Working Group for Depleted Uranium, *Special Report: Medical and Environmental Evaluation of Depleted Uranium*, Vol. 1 (1974) pp. 1, 2.

After DU became controversial in the late 1990s, one Pentagon spokesman claimed the decision to use DU took place 'when it became clear tungsten carbide rounds could not defeat the latest generation of Soviet armour.'[25]

In announcing the switch to DU, the DoD acknowledged some potential adverse consequences of using DU weapons, but it downplayed their significance:

> 'Overall, implementation of the proposed action [use of DU in munitions] is expected to have no significant medical and environmental impact. Depending on conditions locally, significant impact can occur in the event of uncontrolled release of DU.'[26]

Since the use of DU weapons in combat results in an uncontrolled release of DU, this statement appears to contradict itself about the potential for significant impacts from the use of DU weapons. The report goes on to state:

> 'In combat situations involving the widespread use of DU munitions, the potential for inhalation, ingestion, or implantation of DU compounds may be locally significant. However, it should be noted that problems from the use of DU on the battlefield or at sea are insignificant when compared to other dangers of combat.'[27]

Following the release of this report, production and testing of DU weapons expanded, and DU weapons were fielded in the US arsenal by the late 1970s.

British development of DU weapons did not begin in earnest until the 1970s. A 1979 report from the Ministry of Defence noted that the Soviet Union had introduced large numbers of 'sophisticated, heavily armoured vehicles' (T-72 tanks) in Central Europe. The report also noted: 'sources indicate that the USSR is also working on depleted uranium ammunition'.[28] The Soviet actions and the US incorporation of DU ammunition into its arsenal prompted the Ministry of Defence to initiate a 'development and proof firings programme', which was closely followed by a production program.[29]

3.2   Weapons systems using depleted uranium, plutonium contamination, and the alternative of tungsten

Depleted uranium is currently used by a number of countries in ammunition designed to attack armoured targets. Large calibre DU tank rounds are proliferating

---

[25] The Office of the Special Assistant to the Deputy Secretary of Defense for Gulf War Illnesses, 'Remarks by Dr Bernard Rostker at the American Legion Washington Conference' (Washington, D.C., 23 March 1998) pp. 4, 5, <http://www.gulflink.osd.mil/DU_speech.html>.

[26] JTCG/ME, *supra* n. 24, p. vi.

[27] Ibid., p. 96.

[28] UK Ministry of Defence, Memorandum by the Ministry of Defence, 'Anti-Armour Ammunition with Depleted Uranium Penetrators' (March 1979) p. 2.

[29] Ibid., p. 3.

rapidly, but only the US military appears to use smaller calibre DU ammunition shot by fighting vehicles and jets. Following is a non-exhaustive list of countries possessing DU rounds in their arsenals:

- Bahrain – 105 mm (M60 tank)[30]
- China – 100 mm (Type 69 tank), 105 mm (Type 59-II tank, Type 59D tank, Type 63A-1 amphibious tank, Type 80 tank), 125 mm (Type 85, Type 98 tanks)[31]
- Egypt – 120 mm (Abrams tank)[32]
- France – 120 mm (Leclerc tank), 105 mm (AMX-30B2 tank)[33]
- Israel – unknown[34]
- Kuwait – 120 mm (M1A2 tank)[35]
- Oman – 120 mm (Challenger II tank)[36]
- Pakistan – 105 mm, 125 mm (Al-Khalid, T-85, T80-UD tanks)[37]
- Russia – 100 mm (T-55 tank), 115 mm (T-62 tank), 125 mm (T-90, T-84, T-80, T-72, T-64B tanks)[38]

---

[30] US Presidential Determination 94-37 of 19 July 1994, 'Military Sales of Depleted Uranium Ammunition', 59 *Federal Register* (19 July 1994). This is the M833 round purchased from the United States.

[31] US Army, *Worldwide Equipment Guide* (Leavenworth, KS, US Army Training and Doctrine Command) 7 November 2000, pp. 5-6, pp. 4-25; R.D. Fisher, Jr. 'Evolving ground force threat to Taiwan' (Washington, The Jamestown Foundation, 11 March 2003); R.D. Fisher, Jr., 'The Impact of Foreign Weapons and Technology on the Modernization on China's People's Liberation Army', A Review Report for the US-China Economic and Security Review Commission, January 2004, <http://www.uscc.gov/researchpapers/2004/04fisher_report/part3.htm>; 'Type 59 main battle tank', *China Defence Today*, <http://www.sinodefence.com/army/tank/type59.asp>; J. Warford, 'The new Chinese type 98 MBT: A second look reveals more details', *Armor* (May/June 2001), pp. 22-23.

[32] A. O'Sullivan, 'Egypt – the new enemy?', *The Jerusalem Post* (25 August 1999). Egypt has approximately 755 M1A1 tanks. K. Burger, 'More M1A1 Abrams MBTs for Egypt', *Jane's Defence Weekly* (10 August 2001).

[33] T. Gander and C. Cutshaw, eds., *Jane's Ammunition Handbook*, 9th edn., 2000-2001 (Surrey, Jane's Information Group Limited 2000) pp. 226-227; Anonymous, 'DU ammunition types taken into service (non-exhaustive)', *Jane's Defence Weekly* (11 January 2001). Giat Industries manufactures two 120 mm rounds – the OFL 120 F2 APFSDS-T and PROCIPAC APFSDS-T. Giat also manufactures the 105 mm OFL 105 F2 DU round.

[34] US Army Environmental Policy Institute (AEPI), *Health and Environmental Consequences of Depleted Uranium Use by the US Army, Technical Report* (Atlanta, AEPI 1995) p. A-11. Israel shot 20 mm DU ammunition from a shipboard Phalanx gun in 1985.

[35] US Presidential Determination 94-37 of 19 July 1994, 'Military Sales of Depleted Uranium Ammunition', 59 *Federal Register* (19 July 1994). This is the M829 round purchased from the United States. Kuwait has 218 M1A2 Abrams tanks. Burger, *supra* n. 32.

[36] A.H. Cordesman, *The Military Balance in the Middle East – The Southern Gulf by Country: Part XII* (Washington, D.C., Center for Strategic and International Studies, 30 December 1998) p. 14. This is the L26A1 (CHARM 1) round, purchased from the United Kingdom.

[37] Anonymous, 'Pakistan joins DU producer nations', *Jane's Land Forces* (9 May 2001), <http://www.janes.com/defence/land_forces/news/>; Anonymous, 'Pakistan ordnance factories launches Rs 4 billion upgrade plan,' *South-Asian Defence News* (December 2002). The 125 mm round is called NAIZA.

[38] The ammunition manufactured by the Soviet Union and/or Russia containing DU include: 3BM32 125 mm round; 3VBM10/3BM29/30 125 mm round; 3BK21B 125 mm HEAT round with DU liner. US Central Intelligence Agency (CIA), Directorate of Intelligence, 'Science and Weapons Review,' 22 December 1994, p. 2; C. Foss, ed., *Jane's Armour and Artillery, 2000-2001,* 21st edn. (Surrey, Jane's

- Saudi Arabia – 105 mm (M60A3 tank), 120 mm (M1A2 tank)[39]
- Taiwan – 105 mm (M60A3 tank)[40]
- Thailand – calibre unknown[41]
- Turkey – 105 mm (M60A1, M60A3 tanks)[42]
- Ukraine – calibre unknown (possibly 30 mm rounds)[43]
- United Arab Emirates – 120 mm (Leclerc tanks)[44]
- United Kingdom – 20 mm (Phalanx gun), 120 mm (Challenger II tank)[45]
- United States – 20 mm (Phalanx gun), 25 mm (AV-8B aircraft, Bradley Fighting Vehicle), 30 mm (A-10 aircraft), 105 mm (M60A3 tank), 120 mm (M1A1, M1A2 tanks)[46]

In addition, India is reportedly developing DU ammunition.[47] In 2001 US forces discovered DU ammunition among captured Al Qaeda munitions in Afghanis-

---

Information Group Limited 2000) p. 76; Gander and Cutshaw, *supra* n. 33, pp. 231-232. The Sprut-B 125 mm anti-tank gun shoots the BM-42M APFSDS-T round, although the author was not able to confirm if this is a DU round. US Army, *supra* n. 31, pp. 5-6.

[39] US Presidential Determination 94-37 of 19 July 1994, 'Military Sales of Depleted Uranium Ammunition', 59 *Federal Register* (19 July 1994). The 105 mm round is the M833 and the 120 mm round is the M829, both purchased from the United States. Saudi Arabia has 150 M-60A3 and 315 M1A2 Abrams tanks. See also President George H.W. Bush, Address before the United Nations General Assembly, 1 October 1990; Cordesman, *supra* n. 36, p. 38; Burger, *supra* n. 32.

[40] Gander and Cutshaw, *supra* n. 33, p. 190. The United States sold 1,000 rounds of M774 ammunition to Taiwan, but Taiwan has also recently sought to purchase M833 rounds. Commerce Business Daily Issue of 27 February 1995, PSA #1291, <http://www.fbodaily.com/cbd/archive/1995/02(February)/27-Feb-1995/13sol003.htm>. Taiwan has also produced its own 105 mm DU round. US Central Intelligence Agency (CIA), Directorate of Intelligence, 'Science and Weapons Review', 7 October 1993, p. 9.

[41] AEPI, *supra* n. 34, p. A-11.

[42] Gander and Cutshaw, *supra* n. 33, p. 190. Turkey purchased 85,451 M774 rounds from the United States.

[43] W.S. Andrews, 'Depleted uranium on the battlefield: Part 1 – ballistic considerations,' *Canadian Military Journal* (Spring 2003) p. 44. A 2001 interview with Ukrainian Col. Ihor Mazor indicates that Ukraine inherited DU rounds after the breakup of the Soviet Union 'that were made back in the sixties and are now very obsolete…. According to documents, our troops have used such ammunition, and this is beyond a shadow of a doubt.' Interestingly, the Colonel stated that the rounds were 'small caliber air-to-surface projectiles'. D. Tymchuk, 'Ukraine has uranium projectiles, but not in the Army', *The Day* (Ukraine) (23 January 2001).

[44] Gander and Cutshaw, *supra* n. 33, p. 226. The United Arab Emirates purchased 388 Leclerc tanks from France in 1993 and was allowed to purchase the French OFL F2 DU round as well.

[45] Gander and Cutshaw, *supra* n. 33, p. 230. Royal Ordnance Defense manufactures two 120 mm rounds – the L27A1 (CHARM 3) and L26A1 (CHARM 1).

[46] Primex Technologies manufactures the 105 mm M900 round and the 120 mm M829A2 round. Alliant Techsystems manufactures the M829A3 round (10 kg DU penetrator), which will eventually replace the M829A2 round. Gander and Cutshaw, *supra* n. 33, pp. 191, 218-219; Primex Technologies, *1999 Annual Report* (St. Petersburg, Florida 1999) p. 1. Under §620G of the Foreign Assistance Act of 1961 (as amended), the United States is prohibited from selling DU munitions to all countries except NATO members, major non-NATO allies (including Argentina, Australia, Bahrain, Egypt, Israel, Japan, Jordan, Pakistan, Republic of Korea and New Zealand), Taiwan and other countries declared by Presidential Directive.

[47] Anonymous, 'Depleted Uranium (DU) Hazards in Post-Conflict Environments', Geneva International Centre for Humanitarian Deming, GICHD Advisory Note, 25 February 2003.

tan,[48] although the rounds may have been old Soviet munitions left behind from the 1979-1989 war.[49] The Iraqi government under Saddam Hussein started a program for the development of DU ammunition; however, this program did not result in large-scale production or combat use of DU weapons.[50] South Korea produced DU ammunition between 1983 and 1987, but reportedly destroyed its stockpiles in 1989.[51]

Other military uses of DU include tank armour (for the US M1A1 and M1A2 tanks), balance weights used in some aircraft and helicopters,[52] and about 0.1 g is used as a catalyst in certain anti-personnel mines.[53] The US Department of Defense also uses a DU casing in the 'bunker busting' B61-11 nuclear weapon; the DU casing is designed to enable the nuclear warhead to penetrate the ground before detonating.[54] Although other missiles may contain DU counterweights,[55] the US military denies that DU is used in operational Tomahawk cruise missiles, Air Launched Cruise Missiles, Advanced Cruise Missiles, or Conventional Air Launched Cruise Missiles.[56] In addition, a US military spokesman has denied that the Apache helicopter shoots DU rounds.[57]

For at least 20 years the United States used DU contaminated with plutonium and other transuranics in the manufacture of ammunition and tank

---

[48] US Department of Defense News Briefing, 'Sec. Rumsfeld and Gen. Myers', 16 January 2002, <http://www.defenselink.mil/news/Jan2002/t01162002_t0116sd.html>.

[49] CNN, 'Mornings with Paula Zahn', transcript, 25 December 2001; CNN, 'Explosive ordnance disposal at Kandahar airport,' transcript, 26 December 2001.

[50] Melissa Fleming, Senior Information Officer, International Atomic Energy Agency, e-mail to Dan Fahey, 28 July 2003. Iraq produced about ten DU rounds, two or three of which were test fired. See also International Atomic Energy Agency, Iraq Nuclear Verification Office, 'Fact Sheet: Iraq's Nuclear Weapon Programme', 27 December 2002, <http://www.iaea.org/worldatom/Programmes/ActionTeam/nwp2.html>.

[51] Anonymous, 'S. Korea made depleted uranium shells without reporting to IAEA', Japanese Economic Newswire (22 October 2004).

[52] R.C. Magness, *Environmental Overview for Depleted Uranium*, CRDC-TR-85030 (Aberdeen Proving Ground, MD, October 1985) pp. 10-12. In a response to Mr Duncan Smith in the UK Parliament on 2 February 2001, Mr John Spellar, UK Minister of Transport, stated that DU is used in balance weights in the Tristar and Wessex helicopters and C-130 aircraft.

[53] US Army Center for Health Promotion and Preventive Medicine (CHPPM), *Radiological Sources of Potential Exposure and/or Contamination* (Aberdeen Proving Ground, CHPPM, 10 December 1999) pp. 114-120.

[54] P. Richter, 'Old-fashioned hide-outs fuel high-tech weaponry', *The Los Angeles Times* (17 March 2002), p. A1; M.L. Wald, 'US refits a nuclear bomb to destroy enemy bunkers', *The New York Times* (31 May 1997) p. A1.

[55] AEPI, *supra* n. 34, p. 25. The earliest known operational use of DU in munitions was as a ballistic weight in the spotting round for the W54 Davy Crockett missile warhead, deployed with the US Army from 1961 to 1971. See the Office of the Special Assistant to the Deputy Secretary of Defense for Gulf War Illnesses (OSAGWI), *Depleted Uranium in the Gulf (II)* (Washington, DC 2000) p. 94.

[56] Chief of the Radiation Protection Division, Air Force Medical Operations Agency, e-mail message, Subject: 'Cruise missiles', 6 May 1999; Head of Radiological Controls and Health Branch, Chief of Navy Operations, e-mail message, Subject: 'No DU in navy cruise missiles', 4 August 1999.

[57] M.E. Kilpatrick, 'No depleted uranium in cruise missiles or Apache helicopter munitions – comment on an article by Durante and Publiese', 82(6) *Health Physics* (2002) p. 905.

armour.[58] DU rounds recovered by the United Nations Environment Programme in Kosovo, Serbia and Montenegro contained trace amounts of plutonium and other transuranics considered insignificant in terms of the overall radioactivity of the penetrators.[59] However, the United States has not disclosed the amounts of plutonium and other elements in the vast amounts of DU ammunition currently in its arsenal, or used in combat since 1991.[60]

Although a growing number of countries manufacture and possess DU ammunition, concern about the health, environmental and political effects of DU is leading to a re-examination of its use in munitions. In March 2007 Belgium became the first nation to ban the use of DU 'inert ammunitions and armour plates on Belgian territory', although Germany, Switzerland and Canada had earlier foresworn the use of DU weapons as a matter of policy.[61] Many other countries use tungsten alloy munitions instead of DU.[62] Interestingly, the US Navy is currently phasing out its use of DU ammunition for the Phalanx gun and switching to tungsten alloy ammunition 'based on live fire tests showing that tungsten met their performance requirements while offering reduced probabilities of radiation exposure and environmental impact.'[63] The US Marine Corps has decided to forgo the use of DU rounds in favour of tungsten alloy ammunition for its Advanced Amphibious Assault Vehicle, which will be fielded in 2008. A Marine Corps spokesman stated: 'We're not considering depleted uranium anymore because of the environmental problems associated with it, be them real or perceived' [sic].[64]

The US and UK militaries still strongly defend their use of DU weapons but recent advances in the development of tungsten alloy ammunition and tank guns may lead both to reduce their use of DU. In general, DU penetrators can achieve penetrations of 10 to 15 percent in excess of comparable tungsten alloy penetrators.[65] In addition, 'DU rounds can achieve the same penetration as [tung-

---

[58] US Army Test, Measurement and Diagnostic Equipment Activity (USATA), Project Development and Radiation Research Office, 'Tank-Automotive and Armaments Command (TACOM) and Army Materiel Command (AMC) Review of Transuranics (TRU) in Depleted Uranium (DU) Armor', Fort Belvoir, Virginia, 19 January 2000.

[59] United Nations Environment Programme/United Nations Centre for Human Settlements (Habitat), Balkans Task Force, *Depleted Uranium in Kosovo, Post-Conflict Environmental Assessment* (Geneva, March 2001) p. 10; United Nations Environment Programme, Post-Conflict Assessment Unit, *Depleted Uranium in Serbia and Montenegro: Post Conflict Environmental Assessment in the Federal Republic of Yugoslavia* (Geneva, 27 March 2002) p. 153; J.P. McLaughlin, et al., 'Actinide analysis of a depleted uranium penetrator from a 1999 target site in southern Serbia', 64 *Journal of Environmental Radioactivity* (2003) pp. 155-165.

[60] OSAGWI 2000, *supra* n. 55, p. 91. See also J.R. Hightower, et al., *Strategy for Characterizing Transuranics and Technicium Contamination in Depleted UF6 Cylinders*, ORNL/TM-2000/242 (Oak Ridge National Laboratory, October 2000).

[61] W. van den Panhuysen, 'Belgium bans uranium weapons and armour', 10 March 2007, <http://www.bandepleteduranium.org/en/a/118.html>, Andrews, *supra* n. 43, p. 44.

[62] Anonymous, 'Depleted Uranium – FAQs,' *Jane's Defense Weekly* (8 January 2001).

[63] OSAWGI 2000, *supra* n. 55, p. 96.

[64] P. Eisler, 'Military study finds fouled weapons safe', *USA Today* (24 June 2001), <http://www.usatoday.com/news/poison/2001-06-25-hotnukes-side.htm>.

[65] Andrews, *supra* n. 43, p. 44.

sten alloy] rounds at significantly lower velocities, meaning that the DU round remains effective against any given target to significantly greater ranges (up to about 50 to 70 percent greater).'[66] Nonetheless, many militaries have found that tungsten alloy ammunition is sufficient to destroy enemy tanks, and the United States, Germany, Israel and the Republic of Korea are the lead nations currently developing long-rod tungsten alloy ammunition.[67] In the United States, research is underway on an advanced tungsten alloy round that could replace the use of DU in 30 mm rounds shot by the A-10 aircraft,[68] and in the UK development of a new smooth bore tank gun could enable tungsten alloy ammunition to perform comparably with and replace the use of DU ammunition.[69]

Tungsten alloy ammunition also presents environmental and health risks, although perhaps less severe than the effects of DU weapons. In laboratory experiments, tungsten alloy caused DNA and chromosomal damage, as well as tumour formation around implanted pellets.[70] A study of tungsten alloy particles in soil found several effects, including soil acidification and uptake of tungsten alloy particles in earthworms and plants.[71]

4.   THE USE OF DEPLETED URANIUM WEAPONS IN ARMED CONFLICT

Since at least 1985 DU weapons have been shot on land, at sea and in the air during various conflicts (see Table 1). It is possible that the armed forces of the Soviet Union used DU ammunition during the 1980s in Afghanistan;[72] however, the earliest confirmed use of DU ammunition is 1985 when the Israeli Navy shot DU rounds from a Phalanx gun at a boat carrying Palestinian commandos off the coast of Israel.[73] The 1991 Gulf War was the first conflict in which US and British forces shot

---

[66] Ibid.

[67] US Army, 'Processing of tungsten alloys for penetrators (South Korea)', in *Army Science and Technology Master Plan* (21 March 1997) p. II(O) (Annex E). See also Soon Hong, et al., 'Matrix pools in partially mechanically alloyed tungsten heavy alloy for localized shear deformation', *Materials Science and Engineering A* 333(1-2) (2002) pp. 187-192.

[68] D. Hambling, "Safe' alternative to depleted uranium revealed', *New Scientist* (30 July 2003).

[69] S. Rayment, 'Army's new tank gun will end the use of controversial uranium-tipped shells', *The Telegraph* (21 September 2003); Anonymous, 'Challenger 2 smoothbore technology demonstrator contract expected', *Jane's International Defence Review* (21 October 2003).

[70] D.E. McClain, A.C. Miller and J.F. Kalinich, 'Status of health concerns about military use of depleted uranium and surrogate metals in armor-penetrating ammunition', Paper presented at the NATO Human Factors and Medicine Panel Research Task Group 099 'Radiation Bioeffects and Countermeasures' meeting, held in Bethesda, Maryland, USA, 21-23 June 2005, and published in AFRRI CD 05-2; A.C. Miller, et al., 'Effect of the militarily-relevant heavy metals, depleted uranium and heavy-metal tungsten alloy on gene expression in human liver carcinoma cells (HepG2)', 255 *Molecular and Cellular Biochemistry* (2004) pp. 247-256.

[71] N. Strigul, et al., 'Effects of tungsten on environmental systems', 61(2) *Chemosphere* (2005) pp. 248-258.

[72] CNN, *supra* n. 49.

[73] Anonymous, 'Israel military used depleted uranium shells: newspaper', Agence France Presse (11 January 2001); Nuclear Age Peace Foundation, 'IDF confirms possession of DU ammunition', *The Sunflower*, No. 45 (February 2001). There is no credible evidence that the IDF has used weapons

DU ammunition in combat (section 4.1). US aircraft shot DU during combat in Bosnia-Herzegovina (1994-1995), and in Kosovo and Serbia and Montenegro (1999) (section 4.2). The use of DU weapons by US forces since 2001 in Afghanistan is probable but unconfirmed (section 4.3). US and British forces shot DU in Iraq during the 2003 invasion, but the use of DU weapons since 2003 is unknown (section 4.4). Although it is possible that DU weapons have been used in other conflicts, such use has not been confirmed. Claims that Israeli Defense Forces used DU during the 1973 Yom Kippur War,[74] or more recently in Palestinian territories or Lebanon[75] lack credible supporting evidence.

Table 1. Known and Suspected Uses of DU Weapons in Warfare

Location	Armed Force Shooting DU	Year(s)	Number of Rounds	Quantity of DU (kg)
At sea off the Israeli coast	Israeli Navy[76]	1985	Unknown	Unknown
Iraq, Kuwait	US Air Force US Army US Marine Corps UK Royal Army[77]	1991	Tanks: >9,640 Jets: 850,950	Tanks: >39,631 Jets: 246,602 Total: >286,233
Bosnia-Herzegovina	US Air Force[78]	1994–1995	Jets: 10,800	Jets: 3,260
Kosovo, Serbia, Montenegro	US Air Force[79]	1999	Jets: 31,300	Jets: 9,450
Afghanistan	US[80] (use not confirmed)	2001–	Unknown	Unknown

containing DU (or natural uranium or enriched uranium) in the Occupied Palestinian Territories or Lebanon.

[74] See D. Fahey, 'Science or Science Fiction? Facts, Myths and Propaganda in the Debate Over Depleted Uranium Munitions', 12 March 2003, <www.danfahey.com>.

[75] United Nations Environment Programme, *Desk Study on the Environment in the Occupied Palestinian Territories*, UNEP/GP.22/INF/31 (Geneva, 23 January 2003) pp. 88-89; United Nations News Centre, 'Israel did not use depleted uranium during conflict with Hizbollah, UN agency finds', 8 November 2006.

[76] An Israeli Shar class gunboat with a Phalanx gun reportedly 'intercepted and sank a boat carrying a Palestinian commando group heading for Israel'. The 20 mm rounds shot by the Phalanx gun each have a DU penetrator weighing 70 grams. Anonymous, 'Israel military used depleted uranium shells: newspaper', Agence France Presse (11 January 2001); Nuclear Age Peace Foundation, 'IDF confirms possession of DU ammunition', *The Sunflower*, No. 45 (February 2001).

[77] OSAGWI 2000, *supra* n. 55, pp. 102-106. M1 tanks shot the M900 model DU round (3.83 kg); M1A1 tanks shot 6,700 M829 rounds (3.94 kg/DU) and 2,348 M829A1 rounds (4.64 kg/DU). Each 30 mm GAU-8 (PGU-14) round shot by an A-10 contains 302 grams of depleted uranium. The US Marine Corps has not yet publicly announced how much ammunition its tanks shot during the war.

[78] US Department of Defense, news briefing by Mr Kenneth Bacon, 4 January 2001; United Nations Environment Programme (UNEP), *Depleted Uranium in Bosnia and Herzegovina* (Geneva, UNEP, 25 March 2003) p. 264.

[79] Angela Ashton-Kelley, US Air Force 11th Wing, letter to Dan Fahey (31 January 2000) (in author's files). A-10s conducted 112 strikes with DU rounds against 85 targets in Kosovo, ten targets in Serbia and one target in Montenegro. UNEP Serbia and Montenegro, *supra* n. 59, p. 168.

[80] According to news reports, the A-10 and AV-8B aircraft, which shoot DU rounds, on numerous occasions shot small calibre ammunition during combat in Afghanistan. In addition, on three occa-

Table 1. Cont.

Location	Armed Force Shooting DU	Year(s)	Number of Rounds	Quantity of DU (kg)
Iraq	US Air Force US Army US Marine Corps UK Royal Army[81]	2003–	Tanks: >2,650 Bradleys: ~121,000 Jets: ~309,000	Tanks: >12,000 Bradleys: ~10,300 Jets: ~93,400 Total (estimated): 118,000 to 136,000

Table compiled by Dan Fahey

4.1   1991 Gulf War

During the Gulf War American aircraft and American and British tanks shot more than 286,000 kg of DU in north-eastern Saudi Arabia, Kuwait and southern Iraq (Table 2). The A-10 aircraft alone accounted for 83 percent (by weight) of the total DU shot during the war. British, Canadian and American warships were also deployed with DU ammunition for their Phalanx guns,[82] but only one American warship shot DU in anger during the war.[83] The Iraqi military did not possess or use DU weapons during the 1991 conflict.[84]

Table 2. DU Ammunition Used in the 1991 Gulf War

Branch	Weapon System	Ammo Size	Quantity of DU Rounds	Weight of DU (kg)
US Army	M1 tank M1A1 tank	105 mm 120 mm	504[85] 9,048[86]	1,930 37,293
US Air Force	A-10 jet A-16 jet	30 mm 30 mm	782,514[87] 1,000[88]	236,319 302

---

sions, US Secretary of Defense Donald Rumsfeld confirmed that DU munitions were found in December 2001 among captured Al Qaeda weapons near Kandahar. See WISE Uranium Project, 'Current Issues – Depleted Uranium Weapons in Afghanistan', <http://www.wise-uranium.org/dissaf.html>

[81] See D. Fahey, 'Unresolved Issues Regarding Depleted Uranium and Veterans of Operation Iraqi Freedom and Operation Enduring Freedom', 24 March 2004, <www.danfahey.com>.

[82] Andrews, *supra* n. 43, p. 44; UK Ministry of Defence, *Testing for the Presence of Depleted Uranium in UK Veterans of the Gulf Conflict: The Current Position* (London, 19 March 1999) p. 9, fn. 1.

[83] OSAGWI 2000, *supra* n. 55, p. 105.

[84] US General Accounting Office (GAO), *Army Not Adequately Prepared to Deal with Depleted Uranium Contamination*, GAO/NSIAD-93-90 (Washington, D.C., GAO, January 1993) p. 14.

[85] M1 tanks shot the M900 model DU round, which contains a DU penetrator weighing 3.83 kg. OSAGWI 2000, *supra* n. 55, p. 104.

[86] M1A1 tanks shot 6,700 M829 rounds (3.94 kg/DU) and 2,348 M829A1 rounds (4.64 kg/DU). OSAGWI 2000, *supra* n. 55, p. 104.

[87] Each 30 mm GAU-8 (PGU-14) round contains 302 grams of depleted uranium. Bernard Rostker, letter to Dan Fahey, 'Technical Response to FOIA Case Number 97-F-1524, Question Eleven', 11 February 1998 (in author's files); OSAGWI 2000, *supra* n. 55, p. 104.

[88] The F-16 can be modified to an A-16 ('A' signifying 'Attack') with the addition of the GPU30 gun pod for close air support. Flown only by the New York National Guard's 174th Tactical Fighter

Table 2. Cont.

Branch	Weapon System	Ammo Size	Quantity of DU Rounds	Weight of DU (kg)
US Marine Corps	AV-8B Harrier	25 mm	67,436[89]	9,981
	M60A3, M1A1	105, 120 mm	Unknown[90]	Unknown
US Navy	Phalanx gun	20 mm	Unknown[91]	Unknown
UK Army	Challenger tank	120 mm	88[92]	408
Totals			Tanks – 9,640 Jets – 850,950	Tanks – 39,631 Jets – 246,602 Total – 286,233

Table compiled by Dan Fahey

Neither the US Department of Defense nor the UK Ministry of Defence has ever publicly released any estimate of the quantity of Iraqi tanks or other equipment destroyed by DU rounds.[93] In fact, of the approximately 3,700 Iraqi tanks destroyed during the conflict, DU rounds probably destroyed only about 500:

- A-10s destroyed 900 Iraqi tanks with Maverick missiles but just 100 with 30 mm DU ammunition;[94]
- US tanks destroyed approximately 400 Iraqi tanks,[95] mainly with DU rounds; and
- AV-8Bs primarily targeted Iraqi artillery with Rockeye missiles, but artillery as well as some tanks and other targets were likely targeted by DU ammunition.[96]

Wing, the A-16 flew only one Gulf War mission (on 26 February 1991), firing approximately 1,000 30 mm DU rounds. OSAGWI 2000, *supra* n. 55, pp. 99-100.

[89] Each 25 mm GAU/12 (PGU/20) round contains 148 grams of DU. OSAGWI 2000, *supra* n. 54, p. 105.

[90] 'Initially, these tanks used pre-positioned, shipboard munitions stocks, which included DU ammunition. As the Marine M1A1s used up the shipboard stocks, they drew resupply rounds from Army munitions stocks.' OSAGWI 2000, *supra* n. 55, p. 105. The US Marine Corps used 210 M60A3 tanks and 76 M1A1 tanks during Operation Desert Storm. US Department of Defense (DoD), *Conduct of the Persian Gulf War, Final Report to Congress* (Washington, D.C., April 1992) p. 750.

[91] Ships fired DU rounds during testing into the Persian Gulf and a Naval frigate accidentally shot four or five DU shells in response to the launch of a shore-based anti-ship missile. OSAGWI 2000, *supra* n. 55, p. 105.

[92] Rostker 1997, *supra* n. 87.

[93] See, e.g., OSAGWI 2000, *supra* n. 55, Tab F, 'DU Use in the Gulf War', <http://www.gulflink. osd.mil/du_ii/du_ii_tabf.htm>.

[94] The A-10 aircraft destroyed 1,000 tanks during Operation Desert Storm. J.F. Dunnigan and A. Bay, *From Shield to Storm* (New York, William Morrow 1992) p. 284. The DoD report to Congress notes: 'In fact, more than 90 percent of the tank kills credited to the A-10 were achieved with IR Mavericks and not with its 30 mm GAU-8 gun.' DoD 1992, *supra* n. 90, p. 139.

[95] Dunnigan and Bay, *supra* n. 94, pp. 284-286.

[96] 'AV-8B targets included artillery, tanks, armor vehicles, ammunition storage bunkers, convoys, logistics sites, troop locations, airfields and known antiaircraft artillery/surface-to-air missile (SAM) locations. AV-8Bs expended 7,175 Mk-20 Rockeye cluster bombs, 288 Mk-83 bombs, 4,167 Mk-82 bombs, and 83,373 rounds of 25-mm machine gun ammunition.' DoD 1992, *supra* n. 90, p. 672. Note that the AV-8B fired a 4/5 mix of DU/high explosive rounds, which resolves the discrepancy between the amount above and the amount listed in Table 1.

In addition, a total of 31 Abrams tanks and Bradley Fighting Vehicles were contaminated by DU; 21 of these vehicles were accidentally shot by DU rounds during friendly fire incidents.[97]

Interestingly, more than 89 percent of the crewman who were in US vehicles struck by DU rounds survived.[98] Design features, such as the blast doors on tank ammunition compartments, saved the lives of many American tank crewmen. In many cases, DU rounds passed completely through the vehicles, causing minimal damage and leaving the vehicle operable. Only when the DU rounds struck a key component (such as the engine or wheels), or hit ammunition or fuel and ignited a fire, was a vehicle rendered inoperable. Conversely, Iraqi tank crewmen in Soviet-model T-55 and T-72 tanks struck by DU weapons likely had a very low survival rate.

Several pieces of evidence indicate that the vast majority of the DU rounds shot during the war probably missed their targets and deposited relatively intact in the local environment:

- Aircraft accounted for approximately 86 percent (by weight) of the DU shot during the war;
- A strafing attack from an aircraft under optimal conditions typically results in few DU rounds (5-10 percent) hitting a target, with only about 2 percent of the rounds shot actually penetrating the target;[99]
- Tank rounds accounted for approximately 14 percent (by weight) of the total DU released, but more than half this quantity was shot on practice ranges in Saudi Arabia.[100] In combat 'eighty to ninety percent of the tank rounds fired will hit the target and remain in or near it';[101]
- Rounds that hit a soft target or the ground tend to stay intact or break into a few large fragments.[102]

In the absence of evidence to the contrary, it is plausible that well over 80 percent (by weight) of the DU shot during the war did not hit a hard target,[103] thereby minimising the creation of respirable-size DU dust.

---

[97] OSAGWI 2000, *supra* n. 55, pp. 108-109.

[98] Ibid., p. 108.

[99] European Commission, Directorate General, Environment (EURATOM), 'Opinion of the Group of Experts Established According to Article 31 of the Euratom Treaty, Depleted Uranium' (Luxembourg, 6 March 2001) p. 2. In US Air Force tests prior to the Gulf War, ammunition shot from A-10 aircraft had an approximate miss rate of 90%, an approximate hit rate of 10%, and a kill rate of just 2%. US Army Center for Health Promotion and Preventative Medicine (CHPPM), *Depleted Uranium – Human Exposure Assessment and Health Risk Characterization*, No. 26-MF-7555-00D (15 September 2000) R-4. In the Gulf War, the miss rate was likely in excess of 90% because 'of the Iraqi AAA threat, which forced the aircraft to operate at altitudes where the gun was less effective'. DoD 1992, *supra* n. 90, p. 139.

[100] AEPI, *supra* n. 34, p. 79.

[101] Ibid., p. 80.

[102] UNEP Bosnia and Herzegovina, *supra* n. 78, pp. 247-248; AEPI, *supra* n. 34, pp. 42-43.

[103] See, e.g., H. Beach, *The military hazards of depleted uranium*, ISIS Briefing Paper No. 78, January 2001, paras. 18, 19, <http://www.isisuk.demon.co.uk/0811/isis/uk/regpapers/no78long_paper.html#16>.

The US Department of Defense has released a map of the general areas where DU weapons were shot, but there is little information about specific locations where DU was released.[104] According to one US Army report, the major air strikes and tank battles involving DU took place in Iraq.[105] A 2000 Iraqi report claims that most of the DU was shot in Iraq near the southern city of Basra:

'According to the personal communication with number [sic] of Iraqi Army Field Commanders, it was estimated that about 65% of the hit targets by these weapons [sic] were in the Iraqi side of the conflict and 75-80% of the above ratio were found in Al-Basrah War Zone.'[106]

In Kuwait DU has been found at the sites of tank battles,[107] as well as in the Udairi 'boneyard' where destroyed equipment was gathered after the war.[108] Between February 2003 and June 2004, the US military contracted with US-based MKM Engineers to recover DU from the Udairi training range in northern Kuwait,[109] site of a 1991 tank battle.[110] The firm reportedly recovered 22 tons of DU fragments, and shipped DU-contaminated equipment back to the United States.[111]

During 1991 the US Army recovered and disposed of its own vehicles contaminated by DU during fires and friendly fire incidents. Six Bradley Fighting Vehicles were buried in Saudi Arabia due to 'the observed damage and radiological measurements which indicated substantial non-removable depleted uranium contamination within the hull, turret, and crew compartment.'[112] Twenty-five Abrams and Bradleys were wrapped in tarps and shipped to a new $4 million decontamination facility in Barnwell, South Carolina (USA). Workers wearing protective suits decontaminated some vehicles, but the more heavily contaminated equipment was cut up and buried in a nearby radioactive waste dump.[113] Additionally, at least one

---

[104] See a copy of the map at <http://www.ngwrc.org/Dulink/DU_Map.htm>.

[105] CHPPM, *supra* n. 99, pp. R-6, R-7.

[106] Republic of Iraq, Ministry of Higher Education and Scientific Research, 'Conference on the Effects of the Use of Depleted Uranium Weaponry on Human and Environment [sic] in Iraq' (26-27 March 2002) p. 8, posted at the website of the International Depleted Uranium Study Team, <http://www.idust.org/>.

[107] International Atomic Energy Agency (IAEA), *The Radiological Conditions in Areas of Kuwait with Residues of Depleted Uranium* (Vienna, IAEA 2003) pp. 30-31.

[108] See D. Fahey, 'Don't Look, Don't Find: Gulf War Veterans, the US Government and Depleted Uranium, 1990-2000', Military Toxics Project, 30 March 2000, pp. 25-27, <www.danfahey.com>. After the war, the Kuwaiti government awarded contracts to private companies from the United States, France, United Kingdom, Bangladesh, Pakistan, Turkey and Egypt for battlefield clean-up. AEPI, *supra* n. 34, p. 83.

[109] P. Patel, 'Where others fear to tread; firm cleans up with dirty work,' *The Houston Chronicle* (17 August 2004), p. B1.

[110] Fahey 2000, *supra* n. 108, pp. 25-27.

[111] T.D. Williams, 'Weapons dust worries Iraqis', *The Hartford Courant* (1 November 2004) p. A1.

[112] US Army, Depleted Uranium Recovery Team, memorandum to Senior Command Representative, US Army Armament, Munitions and Chemical Command (AMCCOM), 'Vehicle Assessment Report Depleted Uranium Contamination', 14 May 1991.

[113] AEPI, *supra* n. 34, p. 87.

captured Iraqi tank contaminated with DU was rejected for shipment to the United States.[114]

Depleted uranium was also used to reinforce the tank armour on some US M1A1 tanks used in the Gulf War. Of the 2,058 US tanks involved in the February 1991 ground war, 654 (32 percent) were equipped with DU heavy armour on the front of the gun turret.[115]

A few months after the ceasefire, a large munitions fire at the US Army base at Doha, Kuwait resulted in an additional release of DU. On 11 July 1991 a fire in the motor pool and ammunition storage area at Doha consumed dozens of vehicles and large quantities of ammunition, including 660 120 mm DU tank rounds.[116] Approximately 111 of the DU tank rounds destroyed by the fire were located inside Abrams tanks. The other 549 rounds were stored in ammunition storage containers that 'exploded in fires that were of such sustained intensity that steel howitzers and other equipment had melted, making it likely that many DU rounds had been damaged by oxidation'.[117] During post-fire clean-up operations, only approximately 250 DU penetrators from the containers were accounted for, leaving the fate of approximately 300 rounds (1,450 kg/DU) unknown.[118] Depleted uranium has since been detected at a nearby bird sanctuary (in 1995) and in soil and water samples at the Doha site (in 2002).[119]

4.2   The Balkans

In 1994-95 US A-10 aircraft shot approximately 10,800 DU rounds in Bosnia and Herzegovina, releasing 3,260 kg of DU into the environment.[120] On seven occasions between August 1994 and September 1995, A-10 aircraft shot DU weapons either within the 20km exclusion zone around Sarajevo or near Han Pijeak, the headquarters of the Bosnian Serb army,[121] at one tank, artillery and mortar pieces and buildings. The only reported incidence of DU clean-up was the removal by NATO forces to the United States of a box of DU penetrators, fragments and casings from the Hadzici Tank Repair Facility.[122]

In 1999, during the North Atlantic Treaty Organization's (NATO) Operation Allied Force against what was then Serbia and Montenegro, A-10 aircraft shot

---

[114] OSAGWI, *supra* n. 55, p. 115.

[115] DoD 1992, *supra* n. 90, p. 750; D. Fahey, *Case Narrative, Depleted Uranium (DU) Exposures*, 3rd edn. (The Military Toxics Project, the National Gulf War Resource Center, Swords to Plowshares, 20 September 1998) p. 34.

[116] US Army Safety Center, Army Accident Report 910711001 (Fort Rucker, AL, 20 September 1991).

[117] OSAGWI 2000, *supra* n. 55, p. 102.

[118] Ibid., p. 117.

[119] IAEA, *supra* n. 107, pp. 15-17.

[120] Bacon, *supra* n. 78.

[121] UNEP Bosnia and Herzegovina, *supra* n. 78, p. 264; North Atlantic Treaty Organization, 'Briefing by NATO Acting Spokesman Mark Laity and Statement by Ambassador Daniel Speckhard, Chairman Ad Hoc Committee on Depleted Uranium' (Brussels, Belgium, 24 January 2001).

[122] UNEP Bosnia and Herzegovina, *supra* n. 78, p. 104.

approximately 31,300 DU rounds,[123] containing 9,450 kg of DU, at a variety of targets, including armoured vehicles, anti-aircraft artillery and barracks.[124] In total, A-10s conducted 112 strikes with DU rounds against 85 targets in Kosovo, ten targets in Serbia, and one target in Montenegro.[125] Aside from penetrators and contaminated soil recovered by United Nations Environment Programme (UNEP) field missions, the only reported clean-up of DU took place at Cape Arza in Montenegro, where government authorities removed two tons of rock, soil and humus from DU impact sites.[126]

4.3   Afghanistan

The use of DU weapons by US or allied forces since 2001 in the war in Afghanistan remained uncertain when this book went to press. Claims about the use of DU weapons in Afghanistan have neither been confirmed by the US military nor verified by credible investigations. Nonetheless, it appears likely that US forces have used some DU weapons and that Taliban and/or Al Qaeda forces may have possessed DU rounds.

According to news reports, several US weapons platforms that shoot DU rounds have been used in combat in Afghanistan. The Air Force's A-10 aircraft shot 30 mm ammunition at combatants, mortar pieces and vehicles in Afghanistan on at least eight occasions between March 2002 and April 2003.[127] Military press releases report that A-10s were called upon to strafe enemy forces (and involved in a friendly fire incident) as recently as September 2006.[128] However, it is not clear if A-10s are shooting DU ammunition given the lack of armoured targets. The Marine Corps' AV-8B aircraft, another DU shooter, reportedly fired its cannons at combat-

---

[123] Ashton-Kelley, *supra* n. 79.

[124] UNEP Kosovo, *supra* n. 59, pp. 28-29, 36, 48.

[125] UNEP Serbia and Montenegro, *supra* n. 59, p. 168.

[126] Ibid., p. 10.

[127] The reported dates of A-10 attacks are 3-6 March, 21 May, 25 August, 20 September, 15 November and 20 December 2002, and 12 February and 2 April 2003. US Department of Defense News Transcript, 'DoD News Briefing – ASD PA Clarke and Brig. Gen. Rosa', (5 March 2002), <http://www.defenselink.mil/news/Mar2002/t03052002_t0305asd.html>. E. Thomas, 'Leave no man behind,' *Newsweek* (18 March 2002) p. 26; T. Shanker, 'US tells how rescue turned into fatal firefight', *The New York Times* (6 March 2002) p. A1; P. Baker, 'Afghans strengthen US force', *The Washington Post* (8 March 2002) p. A1; E. Schmitt, 'American planes foil an attack on an airfield in Afghanistan', *The New York Times* (22 May 2002) p. A9; C.G. Soriano, 'US to stay in Afghanistan indefinitely', *USA Today* (25 August 2002); US base in Afghanistan attacked', Associated Press (20 September 2002); 'US bases under fire,' Associated Press (15 November 2002); E. Schmitt, 'Paratrooper from New Jersey dies in Afghan firefight near Pakistan border', *The New York Times* (22 December 2002); C. Gall, 'Afghans report 17 civilian deaths in US-led bombing', *The New York Times* (12 February 2003); 'Green berets, allies fight Afghan Taliban,' Associated Press (2 April 2003). In the 2 April 2003 attack on Sikai Lashki, Afghanistan, 'two A-10 fighter jets fired seven white phosphorus rockets and 520 30 mm rounds'.

[128] J.D. Banusiewicz, 'Officials express regret after friendly fire incident,' American Forces Press Service, 5 September 2006, <http://www.defenselink.mil/news/NewsArticle.aspx?ID=697>; cf., Anonymous, 'Coalition forces kill, capture militants in Afghanistan', American Forces Press Service, 7 December 2005, <http://www.defenselink.mil/news/Dec2005/20051207_3568.html>.

ants in April 2003.[129] However, there is no credible evidence to substantiate claims that US forces have used missiles and/or bombs in Afghanistan that contain DU or other forms of uranium.[130] A 10 January 2005 information paper from then-Chairman of the US Joint Chiefs of Staff General Richard Myers states that 'munitions containing DU are not being used in the current stability and support operations in Iraq or Afghanistan.'[131]

The use of DU weapons by Al Qaeda, Taliban, Northern Alliance or other Afghan forces is unknown given currently available public information, although the US Department of Defense has stated that DU weapons were found in December 2001 among captured Al Qaeda weapons near Kandahar. On three occasions, then-US Secretary of Defense Donald Rumsfeld confirmed the discovery of DU ammunition,[132] but the quantity, calibre, and origin of the rounds remain unclear.

4.4   Iraq

The US and UK militaries confirmed that they used DU ammunition during the 2003 invasion of Iraq.[133] Tanks, fighting vehicles and aircraft shot DU ammunition at a wide range of targets; however, there is no evidence that missiles or bombs containing DU (or other forms of uranium) were used during the war. The US weapons systems that are known or suspected to have shot DU ammunition in Iraq include the M1A1 and M1A2 tanks (Army and Marine Corps), M2 Bradley Fighting Vehicle (Army), A-10 aircraft (Air Force), and AV-8B Harrier jet (Marine Corps).[134] In August 2004 a US Department of Defense official stated that 'none of the guided bombs or cruise missiles that the US used in Iraq and Afghanistan contained uranium of any type.'[135] Aside from the United States, the only other armed force known to have shot DU ammunition in Iraq is the UK Army, whose Challenger II tanks shot 120 mm rounds.[136]

---

[129] Gall, *supra* n. 127.

[130] G. Lamartin, US DoD, letter to Hon. John Kyl, 25 August 2004, <http://www.wise-uranium.org/pdf/lamiq04.pdf>; cf., D. Fahey 2003, *supra* n. 74.

[131] R.B. Myers, General, letter to S. Silver, President, Women's International League for Peace and Freedom (WILPF), 27 January 2005, Enclosure 'Information Paper, Subject: Depleted Uranium (DU) Information Summary and Response', 10 January 2005.

[132] DoD News Briefing, *supra* n. 48; US Department of Defense News Transcript, 'Secretary Rumsfeld Roundtable with Radio Media' (15 January 2002), <http://www.defenselink.mil/news/Jan2002/t01152002_t0115sdr.html>; US Department of Defense News Transcript, 'Secretary Rumsfeld interview with Baltimore Sun' (27 December 2001), <http://www.defenselink.mil/news/Dec2001/t12282001_t1227sun.html>; see also 'Current Issues – Depleted Uranium Weapons in Afghanistan (10 February 2002), <http://www.wise-uranium.org/dissaf.html>.

[133] Statement of Dr M. Kilpatrick, 'Depleted Uranium', National Public Radio, Science Friday, 18 April 2003, <http://www.sciencefriday.com/pages/2003/Apr/hour1_041803.html>; UK Ministry of Defence, 'Depleted Uranium – Middle East 2003', 6 June 2003, <http://www.mod.uk/issues/depleted_uranium/middle_east_2003.htm>.

[134] See D. Fahey, 'The Use of Depleted Uranium in the 2003 Iraq War – An Initial Assessment of Information and Policies' (24 June 2003), <www.danfahey.com>.

[135] Lamartin, *supra* n. 130. Thanks to J. Cohen-Joppa for encouraging US Senator Kyl to investigate this matter.

[136] MoD 2003, *supra* n. 133.

The most recent release of information about the quantities of DU shot in Iraq took place during a 6 March 2004 conference on DU at the Massachusetts Institute of Technology.[137] During a debate with the author, Dr Michael Kilpatrick from the Deployment Health Support Directorate stated that the US Air Force had released approximately 93,400 kg/DU (103 tons), and the US Army had released 21,800 kg/DU (24 tons).[138] Together with previously released information, these new numbers indicate that Abrams tanks shot just over half of the Army's DU by weight (11,442 kg/DU in 2,466 rounds of 120 mm ammunition),[139] and Bradley Fighting Vehicles shot the rest (approximately 10,300 kg/DU in approximately 121,000 rounds of 25 mm ammunition).[140] The US Marine Corps has not yet released information about its use of DU weapons, but the Marine Corps is unlikely to have exceeded the amount shot by the Army. The British Ministry of Defence has acknowledged that its Challenger II tanks shot approximately 870 kg/DU (1.9 tons) in 185 rounds of 120 mm ammunition.[141]

There is little information available about the quantity and type of equipment targeted by DU rounds during the 2003 invasion,[142] although press reports indicate that DU rounds were shot at a wide range of non-armoured targets, including Iraqi combatants,[143] trucks[144] and buildings.[145] At least one US Army tank, one British Army tank, two British Army Scimitar armoured vehicles and one US Marine Corps armoured assault vehicle were shot at by DU rounds in accidental friendly fire incidents.[146] Depleted uranium was also apparently released when US ammu-

---

[137] 'Depleted Uranium Weapons: Toxic Contaminant or Necessary Technology', Massachusetts Institute of Technology, 6 March 2004, <http://web.mit.edu/pugwash/du/>.

[138] See additional information and a link to an audio recording of the conference proceedings at <http://www.wise-uranium.org/diss.html#MITDU04>.

[139] Hon. J. Kyl, US Senate, letter to Mr J. Cohen-Joppa, 14 July 2003. The letter from Senator Kyl states that the tank rounds shot by Army tanks were M289A1, which may be inaccurate given that M829A2 rounds are currently in the US arsenal. The M829A1 has a DU penetrator weight of 4.64 kg; the M829A2 has a penetrator weight of 4.74 kg. A.H. Passarella, Director, Freedom of Information and Security Review, Office of the Assistant Secretary of Defense, letter to Mr D. Fahey, 'Technical Response to FOIA Case Number 97-F-1524, Question Eleven', 11 February 1998.

[140] Bradley Fighting Vehicles shoot the 25 mm M919 round, which has a DU penetrator weight of 0.0855 kg. While the exact number of DU tanks rounds was previously reported, the exact number of 25 mm DU rounds shot has not been released. The figures for the Bradley were calculated based on the total amount of DU shot by the Army, as reported by Dr Kilpatrick, minus the quantity of DU shot by Abrams tanks, as reported in July 2003. Consequently, these numbers should be taken as approximations until the DoD releases more accurate figures.

[141] Lord Bach, Under Secretary of State and Minister for Defense Procurement, response to Baroness Miller of Chilthorne Domer, UK Parliament (London, 12 June 2003), <www.publications.parliament.uk/pa/ld199900/ldhansrd/pdvn/lds03/text/30612w03.htm#30612w03_sbhd>. If each round is approximately 4.7 kg, this would equate to approximately 185 rounds shot in combat.

[142] See D. Fahey 2003, supra n. 134.

[143] S.D. Naylor, 'Iraqis ambush armored column; two Abrams tanks destroyed', The Army Times (25 March 2003).

[144] D. Brown, '3rd Infantry: The dash around Karbala', Knight Ridder News Service (2 April 2003).

[145] S. Peterson, 'Remains of toxic bullets litter Iraq', The Christian Science Monitor (15 May 2003); D. Brown, 'At Baghdad airport, a scene of mayhem,' Knight Ridder News Service (4 April 2003).

[146] P. Pae, 'Friendly fire still a problem', The Los Angeles Times (16 May 2003); P. Wachter, 'Fought for freedom with Marines he loved', The State (South Carolina, 25 May 2003); P. Hess, 'Iraq war-

nition trucks containing DU rounds were destroyed[147] and when DU armour on an Abrams tank was breached.[148] Iraqi forces destroyed several Abrams tanks in combat, triggering ammunition explosions that may have resulted in releases of DU fragments and dust.[149]

Battles involving DU shooters took place in cities and towns on the roads along the Tigris and Euphrates rivers from Kuwait to Baghdad, and in Baghdad itself. As of October 2007 the US military had not publicly disclosed the exact locations where American forces shot DU weapons. By contrast, in June 2003 the UK Ministry of Defence (MoD) 'provided details of UK DU firing locations to the United Nations Environment Programme (UNEP) in support of its post-conflict environmental assessment, and directly to recognised non-government organisations, such as aid agencies, in response to ad hoc enquiries'.[150] In addition, the MoD reported that it was cleaning up DU penetrators found lying on the ground and identifying contaminated vehicles so that risk assessments can be conducted to determine whether to decontaminate or dispose of contaminated equipment.[151] On 2 February 2004 the UK Secretary of State for Defence stated:

> 'To date eight military vehicles have been identified as having been hit by depleted uranium (DU) munitions within the southern sector of Iraq under British military control. All these vehicles have been clearly marked. Arrangements are currently being negotiated with the US for a contractor to collect and securely store these military vehicles.'[152]

The MoD has issued a card to its servicemen and women in Iraq that states: 'You may have been exposed to dust containing DU during your deployment…You are eligible for a urine test to measure uranium.'[153]

Clean-up and assessment activities have taken place since 2003, despite the persistence of armed conflict. US-based company MKM Engineers managed a collection site in Iraq where contaminated equipment was gathered,[154] but the fate

---

related US deaths now 147', United Press International (12 May 2003); R. McCarthy, 'Friendly fire kills two UK tank crew', *The Guardian* (26 March 2003); S.D. Naylor, 'Abrams destroyed by friendly, not Iraqi, fire', *The Army Times*, 30 May 2003; US Army Tank Automotive and Armament Command, Abrams Program Manager Office, 'Abrams Tank Systems: Lessons Learned, Operation Iraqi Freedom, 2003', undated; T. Ripley, 'Abrams tank showed 'vulnerability' in Iraq', *Jane's Defence Weekly* (20 June 2003); R. McCarthy, 'British soldier killed by US jet', *The Guardian* (29 March 2003); A. Gillan, 'I never want to hear that sound again', *The Guardian* (31 March 2003).

[147] M.R. Gordon, 'A Soldier's Story', *The New York Times* (9 May 2003); Peterson, *supra* n. 145.

[148] Naylor, *supra* n. 146.

[149] Naylor, *supra* n. 143; S.D. Naylor, '3rd Infantry Division armored unit prevails in 30-hour firefight', *The Army Times* (26 March 2003); S. Christenson, 'Impressions of war', *San Antonio Express-News* (TX) (26 May 2003).

[150] MoD 2003, *supra* n. 133.

[151] Ibid.

[152] UK Parliament, Question 150356, Mr Ingram, Secretary of State for Defence, response to Mr. Hancock, 2 February 2004, Column 747W.

[153] N. Mackay and A. Wilson, 'MoD 'lied' over depleted uranium', *The Sunday Herald* (UK) (29 February 2004).

[154] Patel, *supra* n. 109; Williams, *supra* n. 111.

of this equipment or other contamination, such as penetrator fragments, could not be determined by the author when this book went to press. Field studies conducted during 2004-2005 by the Iraqi Ministry of Environment and UNEP's Post-Conflict Branch identified 311 sites with DU contamination,[155] but it is not clear how many of these sites are from the 1991 war or more recent combat. The Post-Conflict Branch has warned that DU in tanks used for scrap metal recovery presents health and environmental risks, although a study of one scrap metal recovery site found no evidence of DU contamination.[156]

The use of DU weapons since 2003 in Iraq remains unclear. Since DU is primarily an armour-piercing weapon, and because the forces fighting against American forces lack tanks and armoured vehicles, it is unlikely that US forces have shot large amounts of DU weapons since 2003. Nonetheless, US military officials are not publicly releasing information about DU weapons, leading to speculation about ongoing use. Moreover, weapons platforms capable of shooting DU (Bradley Fighting Vehicles, Abrams tanks and A-10 aircraft) continue to be used in combat. One indication of ongoing use of DU ammunition comes from the reports on testing of US service-members for DU exposure.[157] For example, the DoD tested 41 individuals between 1 April and 30 September 2006; two individuals with fragment injuries tested positive for DU.[158] Ongoing DU testing suggests that DU weapons are still being used in combat, although the lack of detail in DoD reporting makes it unclear how long after their suspected exposures these veterans were tested (see also discussion of testing in Chapter 2).

As noted above, however, in January 2005 the DoD denied the ongoing use of DU weapons in Iraq.[159] Given the DoD's history of reporting false or misleading information about the use of DU weapons, it is possible that limited use of DU weapons continues. Another explanation for ongoing testing could be that service-members were potentially exposed to DU in smoke from munitions fires, or during the clean-up of equipment and soil in battle zones.

---

[155] AFX News, 'Iraq faces 40 mln usd bill to clean up toxic, radioactive waste – UN', 11 October 2005, <http://www.forbes.com/business/feeds/afx/2005/11/10/afx2329437.html>.

[156] United Nations Environment Programme (UNEP), Post-Conflict Branch, 'Assessment of Environmental 'Hot Spots' in Iraq' (Geneva, UNEP 2005) pp. 115-116, 119.

[157] For the period 1 October 2004 to 31 March 2005, DoD tested 20 individuals with suspected Level I exposure (in, on or near a vehicle at the time of DU impact), plus 25 individuals with suspected Level II exposure (entry into contaminated vehicles). There was one positive exposure for an Air Force servicemember who had a small fragment of DU removed from his eyelid. The number of Level I and II tests for this period may be substantially higher; fully 1,103 (56%) of the 1,970 service-members tested between June 2003 and 31 March 2005 were not coded for level of exposure. W. Winkenwerder, Deputy Secretary of Defense for Health Affairs, 'Subject: Operation Iraqi Freedom Depleted Uranium Bioassay Results and Semi-Annual Data Submission', Memorandum for Assistant Secretaries of the US Army, Navy and Air Force, 28 July 2005.

[158] W. Winkenwerder, 'Subject: Operation Iraqi Freedom Depleted Uranium Bioassay Results – Sixth Semiannual Report and Request for Data Submission', Memorandum for Assistant Secretaries of the US Army, Navy and Air Force, 5 February 2007.

[159] Myers, *supra* n. 131.

5.   NON-COMBAT USE OF DEPLETED URANIUM WEAPONS

Outside combat, DU weapons have been shot during their development and testing, during training exercises, and as a result of accidental discharges during weapon maintenance.[160] Each country that develops and/or possesses DU weapons has probably shot DU in its territory. In the United States, the Nuclear Regulatory Commission regulates the possession and use of DU weapons,[161] restricting test firing of DU rounds to a few locations, the most heavily contaminated being Aberdeen Proving Ground (Maryland), the now-closed Jefferson Proving Ground (Indiana), and Yuma Proving Ground (Arizona). The UK Ministry of Defence has overseen the testing of DU weapons at the firing range at Eskmeals and the range at Dundrennan, Kirkcudbright in Dumfries and Galloway,[162] although it has also conducted one DU weapons test in France and two in the United States.[163]

Little is known about the fate of DU on many testing and training ranges, but several locations have been extensively studied. The accidental uses of DU weapons at training ranges at Tori Shima, Japan and Vieques, Puerto Rico generated a remarkable amount of political pressure and press attention considering that so little DU was shot at each remote location. In addition, the test firing of DU weapons at the Nevada Test and Training Range is significant for the fact that the US Air Force has stated that it will conduct regular soil and target decontamination. These three examples of non-combat uses of DU weapons are noteworthy because, in each case, political considerations and public pressure prompted the US military to assess the hazards and engage in remedial activities.

5.1   **Tori Shima, Japan**

In December 1995 and January 1996 US AV-8B aircraft shot 1,500 rounds of 25 mm DU ammunition (225 kg/DU) at Tori Shima, an uninhabited island training range located 100 km west of Okinawa in the East China Sea.[164] Though the release of such a small amount of DU so far from the Japanese mainland posed no risk to the Japanese people as such, the US Air Force quickly sent a team of scientists to the island to gather DU penetrators and take soil and water samples.[165]

---

[160] On 4 May 1994, the USS Lake Erie (CG 70) accidentally shot two 20 mm DU rounds from a Phalanx Close-In Weapons System (CIWS) gun during gun maintenance. US Navy message 132025Z JUL 94, 'Explosive mishap report', 13 July 1994. Thanks to K. Kajihiro from the American Friends Service Committee – Hawaii for providing the author with this document.

[161] AEPI, *supra* n. 34, p. 37.

[162] Dr T. Carter, *Comparison of Kirkcudbright and Eskmeals Environmental Monitoring Data for Generalized Derived Limits for Uranium* (London, Ministry of Defence June 2002) p. vii.

[163] Mr J. Spellar, response to Mr L. Smith, UK Parliament (London, 12 June 2003), <http://www.parliament.the-stationery-office.co.uk/pa/cm200001/cmhansrd/vo010123/text/10123w11.htm>.

[164] US Air Force, Detachment 3, Armstrong Laboratory, memorandum for 18th Medical Group, 'Consultative Letter (CL), AL/OE-CL-1996-0004, Site Assessment at Aerial Gunnery Range Whiskey 176, Tori Shima, Japan', 18 March 1996.

[165] Ibid.

A total of 169 DU penetrators (11 percent) were found and recovered during two trips to Tori Shima. Most of the remaining 1,351 DU penetrators (89 percent) were presumed to have landed in the East China Sea.[166] The use of DU weapons on Japanese territory was seized upon by groups opposing the US military presence on Okinawa as evidence of the military's disregard for the environment,[167] a fact that may have prompted the Air Force to send teams of scientists to the remote island.

5.2   Vieques, Puerto Rico

On 19 February 1999 two US AV-8B jets shot a total of 263 rounds of 25 mm DU ammunition (39 kg/DU) during US military training exercises at Vieques, Puerto Rico.[168] The live fire area at Vieques is located on the eastern tip of the island, approximately nine miles from the nearest civilian population. After notifying the Nuclear Regulatory Commission (NRC) that it had violated its license for the possession and use of DU weapons, the Navy dispatched scientists to recover the DU rounds.[169] The small amount of DU released and its remote location minimise the risks to the island's inhabitants,[170] but Navy scientists nonetheless recovered dozens of intact penetrators.[171]

The Vieques DU incident came at a time when the local population was actively campaigning for the Navy to stop using the eastern half of the island as a bombing range.[172] Some residents of Vieques claimed that the Navy's use of DU ammunition was polluting the environment on Vieques and endangering the health of the local population. Public opposition combined with the Navy's violation of its NRC license apparently prompted the Navy to assess and clean-up DU at Vieques.

5.3   Nevada Test and Training Range

The Nevada Test and Training Range (NTTR) contains the only approved air-to-ground target area in the United States for the use of DU weapons. Between 1976-1977 and 1982-1993, A-10 aircraft fired approximately 27,800 kg of DU at rows of

---

[166] Ibid.

[167] D. Allen, 'Okinawans have been concerned about depleted uranium ammo', *Stars and Stripes* (Pacific Edition) (16 January 2001).

[168] US Navy, letter to the US Nuclear Regulatory Commission, 'Subject: Improper Expenditure of Depleted Uranium Munitions', Washington, DC, 1 June 1999.

[169] US Nuclear Regulatory Commission (NRC), 'Technical Evaluation Report: US Navy – Vieques Island Review of Survey Work Plan' (Atlanta, GA, NRC, 21 March 2000).

[170] Agency for Toxic Substances and Disease Registry (ATSDR), 'Focused Petitioned Public Health Assessment, Drinking Water Supplies and Groundwater Pathways Evaluation, Isla de Vieques Bombing Range, Vieques, Puerto Rico', Atlanta, GA, 20 February 2001, p. 10.

[171] US Nuclear Regulatory Commission, 'Preliminary Notification of Event or Unusual Occurrence, PNO II-99-020, Subject: Press Interest in Event Involving Depleted Uranium Ammunition', 1 June 1999.

[172] M. Chapman, 'Uranium scare on enchanted island', *The Telegraph* (UK) (4 February 2001); T. Eglund, 'Depleted uranium: The Vieques-Kosovo connection,' *The Gully* (12 February 2001), <http://www.thegully.com/essays/puertorico/010212depleted_uranium.html>.

tanks.[173] In 1996 pressure from environmental groups and the US Fish and Wildlife Service prompted the Air Force to stop shooting DU at Nellis, but in 1998 the Air Force announced its intention to resume the use of DU after stating that it would periodically remove solid DU penetrators and fragments from the soil and either recycle or dispose of them as low-level radioactive waste.[174] In 2002 the Air Force resumed test firing of DU weapons at the NTTR.[175]

There are approximately 180 tank and vehicle targets contaminated with DU in a storage area near the target area; the target area itself contains six tanks.[176] Due to public pressure, the Air Force announced plans to clean-up DU-contaminated targets and debris from the NTTR. In October 2005 the US Nuclear Regulatory Commission (NRC) approved US Air Force plans to transport four contaminated tanks (each with approximately 12 kg/DU) to the US Ecology facility in Grand View, Idaho.[177] According to the Air Force, 'for these tanks, the cost for disposal by burial was determined to be less than attempting to decontaminate them.'[178]

6. CONCLUSION

Military officials justify the use of DU ammunition by claiming it is necessary to penetrate advanced tank armour, but recent wars suggest that there is widespread use of DU weapons against non-armoured targets such as soldiers, buildings, cars and trucks. Although most of the DU shot in combat has probably deposited intact in the ground, it is clear that DU rounds have hit hundreds of tanks and other equipment, creating localised areas of contamination. Over the course of years or decades, the rounds that deposited in the ground will degrade and contamination will further disperse. The proliferation of DU weapons virtually ensures their use in future conflicts, but public interest in DU weapons has so far been limited to conflicts which involve the American and British militaries.

The US Department of Defense and UK Ministry of Defence have conducted limited remediation of DU contamination after armed conflict. However, the lack of transparency (particularly by the DoD) about battlefield clean-up opera-

[173] US Air Force, Air Combat Command, 'Draft – Nevada Test and Training Range Depleted Uranium Target Disposal Environmental Assessment' (September 2004) pp. 1-5; US Air Force, Air Combat Command, *Final Environmental Assessment for Resumption of Use of Depleted Uranium Rounds at Nellis Air Force Range Target 63-10* (September 1998) pp. 2-4.

[174] US Air Force 1998, *supra* n. 173, pp. 3-8-3-9.

[175] US Air Force 2004, *supra* n. 173, pp. 1-5.

[176] Ibid., pp. 1-6.

[177] US Federal Register, 'Environmental Assessment and Finding of No Significant Impact for Department of the Air Force's Request for 10 CFR 20.2002 Authorization, for Disposal of Four Tanks Containing Depleted Uranium to a Subtitle C RCRA Hazardous Waste Disposal Facility', Vol. 70, No. 205 (25 October 2005), <http://www.epa.gov/fedrgstr/EPA-IMPACT/2005/October/Day-25/i5878.htm>.

[178] US Air Force Institute for Operational Health (AFMC), Memorandum for AFSMA/SGPR (Dr Bhat), 'SUBJECT: Consultative Letter, IOH-SD-BR-CL2004-0054, Radiological Assessment for Burial of Four Tanks Containing Unimportant Quantities of Radioactive Materials at the US Ecology Hazardous Waste Treatment and Disposal Facility in Idaho', 23 June 2004, <http://www.wise-uranium.org/dissti.html#NELLIS>. Thanks to the WISE Uranium Project for this document.

tions has led to speculation about ongoing DU exposures and health effects. Clean-up activities by the DoD at several training ranges demonstrate official acknowledgment of DU's risks and provide a standard against which battlefield remediation can be measured.

Chapter 2
ENVIRONMENTAL AND HEALTH CONSEQUENCES
OF THE USE OF DEPLETED URANIUM WEAPONS

Dan Fahey

1.   INTRODUCTION

In recent years, hundreds of scientific studies have shed new light on the environmental and health effects of the use of depleted uranium (DU) munitions, although many uncertainties remain. Environmental assessments of DU impact sites in the Balkans and Kuwait found localised contamination and corroding penetrators, but the fate of DU released in Iraq in 1991 and since 2003 remains largely unknown. DU has caused cancer and neurological disease in laboratory animals, but conclusive evidence of similar effects in humans is lacking. Absence of evidence should not be interpreted as evidence of absence, however, as there have been few long-term health studies of soldiers or civilians with confirmed DU exposures. Future research and studies will clarify the environmental and health consequences of using DU weapons, but a complete picture of DU's effects may not emerge for several decades.

This chapter describes the methods by which DU is released into the environment during armed conflict (section 2), and assesses air, soil, water and biological contamination (section 3). It discusses human exposure scenarios and estimates (section 4), and reviews possible health effects from intakes of DU (section 5). Special attention is given to the findings and limitations of the US study of Gulf War veterans (section 6). The chapter concludes with a summary of the status of knowledge of the environmental and health effects of the use of DU weapons (section 7).

2.   RELEASE OF DEPLETED URANIUM INTO THE ENVIRONMENT

In warfare, DU can be released into the environment when munitions containing DU are shot from aircraft or vehicles, when DU armour is breached, or as a result of fires consuming DU rounds. The risk to public health and the local environment depends on the quantity, size, distribution and solubility of the DU released, and the extent to which different factors affect its movement in the environment.[1] The re-

---

[1] The Royal Society, *The Health Hazards of Depleted Uranium Munitions, Part II* (London 2002) p. 19 (hereinafter, Royal Society Part II).

McDonald/Kleffner/Toebes (eds.), Depleted Uranium Weapons and International Law
© 2008, T·M·C·ASSER PRESS, *The Hague, The Netherlands and the Authors*

lease of DU may contaminate soil, water, air, plant and animal life, and potentially expose human populations to DU through implantation, inhalation, ingestion or wound contamination.

2.1   Depleted uranium from impacts

The impact of DU penetrators against tank armour or other hard targets creates respirable-size DU dust that contaminates the impact site. The amount of dust created by an impact depends on the energy of the impact and the type of material impacted. The greatest quantities of DU dust are created by impacts against heavily armoured vehicles or other dense objects (e.g., reinforced bunkers, rocks). The impact of DU rounds against soft surfaces (e.g., soil, trees) generally creates significantly smaller amounts of DU dust.

The impact of a DU penetrator against a hard target creates local temperatures as high as 1,800°C.[2] The high speed and thermal softening of the penetrator results in a 'self-sharpening' effect known as 'adiabatic shearing', whereby small particles break off and burn, creating solid aerosol particles.[3] Test data from the United States demonstrate that, normally, about 20 percent of a DU penetrator is aerosolised upon impact with a tank.[4] The impact of one 120mm DU tank round could therefore create approximately 950g of DU dust.[5] During a single attack by an A-10 aircraft shooting a burst of 30mm ammunition, between five and 16 DU bullets will likely hit the target, creating roughly 300g to 960g of aerosol.[6] The majority of the DU particles created by an impact are considered insoluble in lung

---

[2] W.S. Andrews, 'Depleted uranium on the battlefield: Part 1 – ballistic considerations', *Canadian Military Journal* (Spring 2003) p. 44. A French test in which DU rounds were shot at a tank found that 'at the impact point, the temperature was 600°C'. V. Chazel, 'Characterization and dissolution of depleted uranium aerosols produced during impacts of kinetic energy penetrators against a tank', 105 *Radiation Protection Dosimetry* (2003) p. 164.

[3] US Army Center for Health Promotion and Preventative Medicine (CHPPM), 'Depleted Uranium – Human Exposure Assessment and Health Risk Characterization', No. 26-MF-7555-00D (15 September 2000) p. 12.

[4] US Army testing found that normally 10-35% (but up to 70%) of the round oxidises into dust upon impact with a hard target. Twenty percent is commonly used to determine the amount of dust created by an impact. The Office of the Special Assistant to the Deputy Secretary of Defense for Gulf War Illnesses (OSAGWI), 'Depleted Uranium in the Gulf (II)' (Washington, D.C. 2000) p. 203.

[5] The 120 mm M829A2 tank round contains a DU penetrator weighing 4.74 kg. B. Rostker, letter to D. Fahey, 'Technical Response to FOIA Case Number 97-F-1524, Question Eleven', 11 February 1998 (in author's files).

[6] The number of penetrators hitting a target varies with the type of target, but 90 to 95% of the projectiles generally miss the target during air attacks. European Commission, Directorate General, Environment (EURATOM), 'Opinion of the Group of Experts Established According to Article 31 of the Euratom Treaty, Depleted Uranium', (Luxembourg, 6 March 2001) p. 2. 'The weight of one [30 mm] penetrator is approximately 300 g…A typical burst of fire occurs for two to three seconds and involves 120 to 195 rounds. These hit the ground in a straight line, one to three meters apart, depending on the angle of the approach, and cover an area of about 500 m². ' United Nations Environment Programme (UNEP), *Depleted Uranium in Kosovo, Post-Conflict Environmental Assessment* (Geneva, UNEP, March 2001) p. 10 (hereinafter, UNEP Kosovo Report). A typical combat load for an A-10 is 1,100 rounds of 30 mm ammunition mixed at a ratio of five depleted uranium rounds to one high explosive round. OSAGWI, *supra* n. 4, p. 104.

fluid. Other major elements found in DU dust at impact sites include iron, manganese, and silicon.[7] Once the armour has been perforated, the burning fragments and flying debris may ignite fuel, ammunition or other combustible material, disabling or destroying the vehicle and its crew.[8]

About 90 percent of the DU dust created by the impact of a tank round against a hard target falls to the ground within 50 meters of the target,[9] although airborne DU has been found up to 400 metres from the impact site immediately following an impact.[10] The DU dust created by air attacks typically spreads out over a larger area up to approximately 100 metres from the impacted target.[11]

Significant amounts of DU dust and fragments may also be deposited inside a target vehicle. Measurable quantities of DU oxide particles can resuspend when individuals reenter a contaminated vehicle; the resuspended aerosols contain particles of respirable size.[12] The end result of the impact of DU ammunition against a hard target is not only a destroyed or disabled vehicle but also an impact site contaminated with DU dust and fragments.

Depleted uranium penetrators that hit soft targets or miss targets altogether deposit more or less intact in the local environment. These impacts may result in localised points of ground contamination, and over time the penetrator fragments chemically react with soil and moisture to corrode and leach into the soil beneath the fragments. Test range data from the United States indicate that penetrators shot from aircraft will deposit close to the surface in thick soil (e.g., clay), but can go as deep as six to seven metres in soft soil (e.g., sand).[13] One study of DU in Kuwait found few DU rounds on the ground's surface; most rounds shot by aircraft were assumed to lie at least 0.5 m below the ground.[14] Other rounds may ricochet and come to rest on the surface hundreds or thousands of meters from the targeted area.[15]

2.2   Depleted uranium from corroding penetrators

The corrosion rate of intact DU rounds or fragments is determined by several variables, including soil type, the oxygen content of its surroundings, presence of wa-

---

[7] R.E.J. Mitchel and S. Sunder, 'Depleted uranium dust from fired munitions: physical, chemical, and biological properties', 87(1) *Health Physics* (2004) pp. 57-67; cf., S.J. Jette, et al., 'Aerosolization of M829A1 and XM900E1 Rounds Fired Against Hard Targets', PNL-7452 (Richland, W.A., Battelle Pacific Northwest Laboratory, August 1990) p. 4.1.

[8] R. Pengelley, 'The DU debate: what are the risks?', *Jane's Defence Weekly* (15 January 2001).

[9] CHPPM 2000, *supra* n. 3, p. R-2.

[10] R.L. Fliszar, 'Radiological Contamination from Impacted Abrams Heavy Armor', Technical Report BRL-TR-3068 (Aberdeen Proving Ground, MD, Ballistic Research Laboratory December 1989) pp. 12, 37-38.

[11] Jette, *supra* n. 7, p. 4.1.

[12] OSAGWI, *supra* n. 4, pp. 54-55.

[13] UNEP Kosovo Report, *supra* n. 6, p. 119.

[14] H. Bem and F. Bou-Rabee, 'Environmental and health consequences of depleted uranium use in the 1991 Gulf War', 30 *Environment International* (2004) p. 130.

[15] US Air Force, Air Combat Command, 'Final Environmental Assessment for Resumption of Use of Depleted Uranium Rounds at Nellis Air Force Range Target 63-10' (September 1998) pp. 2-4; UNEP Kosovo Report, *supra* n. 6, p. 11; Royal Society Part II, *supra* n. 1, p. 88.

ter, size of metal pieces, presence of protective coatings and the salinity of any water present.[16] In general, the corrosion rate of DU penetrators is greatest in the presence of water (particularly salt water), humidity or moist soil, with slower rates of corrosion observed for penetrators exposed to air.[17] Corrosion rates are highly variable depending on location and environment, but penetrators may completely disintegrate into particulate matter within five to 35 years.[18] One study of penetrators recovered in Kosovo and Serbia found that 'DU projectiles lose as much as 6% weight per year due to humidity and oxidising conditions that prevail in this area.'[19]

As penetrators or fragments corrode, the particles will disperse in the environment. Soil disturbance during agriculture, decontamination or construction work, or the movement of people or vehicles may resuspend DU dust, which could then be inhaled.[20] Some DU dust could dissolve in soil runoff and be transported into plants, surface waters or groundwater.[21] In addition, discovery of corroding penetrators by adults or children could result in ingestion or inhalation of significant amounts of DU.[22]

2.3   Depleted uranium from fires

Fires involving the ignition and dispersal of DU have occurred at munitions storage areas,[23] manufacturing sites,[24] airplane crash sites,[25] and during destruction of tanks

---

[16] United Nations Environment Programme (UNEP), *Post-Conflict Assessment Unit, Depleted Uranium in Serbia and Montenegro: Post Conflict Environmental Assessment* (Geneva, UNEP, 27 March 2002) p. 14 (hereinafter, UNEP Serbia and Montenegro); cf., W. Schimmack, U. Gertsmann, U. Oeh, W. Schultz and P. Schramel, 'Leaching of depleted uranium in soil as determined by column experiments', 44 *Radiation and Environmental Biophysics* (2005) pp. 183-191.

[17] Royal Society Part II, *supra* n. 1, pp. 92-93. See also N. Priest, 'UK Research on DU', presentation to the US Institute of Medicine, Washington, D.C., 28 June 2007 (in author's files).

[18] Royal Society Part II, *supra* n. 1, p. 21; UNEP Serbia and Montenegro, *supra* n. 16, p. 27; United Nations Environment Programme, 'Depleted Uranium in Bosnia and Herzegovina', (Geneva, UNEP, 25 March 2003) pp. 247-248 (hereinafter, UNEP Bosnia); US Army Environmental Policy Institute (AEPI), 'Health and Environmental Consequences of Depleted Uranium Use by the U.S. Army, Technical Report', (Atlanta, AEPI 1995) p. 152 (hereinafter, AEPI Report 1995). In the United Kingdom, a 'DU Corrosion Program' examined corrosion rates of DU penetrators in marine and terrestrial environments. At a 28 June 2007 meeting of the US Institute of Medicine in Washington, D.C., Nick Priest reported that 'interpreted time-scales to complete corrosion' in terrestrial experiments were 103 years at Eskmeals and 14 years at Kirkcudbright. Priest, *supra* n. 17.

[19] M. Esposito, et al., 'Survey of natural and anthropogenic radioactivity in environmental samples from Yugoslavia', 61 *Journal of Environmental Radioactivity* (2002) p. 281; cf., L.A. Di Lella, F. Nannoni, G. Protano and F. Riccobono, 'Uranium contents and 235U/238U atom ratios in soil and earthworms in western Kosovo after the 1999 war', 337 *Science of the Total Environment* (2005) pp. 109-118.

[20] US Army Chemical School, 'Development of Depleted Uranium Support Packages: Tier 1 – General Audience' (October 1995) p. 28; UNEP Serbia and Montenegro, *supra* n. 16, p. 9.

[21] Royal Society Part II, *supra* n. 1, p. 21.

[22] UNEP Serbia and Montenegro, *supra* n. 16, p. 28.

[23] R.I. Scherpelz, et al., 'Depleted Uranium Exposures to Personnel Following the Camp Doha Fire, Kuwait, July 1991' (Richland, Washington, Pacific Northwest National Laboratory, June 2000).

[24] J. Guthrie, 'Factory in leak scare', *Birmingham Evening Mail* (UK) (27 June 2003); UK National Radiation Protection Board, 'Fire at a Royal Ordnance Factory,' 15 February 1999.

[25] R. Sisk, 'Cool rescue, warthog and all,' *New York Daily News* (20 July 2003).

and ammunition trucks.[26] In severe fire conditions, such as large munitions fires or after the crash of an airplane carrying DU rounds, DU penetrators 'slow cook' and may completely oxidise.[27] In fires of less intensity and duration, the amount of oxidation and resultant contamination may be considerably lower.

A large percentage of the DU oxides produced by a fire are respirable in size.[28] Some of these oxides may become entrained in smoke and travel downwind, or spread around the site of the fire by the force of explosions, wind or water from fire fighting efforts.[29] Fires in tanks or other vehicles may detonate fuel or ammunition, resulting in partial or complete oxidation of DU rounds and dispersal of dust and whole or partial penetrators around the site of the fire.[30] During a fire, people may inhale DU entrained in smoke, and after a fire people may inhale or ingest DU during decontamination and cleanup activities.[31]

3. ENVIRONMENTAL CONTAMINATION

The amount and form of DU released in a given area during combat may vary widely, and over time the health and environmental risks will change. On the battlefield, the concentrations of DU in the environment may range from relatively high concentrations near impacted vehicles, to minor levels of localised contamination around penetrators lying on the ground.[32] Over time, the risks to the environment and public health may shift as the potential for airborne DU decreases and the potential for soil, water or biological contamination from corroding penetrators increases.[33]

3.1 Depleted uranium in air

Depleted uranium may contaminate air as a result of ammunition impacts, fires or resuspension of dust. The quantity and concentration of airborne DU depends on a

---

[26] US Army Armament, Munitions and Chemical Command, Memo from Depleted Uranium Recovery Team to Senior Command Representative, AMCCOM – Southwest Asia (SWA), 'Vehicle Assessment Report, Depleted Uranium Contamination', 14 May 1991; M.R. Gordon, 'A soldier's story', *The New York Times* (9 May 2003); S. Peterson, 'Remains of toxic bullets litter Iraq', *The Christian Science Monitor* (15 May 2003).

[27] J. Mishima, et al., 'Potential Behavior of Depleted Uranium Penetrators under Shipping and Bulk Storage Accident Conditions', PNL-5415 (Richland, WA, Battelle Pacific Northwest Laboratory March 1985) p. v.

[28] 'In general, DU aerosols in the respirable range are produced when penetrators are exposed to temperatures greater than 500° C for burn times longer than 30 minutes.' Royal Society Part II, *supra* n. 1, p. 89.

[29] US Army, Guidelines for Safe Response to Handling, Storage, and Transportation Accidents Involving Army Tank Munitions or Armor Which Contain Depleted Uranium, Technical Bulletin 9-1300-278 (28 September 1990) pp. 3-2, 7-1.

[30] OSAGWI, *supra* n. 4, pp. 53, 163-166.

[31] Scherpelz, *supra* n. 23, p. ii.

[32] UNEP Serbia and Montenegro, *supra* n. 16, pp. 24-26.

[33] Royal Society Part II, *supra* n. 1, p. 22; UNEP Serbia and Montenegro, *supra* n. 16, p. 16.

number of variables, including particle size, weather conditions and the quantity and cause of the initial release. In general, the risk of inhalation of airborne DU decreases as the time and distance increase from the point of release.

Most airborne DU falls to the earth within 50-100 meters of an impacted target or within a few kilometres of a large munitions fire,[34] but the potential exists for long-range airborne transport of small amounts of DU during a military conflict.[35] Long-range transport of DU particles was demonstrated by a 1979 study at the Knolls Atomic Power Laboratory in upstate New York, which unexpectedly found DU particles in 16 air filters at three different locations. The source of the DU particles proved to be the National Lead Industries munitions plant in Colonie, NY, which manufactured 30 mm DU rounds for the US Air Force. Three of the 16 air filters containing DU were located 41 km from the National Lead plant.[36]

Resuspension of DU dust may be natural (e.g., wind driven) or anthropogenic (e.g., due to human activity, vehicular movement, or handling corroded penetrators).[37] Entering a heavily contaminated vehicle is also likely to result in resuspension of DU dust. In general, resuspension in confined spaces will result in higher inhalation exposures than resuspension in the natural environment, and resuspension rates decrease with time as DU dust is dispersed in the environment.[38] During field studies in Bosnia, the United Nations Environment Programme's (UNEP) Post Conflict Assessment Unit detected low levels of airborne DU dust at two locations, more than seven years after DU rounds had been shot.[39] In 2001 low levels of DU were detected in the air at three locations in Serbia and one location in Montenegro, two and a half years after DU rounds had been shot in combat.[40] Based on experimental data and field observations, resuspension of DU dust is not a significant exposure pathway, although the many uncertainties in this area have prompted calls for additional research and cleanup of sites where airborne DU is detected.[41]

3.2   Depleted uranium in soil

Although uranium is naturally occurring in soils around the world, the use of DU weapons in armed conflict may increase by several orders of magnitude the con-

---

[34] Royal Society Part II, *supra* n. 1, p. 90.

[35] Ibid., p. 96.

[36] L. Dietz, 'Contamination of Persian Gulf War Veterans and Others by Depleted Uranium', 19 July 1996, <http://www.wise-uranium.org/dgvd.html>.

[37] UNEP Bosnia, *supra* n. 18, p. 187; cf., J.J. Whicker, et al., 'From dust to dose: Effects of forest disturbance on increased inhalation exposure', 368 *Science of the Total Environment* (2006) pp. 519-530.

[38] The Royal Society, 'The health hazards of depleted uranium munitions, Part I' (London, Royal Society, 2001) p. 43 (hereinafter, Royal Society Part I).

[39] UNEP Bosnia, *supra* n. 18, pp. 36, 42.

[40] G. Jia, M. Belli, U. Sansone, S. Rosamilia, S. Guadino, 'Concentration and characteristics of depleted uranium in water, air, and biological samples collected in Serbia and Montenegro', 63 *Applied Radiation and Isotopes* (2005) p. 396.

[41] UNEP Serbia and Montenegro, *supra* n. 16, p. 138; UNEP Bosnia, *supra* n. 18, p. 42.

centrations of uranium in soil. Dust created by impacts or fires will settle on the ground, and DU penetrators may deposit on the ground surface or penetrate soil down to a depth of several metres. Over time, DU fragments and particles created by impacts, fires or corrosion may migrate, contaminating soil and water.

The mobility of DU in soil varies depending on the quantity, form and size of the DU, the soil type and the local geological and weather conditions. DU may bind with soils high in organic matter and be slowly transported in rain or surface waters. However, DU generally experiences greater mobility in semi-arid soils with little organic matter. The migration and dispersal of DU in soil may decrease its concentration but increase the likelihood of groundwater contamination and the costs and technical difficulties of DU remediation.[42]

In the United States, target ranges at Aberdeen Proving Ground (Maryland) are contaminated with approximately 80,000 kg of DU from decades of ammunition tests.[43] Several studies found contamination of soil down to 20 cm beneath corroding penetrators.[44] The US Army is overseeing the removal of several tons of DU fragments and contaminated soil from the target areas.[45]

During the 1991 Gulf War the major air strikes and tank battles involving DU reportedly took place in Iraq.[46] American tanks and aircraft sometimes engaged dozens of Iraqi tanks in a single location, potentially resulting in large concentrations (hundreds of kg) of DU dust and debris in the soil at battle sites. Although representatives of the Iraqi government under Saddam Hussein claimed high levels of radioactivity in the soil in southern Iraq,[47] these claims have not been independently verified.

There have been several studies of soil and equipment in battlefield areas in Kuwait. During studies conducted in the 1990s, Kuwaiti researchers found DU in soil in battlefield areas and near the site of the Doha munitions fire.[48] In 1995 US Army tests found small amounts of DU in the soil next to contaminated tanks in a containment area in western Kuwait known as the 'boneyard'.[49] Some of the entry and exit holes on these tanks exhibited radioactivity levels 20 to 24 times above background. The test results demonstrated that there was little hazard to US troops

---

[42] Royal Society Part II, *supra* n. 1, p. 22.

[43] AEPI Report 1995, *supra* n. 18, pp. 43, 173.

[44] W. Dong, et al., 'Sorption and bioreduction of hexavalent uranium at a military facility by the Chesapeake Bay', 142 *Environmental Pollution* (2006) pp. 132-142; AEPI Report 1995, *supra* n. 18, p. 147.

[45] Allied Technology Group, 'Aberdeen Proving Ground: Transonic Range Depleted Uranium Study Area, Detailed Work Plan' (Oak Ridge, T.N., September 1999) p. 1; cf., C.C. Choy, et al., 'Removal of depleted uranium from contaminated soils', 136 *Journal of Hazardous Materials* (2006) pp. 53-60.

[46] CHPPM, *supra* n. 3, pp. R-6, R-7.

[47] See, e.g., O. Ward, 'Iraq casualties mounting after eight years of exposure to depleted uranium', *The Toronto Star* (Canada) (12 July 1999); R. Norton-Taylor, 'Doctor blames west for deformities,' *The Guardian* (UK) (30 July 1999).

[48] Bem and Bou-Rabee, *supra* n. 14, pp. 123-134.

[49] These tanks had been moved to the boneyard from the battlefield, therefore the soil contamination was due to runoff from the tank surface or interior.

'as long as there are no ongoing operations within the boneyard'.[50] Subsequent US Army tests found small amounts[51] of DU in soil within the boneyard and at Doha, Kuwait, but at levels considered to pose little threat to human health.[52]

In September 2002 scientists from the International Atomic Energy Agency visited several sites in Kuwait where DU was known or suspected to have been released. Investigators found complete DU penetrators and fragments at some locations, but concluded that 'depleted uranium does not pose a long-term radiological hazard to the general population of Kuwait.'[53] The IAEA scientists did not examine the Udairi Training Range in northern Kuwait, a 1991 battle site that became a training ground for US and allied troops throughout the 1990s and then a staging area for troops deploying to Iraq since 2003.[54] In 2003-2004, an American contractor removed 22 tons of DU fragments and contaminated soil from the Udairi range.[55]

In Bosnia UNEP's Post Conflict Assessment Unit (renamed the Post-Conflict Branch in 2005) found that penetrators hitting the ground result in contamination of several grams of DU per kilogram of soil.[56] Localised soil contamination near penetrators (within 1-2 m) was found at several locations, as well as limited dispersion to the soil beneath corroding penetrators.[57]

In Kosovo, Serbia and Montenegro US A-10 aircraft fired between 50 and 2,320 rounds of 30mm ammunition at ground targets, representing releases of 15 kg to 700 kg at individual locations.[58] The vast majority of these rounds are believed to have missed their targets and deposited more or less intact in the ground.[59] The UNEP mission to Serbia and Montenegro did not find 'significant, widespread contamination of the ground surface' by DU except at localised points underneath and adjacent to penetrators on and in the ground.[60] At Cape Arza in Montenegro,

---

[50] US Army Medical Research Institute of Infectious Diseases (USAMRIID), 'Vigilant Warrior '94, Forward for the Soldier, Medical Problem Definition and Assessment Team', (Fort Detrick, M.D., 8 May 1995) pp. 17, 20.

[51] The highest level of DU radioactivity measured in a soil sample was 11.7pCi/g, less than the 35 pCI/g screening guideline for unrestricted public access to a contaminated area.

[52] OSAWGI, *supra* n. 4, p. 53.

[53] International Atomic Energy Agency (IAEA), Radiological Conditions in Areas of Kuwait with Residues of Depleted Uranium (Vienna, IAEA 2003) p. 1.

[54] For more information about the Udairi Training Range, see 'Udairi Training Range', GlobalSecurity.org, <http://www.globalsecurity.org/military/facility/udairi.htm>. In May 2004, the base adjacent to the training range, known as Camp Udairi, became Camp Buehring. GlobalSecurity.org, 'Camp Buehring', undated, <http://www.globalsecurity.org/military/facility/camp-buehring.htm>.

[55] P. Patel, 'Where others fear to tread; firm cleans up with dirty work', *The Houston Chronicle* (17 August 2004) B1; T.D. Williams, 'Weapons dust worries Iraqis', *The Hartford Courant* (1 November 2004) A1; D. Fahey, 'Don't Look, Don't Find: Gulf War Veterans, the U.S. Government and Depleted Uranium, 1990-2000' (Lewiston, Military Toxics Project, 30 March 2000) pp. 25-27, <www.danfahey.com>.

[56] UNEP Bosnia, *supra* n. 18, p. 159.

[57] Ibid., p. 32.

[58] UNEP Kosovo Report, *supra* n. 6, p. 133.

[59] M. Mellini and F. Riccobono, 'Chemical and mineralological transformations caused by weathering in anti-tank DU penetrators ('the silver bullets') discharged during the Kosovo war', 60 *Chemosphere* (2005) p. 1250.

[60] UNEP Serbia and Montenegro, *supra* n. 16, p. 24.

where 300 DU rounds (90 kg/DU) were shot, government authorities have removed two tons of rock, soil and humus from DU impact sites.[61] The UNEP mission to Kosovo found lower levels of DU contamination in soil around ground impacts and penetrators than those found in Serbia and Montenegro.[62]

An analysis of soil samples from Kosovo found 'thousands of [DU] particles…indicating that several hundred thousands or a million particles can be present in a few milligrams of DU contaminated soil.'[63] Most of the DU particles were less than five microns in diameter, and more than 50 percent of the particles were less than 1.5 microns in size. 'This means that the DU particles display a large surface area for dissolution processes and have a potential for re-suspension and inhalation under arid conditions.'[64]

3.3   Contamination of water

Contamination of water by DU can be caused by deposition of DU dust or disintegration of fragments or penetrators.[65] Depleted uranium's ability to contaminate water depends upon the quantity, size and solubility of the DU, the soil chemistry and the proximity of contaminated soil to surface and sub-surface water. In general, the greatest potential for human consumption of contaminated water occurs where drinking water is drawn from small lakes or shallow groundwater sources in areas where large amounts of DU are released during a conflict.[66] The corrosion rate of DU penetrators and their mobility in soil will determine the rate of contamination, but contamination of water supplies at levels that may pose a risk to public health might not occur for years or decades, highlighting the need for continued environmental monitoring of water and food supplies over many decades in areas where DU weapons have been used.[67]

Environmental monitoring at several US locations has identified water contamination by DU. Depleted uranium particles were found at the bottom of a reservoir 1 km away from the former National Lead Industries plant in Colonie, NY, where DU rounds were manufactured; the particles were likely transported into the reservoir by groundwater runoff from the plant site, although some may

---

[61] Ibid., p. 10.

[62] 'The range of values from the UNEP Kosovo Mission within the first 1-2 meters of impact holes was 0.5±0.25 MG DU/kg in surface soils (0.5 cm), independent of the hardness of the surface. This picture was not confirmed in Serbia and Montenegro. There, soil samples taken at similar distances from penetrator impacts often contained several mg DU/kg.' UNEP Serbia and Montenegro, *supra* n. 16, pp. 125-126.

[63] P.R. Danesi, et al., 'Depleted uranium particles in selected Kosovo samples', 64 *Journal of Environmental Radioactivity* (2003) p. 152.

[64] Ibid.

[65] A. Bleise, et al., 'Properties, use and health effects of depleted uranium (DU): a general overview', 64 *Journal of Environmental Radioactivity* (2003) p. 9.

[66] UNEP Serbia and Montenegro, *supra* n. 16, p. 131; Royal Society Part II, *supra* n. 1, pp. 27, 98-100.

[67] Royal Society Part II, *supra* n. 1, p. 26.

have also been deposited by wind.[68] Monitoring at US Army testing and training ranges has not detected DU migration out of the impact areas or into sub-surface groundwater, but it has measured limited movement of DU in surface water within the impact areas.[69] At Sandia National Laboratory in New Mexico, a 1997 rainstorm dislodged buried DU and scattered it in an adjacent flood plain, requiring the cleanup of more than 4,000 barrels of contaminated soil.[70] Scientists at Aberdeen Proving Ground monitor the migration of DU from the soil and sediments of testing ranges towards the estuary of Chesapeake Bay; no movement of DU into the bay has been found, but a 2005 article noted 'during extreme climate events such as heavy rainfall, [DU] will be susceptible to erosion and flushing into the estuary even when tightly bound to [natural organic matter] and other particulates present at the site.'[71]

At the Starmet (formerly Nuclear Metals) munitions plant in Concord, Massachusetts, DU dumped in pits during the production of 30mm DU rounds has contaminated underground drinking water supplies.[72] The state of Massachusetts permits drinking water to contain up to 29 micrograms of uranium per litre, but test wells at the Starmet site have measured levels up to 87,000 micrograms of uranium per litre of water.[73] One study found DU in the sapwood and bark of oak trees on the Starmet site; the transfer of DU to the sapwood apparently took place through uptake of contaminated groundwater, while the finding of DU in tree bark may indicate deposition of airborne DU from dust resuspended by nearby remediation activities.[74] Cleanup of the 46-acre site was due to be completed by 2006; as of 23 December 2005, contractors had carted away 2,540 of the 3,650 drums of DU-contaminated soil as well as 100 of the 317 tons of DU metal.[75]

The UNEP missions in the Balkans found limited evidence of water contamination. In Bosnia UNEP found DU in one sample of drinking water at one site, but 'the concentration was low and insignificant from a radiological and chemical-toxicological point of view.'[76] UNEP did not find any signs of DU in water samples from Kosovo, Serbia or Montenegro,[77] but one study reported the possible pres-

---

[68] D. Lo, et al., 'Location, identification, and size distribution of depleted uranium grains in reservoir sediments', 89 *Journal of Environmental Radioactivity* (2006) pp. 240-248.

[69] AEPI Report 1995, *supra* n. 18, p. 147.

[70] B. Murphy, 'New technology cleans up residue from Sandia's early Cold War weapons test program', Sandia Lab News 50(23), 20 November 1998, <http://www.sandia.gov/LabNews/LN11-20-98/du_story.htm>.

[71] Dong, et al., *supra* n. 44, pp. 132-142.

[72] US General Accounting Office (GAO), 'Hazardous Waste: Information on Potential Superfund Sites,' GAO/RCED-99-22 (Washington, D.C., Government Printing Office, November 1998) p. 170.

[73] M. Orey, 'Uranium waste site has a historic New England town up in arms', *The Wall Street Journal* (1 March 2001) p. B1.

[74] J.D. Edmands, et al., 'Uptake and mobility of uranium in black oaks: implications for biomonitoring depleted uranium-contaminated groundwater', 44 *Chemosphere* (2001) pp. 789-795.

[75] Davis Bushnell, 'Uranium cleanup moves ahead', *The Boston Globe* (1 January 2006).

[76] UNEP Bosnia, *supra* n. 18, pp. 35-36; cf., G. Jia, et al., 'Concentration and characteristics of depleted uranium in biological and water samples collected in Bosnia and Herzegovina', 89 *Journal of Environmental Radioactivity* (2006) pp. 172-187.

[77] UNEP Kosovo, *supra* n. 6, p. 24; UNEP Serbia and Montenegro, *supra* n. 16, p. 29.

ence of very low levels of DU in water from a private well in Serbia.[78] Since most penetrators shot in the Balkans are believed to have deposited more or less intact in the soil, there is a risk of future DU contamination of groundwater and drinking water: 'Heavy firing of DU in one area could increase the potential source of uranium contamination of groundwater by a factor of 10 to 100.'[79] Such contamination might pose a chemical toxicity hazard to local populations. UNEP and the Royal Society recommend periodic testing of groundwater used for drinking within and adjacent to areas where DU has been used.[80]

In Kuwait scientists did not find any presence of DU in water from the country's main underground source or in the Persian Gulf; no DU was found in coastal bottom sediments.[81]

3.4   Biological contamination

Depleted uranium particles in the air, water and soil may contaminate soil microbes, plants and animals. The extent and effects of biological contamination depend on a number of variables, including the quantity and form of DU released and the characteristics of the local ecosystem. Although the biological effects of DU contamination have been poorly studied, biological contamination could affect ecosystem function and result in human consumption of food and water contaminated by DU. In addition, certain organisms known to absorb and concentrate DU may aid assessments of the presence and risks of DU in battlefield areas.

Depleted uranium may have adverse effects on the diversity and functioning of microorganisms in soil. Microorganisms lie at the base of many food chains, and they play an important role in influencing the concentration and composition of organic matter in soil, which has been demonstrated to control the mobility of uranium in soil and the amount available for uptake into plants.[82] Adverse effects on microorganisms are only likely in areas adjacent to corroding penetrators, or where DU impacts have resulted in high concentrations of DU in soil.[83] A study of DU at Aberdeen Proving Ground showed that some microorganisms function to bioreduce DU to forms that immobilise uranium in surface soil sediments.[84]

DU may contaminate plants through deposition on plant surfaces or root uptake. Deposition of airborne DU might significantly increase the concentration of DU on foliage and unwashed fruits and seeds.[85] Agricultural crops, plants and

[78] Jia, et al., *supra* n. 40, p. 397.

[79] UNEP Kosovo, *supra* n. 6, p. 26; UNEP Serbia and Montenegro, *supra* n. 16, p. 34.

[80] UNEP Bosnia, *supra* n. 18, p. 44; UNEP Serbia and Montenegro, *supra* n. 16, p. 36; Royal Society Part II, *supra* n. 1, p. xi.

[81] A.Z. Al.-Zamel, F. Bou-Rabee, M. Olszewski and H. Bem, 'Natural radionuclides and 137Cs activity concentration in the bottom sediment cores from Kuwait Bay', 266(2) *Journal of Radioanalytical and Nuclear Chemistry* (2005) pp. 269-276; Bem and Bou-Rabee, *supra* n. 14, p. 130.

[82] Royal Society Part II, *supra* n. 1, p. 23.

[83] Ibid.

[84] Dong, et al., *supra* n. 44, pp. 10-11.

[85] UNEP Bosnia, *supra* n. 18, p. 44; UNEP Serbia and Montenegro, *supra* n. 16, p. 36; Royal Society Part II, *supra* n. 1, p. xi.

trees may also absorb DU through their roots, although the rates of absorption are believed to vary widely among species and in different soils.[86] At Aberdeen Proving Ground in Maryland, grasses and marsh plants in a target area contain DU taken up through their roots.[87] Some agricultural crops, including barley,[88] carrots, alfalfa, chard and sugar beet,[89] are known to bioaccumulate uranium, but for many plants, root uptake of uranium is small and does not result in significant translocation to fruits or seeds. Additionally, contamination of farmland with uranium may have a detrimental effect on the productivity of certain crops, such as wheat.[90]

Of particular interest is the use of lichen and tree bark as bio-indicators of DU contamination. During UNEP-led field missions in Bosnia, Kosovo, Serbia and Montenegro, the presence of DU in lichen was found to be a good indicator of the past presence of airborne DU, most likely from dust created by the impact of penetrators against a hard target.[91] A related study determined that 'the field surveys in Serbia and Montenegro have clearly demonstrated that lichens, mosses, mushrooms, leaves and barks are sensitive bioindicators of past airborne contamination of DU dust or particles, in particular in areas where DU ammunition has been used.'[92] However, one study of lichens at battle sites in Kosovo did not find any evidence of DU contamination.[93]

Animals may be exposed to DU through consumption of contaminated plants, soil or water, or inhalation of airborne DU. Herbivores typically consume considerable quantities of soil while grazing; consequently, they may consume DU in soil, contaminated grasses or on contaminated plant surfaces.[94] DU has been found in deer, kangaroo rats and other animals living within US testing ranges, although no significant health effects from DU have been reported in these animals.[95] An ecological risk assessment for Yuma Proving Ground in Arizona, USA, where DU ammunition is shot in testing, determined that several species could be at

---

[86] Ibid.

[87] US Army Tank-Automotive and Armaments Command, *Environmental Assessment of the Abrams Heavy Armor System, Technical Report* (13 April 1998) pp. 17-18; AEPI Report 1995, *supra* n. 18, p. 147.

[88] A.A. Kasianenko, et al., 'The migration of U-238 in the system 'soil-plant' and its effect on plant growth', 1276 *International Congress Series* (2005) pp. 223-224.

[89] AEPI Report 1995, *supra* n. 18, pp. 101-104.

[90] World Health Organization, 'Depleted uranium: Sources, Exposure and Health Effects' (Geneva, WHO 2001) p. 34 (hereinafter, WHO).

[91] UNEP Bosnia, *supra* n. 18, pp. 36-37; UNEP Kosovo Report, *supra* n. 6, pp. 157-158; UNEP Serbia and Montenegro, *supra* n. 16, pp. 30-31; cf., G. Jia, et al., *supra* n. 76.

[92] Jia, et al. , *supra* n. 40, p. 396; cf., UNEP Serbia and Montenegro, *supra* n. 16, p. 31. See also S. Loppi, et al., 'Lichens as biomonitors of uranium in the Balkan area', 125 *Environmental Pollution* (2003) pp. 277-280; L.A. Di Lella, 'Lichens as biomonitors of uranium and other trace elements in an area of Kosovo heavily shelled with depleted uranium rounds', 37(38) *Atmospheric Environment* (2003) pp. 5445-5449.

[93] L.A. Di Lella, et al., 'Environmental distribution of uranium and other trace elements at selected Kosovo sites', 56 *Chemosphere* (2004) pp. 861-865.

[94] WHO, *supra* n. 90, p. 56.

[95] Jefferson Proving Ground Restoration Advisory Board, meeting minutes, 22 August 2001, Madison, Indiana (in author's files); WHO, *supra* n. 90, p. 51.

risk of adverse effects of DU.[96] One study in the former Yugoslavia did not find any evidence of DU or other radionuclides in 30 samples of honey.[97] A study of contaminated areas of Kosovo did not find any evidence of DU contamination in earthworms.[98]

Contamination of plants and animals may present a hazard to human populations who live in or secure food supplies from areas containing DU dust and debris. The World Health Organization notes:

> 'Data regarding bio-uptake of uranium into plants and animals indicates that bioaccumulation factors, while not being high may be in some cases significant over the longer term, particularly where local consumption patterns indicate a preference for foodstuffs shown to potentially bioaccumulate uranium (i.e. kidneys of cattle).'[99]

In 2002 the IAEA found DU in foodstuffs collected at farms in Kuwait, but it concluded that the levels were low and of 'little radiological significance'.[100] The Royal Society has called attention to the potential for human consumption of crops grown on contaminated lands and food prepared in poor hygienic conditions, and the problems associated with the drying of food on soil potentially contaminated by DU.[101]

A 1995 article in the US Army magazine *Armor* offers practical advice to military officers inhabiting or working in a contaminated environment:

> 'Take care to ensure mess [food], shower, and bivouac sites are not in an area of either known DU contamination or where DU dust may have been carried by recent rains ... Of course, always keep personnel away from contaminated equipment or terrain unless required to complete the mission.'[102]

This guidance for soldiers applies equally to civilians and highlights the need for timely identification, study and, where necessary, decontamination of areas where DU has been released by fires or the use of DU ammunition.

4.   HUMAN EXPOSURE ESTIMATES

The use of DU weapons in combat or the release of DU from fires can result in the exposure of soldiers and civilians to DU particles. Although prolonged external

---

[96] M. Fan, et al., 'Using a probabilistic approach in an ecological risk assessment simulation tool: test case for depleted uranium (DU)', 60(1) *Chemosphere* (2005) pp. 111-125.

[97] Esposito, *supra* n. 19, p. 280.

[98] Di Lella, et al., *supra* n. 19, pp. 109-118.

[99] WHO, *supra* n. 90, p. 41.

[100] IAEA 2003, *supra* n. 53, pp. 18-20.

[101] Royal Society Part II, *supra* n. 1, p. 110.

[102] P. Paulsen, 'Depleted uranium without the rocket science', *Armor* (July-August 1995) p. 34.

exposure to DU metal can be hazardous,[103] DU has the greatest potential to effect health effects when it enters the body. Routes of exposure to DU include:

- Injection of fragments through wounds;
- Inhalation of DU dust;
- Ingestion of DU directly or in contaminated food, soil and water;
- Wound contamination by DU dust;[104]
- Dermal absorption.

Injection of fragments and inhalation of DU dust are considered to be the routes of exposure most likely to potentially cause health effects, although the significance of each type of exposure remains unclear due to a lack of data on exposures to DU during and after armed conflict.

As a result of the use of DU weapons in the Gulf War and the Balkans, several likely exposure scenarios have emerged:

- Soldiers inside a vehicle hit by one or more DU rounds may be wounded by DU fragments, inhale and ingest DU dust particles, and suffer wound contamination by DU dust;
- Rescue soldiers who enter vehicles immediately after they are struck are also likely to inhale or ingest DU dust, and they may suffer cuts or abrasions that become contaminated;
- Soldiers or civilians may inhale resuspended dust during activity (e.g., agriculture, construction, decontamination) in or around a contaminated site;
- Soldiers or civilians downwind of a DU impact site or fire may inhale DU dust entrained in the smoke;
- Soldiers or civilians may ingest DU in contaminated food, soil, or water;
- Civilians may inhale or ingest DU when handling corroding penetrators found in soil or foliage.

Several agencies recommend limits for exposure to uranium compounds to minimise health effects among occupational workers and the public (Table 1).[105] In addition

---

[103] WHO, *supra* n. 90, p. 84; R. Pöllänen, et al., 'Characterisation of projectiles composed of depleted uranium', 64 *Journal of Environmental Radioactivity* (2003) p. 142.

[104] Not much is known about wound contamination by DU dust, but a recent article notes: 'From biokinetic and dosimetric perspectives, contaminated wounds are intrinsically different from inhalation or ingestion intakes because the behaviour of the radioactive contaminant is dependent not only on the physiochemical form of the contaminating material, but also on the pathophysiological response to the tissue in which it is deposited...It is also possible that the radiation risk for the altered wounded tissues may be different (greater?) from that of normal tissue because the inflammatory and repair processes generally result in cell populations that are rapidly proliferating.' R.A. Guilmette and P.W. Durbin, 'Scientific basis for the development of biokinetic models for radionuclide-contaminated wounds', 105 (1-4) *Radiation Protection Dosimetry* (2003) p. 213.

[105] These limits are based upon studies of animals and humans exposed to uranium. Occupational limits are higher because workers assume greater risks than members of the public, and members of the

to these limits, the US Health Physics Society estimates that temporary and permanent kidney effects may occur for single-event inhaled soluble DU intakes above 8 mg and 40 mg, respectively.[106]

Table 1. Recommended Limits on Intake

	United States	**Others**
Members of the Public	0.05 mg/15 minutes[107] 0.5 mg/day[108]	0.035 mg/day[109] 4.5 mg/year[110]
Occupational Workers	0.18 mg/15 minutes[111]	0.18 mg/15 minutes[112]

public are assumed to have a range of sensitivities to toxins. N. Harley, et al. (RAND), *A Review of the Scientific Literature as it Pertains to Gulf War Illnesses, Volume 7: Depleted Uranium* (Washington, D.C., RAND, National Defense Research Institute 1999) pp. 9, 10.

[106] OSAGWI, *supra* n. 4, p. 17.

[107] This limit is for inhalation of insoluble uranium based on a short-term exposure limit of 0.15 mg/m³ based on a breathing rate of 9.6 m³ per eight-hour working day. US National Institute for Occupational Safety and Health (NIOSH), *Pocket Guide to Chemical Hazards* (1994); see also UNEP Bosnia, *supra* n. 18, p. 261.

[108] This limit is for inhalation of insoluble uranium based on chronic exposure limit of 0.05 mg/m³ based on a breathing rate of 9.6 m³ per eight-hour working day. NIOSH, *supra* n. 107; see also UNEP Bosnia, *supra* n. 18, p. 261. Another reference states that the limit for inhalation of DU for members of the public equates to breathing a mass of 0.2 mg/day; Fliszar, *supra* n. 10, p. 18.

[109] The International Commission on Radiological Protection and the World Health Organization prescribe slightly different limits on intake by inhalation for members of the public, based partly on differences in limits based on chemical toxicity and radiation dose. To resolve this discrepancy, a recommendation has been made 'that a unified…daily intake of 35 [micrograms] would be acceptable in most cases. This value would satisfy the constraints imposed by radiation dose and chemical toxicity. However, for protracted exposure to highly insoluble uranium compounds, a further three-fold reduction may be considered appropriate.' N. Stradling, et al., 'Anomalies between radiological and chemical limits for uranium after inhalation by workers and the public', 105 *Radiation Protection Dosimetry* (2003) p. 178.

[110] This refers to an inhalation of type S (insoluble) natural uranium and is based on a 1 micron activity mean aerodynamic diameter (AMAD). As noted above (see *supra* n. 7), the majority of DU particles created by an impact are insoluble. The limit for intake by inhalation of type M (moderately soluble) natural uranium is 13 mg/year; for type F (soluble) it is 75 mg/year. Stradling, et al., *supra* n. 109, p. 176. Another reference states: 'The Annual Limit of Intake for uranium-238, for a member of the public, as specified by the International Committee for Radiological Protection, equates to breathing in a mass of approximately 8 mg of Depleted Uranium.' Lord Gilbert, UK Ministry of Defence, letter to The Countess of Mar, 2 March 1998 (in author's files).

[111] For brief exposures, the American Conference of Governmental Industrial Hygienists (ACGIH) set a short-term exposure limit (STEL) of 0.6 mg/m³ over a 15-minute period. At a breathing rate of 9.6 m³ per eight-hour working day, this equates to a recommended short-term limit on inhalation intake of 0.18 mg. US Agency for Toxic Substances and Disease Registry (ATSDR), 'Toxicological Profile for Uranium' (Washington, D.C., US Public Health Service, September 1999) p. 9 (hereinafter, ATSDR Report). 'The STEL (i.e., less than a 15 minute exposure followed by periods of minimal or no exposure) would apply to the shorter-term exposures occurring in the Gulf War (e.g., entering damaged equipment).' OSAGWI, *supra* n. 4, p. 19.

[112] For brief inhalation exposures, the UK Health and Safety Executive (HSE) set a short-term exposure limit (STEL) of 0.6 mg/m³ over a 15-minute period. At a breathing rate of 9.6 m³, this equates to a recommended short-term limit on intake of 0.18 mg, based on chemical toxicity. UK Health and Safety Executive (HSE), *Occupational Health Exposure Limits 2000*, EH40/2000 (Sudbury, Suffolk, HSE Books 2000).

Table 1. Cont.

	United States	**Others**
Occupational Workers	2 mg/day [113] 10 mg/week [114] 480 mg/year [115]	2 mg/day[116] 130 mg/year[117]

Table compiled by Dan Fahey

The recommended limits on intake provide a basis from which to assess the significance of theoretical exposure estimates in a range of battlefield scenarios (Table 2). The Royal Society has generated a series of estimates intended to be generic for soldiers and civilians in conflicts where DU weapons are used.[118] In 1999 the US Army Center for Health Promotion and Preventive Medicine developed a set of exposure estimates that were subsequently criticised as 'incomplete and misleading' by the Presidential Special Oversight Board on Gulf War Veterans' Illnesses.[119] Consequently, the Army undertook a series of live-fire tests of DU rounds, known as the Capstone Project, and released revised estimates in 2004 (those listed below are the Capstone estimates).[120] In 2005 Sandia National Laboratories (US) published a study that included exposure estimates. The Royal Society, US Army and Sandia estimates are similar in some cases; in others they vary by orders of magnitude.

---

[113] Based on an 8-hour workday, 40-hour workweek maximum air concentration limit of 0.2 mg/m³, with an average breathing rate of 9.6 m³ per 8 hour working day. ATSDR, *supra* n. 111, pp. 322, 329. The 2 mg figure applies for both soluble (type F) and insoluble (type S) compounds; Stradling, et al., *supra* n. 109, p. 177; Dr N. Harley, statement to the Presidential Special Oversight Board for Department of Defense Investigations of Gulf War Chemical and Biological Incidents (Washington, D.C., 19 July 1999); OSAGWI, *supra* n. 4, p. 18.

[114] US Code of Federal Regulations, 10 CFR 20, 'Standards for Protection Against Radiation', Subpart C, 20.1201, 'Occupational Dose Limits for Adults', 1 January 2001.

[115] The annual limit on inhalation intake is based on the volume of air that a worker is assumed to breathe in a year (2,400 m³), and the occupational exposure limit of 0.2 mg/m³. ATSDR, *supra* n. 111, pp. 321, 329; US Code of Federal Regulations (CFR), 29 CFR 1926.55, Appendix A, 'Threshold Limit Values of Airborne Contaminants for Construction', Uranium, 1 July 2000.

[116] This is the occupational exposure limit for soluble natural uranium compounds (0.2 mg/m³), at a breathing rate of 9.6 m³ per 8-hour working day, based on chemical toxicity. Ibid.

[117] This refers to an inhalation of type S (insoluble) natural uranium based on a 5 micron activity mean aerodynamic diameter.

[118] Royal Society Part I, *supra* n. 38, p. 5.

[119] Presidential Special Oversight Board for Department of Defense Investigations of Gulf War Chemical and Biological Incidents, 'Special Oversight Board Analysis (Ver. 2) of OSAGWI's DU Report', (Washington, D.C., 19 February 1999) (in author's files).

[120] M.A. Parkhurst, et al., 'Depleted Uranium Aerosol Doses and Risks: Summary of U.S. Assessments', PNWD-3476 Prepared for the U.S. Army by Battelle (Richland, W.A., Battelle, October 2004) <http://www.deploymentlink.osd.mil/du_library/du_capstone/index.pdf>.

Table 2. Estimated Intakes in Exposure Scenarios, Durations of Exposure

	Royal Society 'Central' Estimate[121] (Time)	Royal Society 'Worst-Case' Estimate[122] (Time)	US Army 'Most Likely' Estimate[123] (Time)	US Army 'Upper Bound' Estimate[124] (Time)	Sandia 'Nominal Exposure'[125] (No duration specified)	Sandia 'Maximum Exposure'[126] (No duration specified)
Soldiers in an armoured vehicle penetrated by a DU round	250 mg (1 minute)	5000 mg (1 hour)	10-280 mg (1 minute) 43-710 mg (5 minutes)	91-970 mg (1 hour) 110-1000 mg (2 hours)	250 mg *inhalation* 330 mg *fragments* 15 mg *ingestion*	4000 mg *inhalation* 1800 mg *fragments* 500 mg *ingestion*
Soldiers who enter vehicles to rescue occupants immediately after a DU impact	250 mg (1 minute)	5000 mg (1 hour)	27-200 mg (10 minutes)	No estimate	250 mg *inhalation* 15 mg *ingestion*	4000 mg *inhalation* 500 mg *ingestion*
People who work in and around DU-impacted equipment	1 mg *inhalation* 0.5 mg *ingestion* (1 hour)	200 mg *inhalation* 50 mg *ingestion* (10 hours)	0.45 mg *inhalation* 10.6 mg *ingestion* (1 hour)	14.5 mg *inhalation* 10.6 mg *ingestion* (1 hour)	40 mg *inhalation* 30 mg *ingestion*	600 mg *inhalation* 300 mg *ingestion*
Child at play	No estimate	No estimate	No estimate	No estimate	54 mg *inhalation* 3000 mg *ingestion*	226 mg *inhalation* 9000 mg *ingestion*
People downwind of DU impacts	0.07 mg *inhalation* (passage of plume)	4.9 mg *inhalation* (passage of plume)	0.00006 mg *inhalation* (passage of plume)	0.04 mg *inhalation* (passage of plume)	0.003 mg *inhalation*	0.1 mg *inhalation*

---

[121] The Royal Society's *central estimate* 'is intended to be representative of the average individual within the group (or population) of people exposed in that situation'. Royal Society Part I, *supra* n. 38, pp. 6, 41-43. See also Annexe C of the Royal Society Report, <http://www.royalsoc.ac.uk/policy/du_c.pdf>.

[122] 'We calculated a 'worst case' estimate using values at the upper end of the likely range, but not extreme theoretical possibilities. The aim is that it is unlikely that the value for any individual would exceed the worst case... If even the worst-case assessment for a scenario leads to small exposures, then there is little need to investigate more closely. If, however, the worst-case assessment for a scenario leads to significant exposures, it does not necessarily mean that such high exposures have occurred, or are likely to occur in a future battlefield, but that they might have occurred, or might occur in future conflicts, and further information and assessment are needed.' Royal Society Part 1, *supra* n. 38, pp. 6, 41-43.

[123] Parkhurst, et al., *supra* n. 120, Chapters 3, 4.

[124] Ibid.

[125] A.C. Marshall, *An Analysis of Uranium Dispersal and Health Effects Using a Gulf War Case Study*, Sandia Report SAND2005-4331 (Albuquerque, NM: Sandia National Laboratories, July 2005) pp. 59-60.

[126] Ibid.

Table 2. Cont.

	Royal Society 'Central' Estimate (Time)	Royal Society 'Worst-Case' Estimate (Time)	US Army 'Most Likely' Estimate (Time)	US Army 'Upper Bound' Estimate (Time)	Sandia 'Nominal Exposure' (No duration specified)	Sandia 'Maximum Exposure' (No duration specified)
Inhalation of resuspended DU from soil	0.8 mg[127] (27 days)	80 mg[128] (27 days)	No estimate	No estimate	0.001 mg *inhalation*	0.003 mg *inhalation*

Table compiled by Dan Fahey

These estimates confirm previous assumptions that the greatest exposures take place among people inside armoured vehicles penetrated by DU rounds. The Royal Society's central estimate of 250 mg/minute indicates that soldiers inside vehicles struck by DU weapons may inhale – in one minute – almost 1,400 times the recommended short-term limit for occupational exposures. The Army's 'most likely' estimate suggests that soldiers inside DU-pierced vehicles could inhale nearly 150 percent of the US's annual occupational exposure limit in a span of just five minutes. The Sandia estimate indicates that 'a child playing in or near vehicles destroyed by DU munitions may inhale or ingest significant quantities of DU.'[129]

For other exposure scenarios, the picture emerging from the estimates is more complex. The Army's estimates for people working in or on contaminated equipment for one hour range from 10 to 322 percent of the annual limit for public exposure. Casual, fleeting exposures (such as being downwind of an impact) may not result in a significant exposure, but one-size-fits-all exposure estimates may not accurately reflect the variety of exposures on a battlefield, or have relevance for children or susceptible individuals. The Royal Society, US Army and Sandia estimates do not imply that health effects will or will not result from DU exposures, but they provide a basis for understanding how much DU people may take into their bodies following the use of DU weapons. As such, they provide a point of comparison against the regulatory limits for DU intake, and highlight the need to inform both civilian populations in battlefield areas and members of the armed forces about the use of DU weapons and appropriate measures to prevent or limit exposure.

UNEP has called attention to the risk posed to people picking up corroding penetrators. Based on its study of DU penetrators found in the soil in Bosnia, Serbia and Montenegro, UNEP's Post-Conflict Assessment Unit (name changed to the

---

[127] It is assumed that 'soldiers spend 4 weeks in an area, starting from the time it is contaminated with 1 g m² DU; all the DU is respirable; and the soldiers' activities cause enhanced resuspension of the DU owing to normal heavy vehicle movements, but the soldiers are not undertaking digging, ploughing or clearance operations'. 'The central estimate is based on UK-like conditions....' Royal Society Part I, *supra* n. 38, p. 43, Annexe C.

[128] '[T]he worst case [estimate] is based on arid, dusty conditions.' Royal Society Part I, *supra* n. 38, p. 43, Annexe C.

[129] Marshall, *supra* n. 125, p. 70.

Post-Conflict Branch in 2005) determined that '[g]ram quantities of corroded ura-nium can easily be removed from the penetrators by mechanical contact'.[130] This DU could contaminate a person's hands and be ingested or become airborne and be inhaled.

The distribution and retention of inhaled or ingested DU depends on the quantity, size and solubility of the particles that enter the body. The characteristics of oxides created by an impact may vary considerably, but the aerosol is generally understood to contain a high percentage of respirable size particles (91 to 96 per-cent),[131] most of which are insoluble in lung fluid (52 to 83 percent).[132] If inhaled, respirable size particles may deposit in the deep lung, while larger particles gener-ally clear the lungs and are either swallowed or expunged from the body. The solu-bility of inhaled particles affects the length of time that DU will remain in the body. If inhaled, insoluble particles tend to remain trapped in the lungs for a period of years to decades.[133] Soluble particles transfer from the lung to the bloodstream, where 90 percent of the DU will leave the body through the kidneys in the first three days after exposure.[134] The remaining ten percent deposits in the bones, lymph nodes, brain, testicles, liver, kidneys and other organs, from where it is excreted over a period of months or years.[135]

Ingestion of DU can occur through consumption of contaminated food, soil or water. People working or playing in contaminated soil may also ingest DU if they fail to carefully wash hands, face and other exposed skin areas prior to eating or drinking.[136] In addition, ingestion of contaminated soil could pose a risk to 'dis-advantaged or tribal communities in which practices such as geophagy [consump-tion of soil to supplement diet] are common.'[137] Ingestion of DU is generally less of a concern than inhalation because the body poorly absorbs ingested DU and excretes most of it (approximately 90 percent) over a period of a few days after an acute exposure.[138]

Children playing in contaminated soil may be particularly susceptible to ingestion of DU through hand-to-mouth transfer or inhalation of resuspended dust.[139] Children consume more calories per kilogram of body weight than adults and may

---

[130] UN Bosnia, *supra* n. 18, pp. 32-35; UNEP Serbia and Montenegro, *supra* n. 16, p. 28.

[131] Andrews, *supra* n. 2, p. 44. See also Danesi 2003, *supra* n. 63, pp. 143-154.

[132] US Army Armament Research, Development and Engineering Center (ARDEC), memoran-dum for Commander, U.S. Army Materiel Command, 'Summation of ARDEC Test Data Pertaining to the Oxidation of Depleted Uranium During Battlefield Conditions', 8 March 1991.

[133] ATSDR, *supra* n. 111, pp. 151, 162.

[134] Harley, et al., *supra* n. 105, p. 8.

[135] T.C. Pellmar, et al., 'Distribution of uranium in rats implanted with depleted uranium pellets', 49 *Toxicological Sciences* (1999) p. 34; WHO, *supra* n. 90, p. 64; M. McDiarmid, et al., 'Health effects of depleted uranium on exposed Gulf War veterans', 82 *Environmental Research* (2000) p. 176.

[136] See WHO, *supra* n. 90, p. 128.

[137] Royal Society Part II, *supra* n. 1, p. 109.

[138] WHO, *supra* n. 90, p. 64.

[139] C. Giannardi and D. Dominici, 'Military use of depleted uranium: assessment of prolonged exposure', 64 *Journal of Environmental Radioactivity* (2003) p. 233; WHO, *supra* n. 90, p. 57; Royal Society Part II, *supra* n. 1, p. 27.

have a higher gastrointestinal absorption of metals, making them potentially more at risk of developing health problems from an intake of DU.[140] In addition, deposition of DU in the bones and organs of growing children and infants may have unanticipated consequences. The Royal Society and the World Health Organization have each highlighted the risk to children playing in soil contaminated by DU.[141]

4.1   Test results

There are large gaps in knowledge about the actual extent and levels of DU exposure among soldiers or civilians. Although testing to date has not revealed widespread exposures, it is far from certain that the people tested include everyone actually at risk of exposure. In addition, in some cases testing took place years after the suspected exposure; the World Health Organization notes that such distant testing may yield results that are highly inaccurate and unreliable.[142] The Royal Society and the World Health Organization recommend measuring urinary uranium of individuals exposed to DU as soon as possible after exposure in order to assess the potential for future health effects.[143] US Army guidelines recommend testing within 180 days of a suspected exposure and note that testing beyond 180 days (with the test being used by DoD) is unlikely to identify inhalation exposures;[144] this means that US veterans tested beyond 180 days of exposure may receive false negative test results. Following is a survey of the principal testing results to date.

4.1.1   *1991 Gulf War*

In 1997 and 1999 the US DU Program collected urine from a total of 45 veterans exposed to DU in friendly fire incidents; six to eight years after the war, 26 (58 percent) had detectable levels of DU.[145] Separately, the DU Program tested 396 veterans who self reported a possible DU exposure; 22 tested positive for elevated uranium in their urine, but isotopic analysis showed that only three of these positive results were for DU.[146] Test results from Canadian veterans do not indicate any DU exposure.[147] As of 9 August 2006, the British military's 'Depleted Uranium Over-

---

[140] Royal Society Part II, *supra* n. 1, p. 72.

[141] WHO, *supra* n. 90, pp. vii, 57, 59; Royal Society Report, Part II, *supra* n. 1, pp. 72, 109, 125, 127.

[142] See WHO, *supra* n. 90, p. 112.

[143] WHO, *supra* n. 90, p. 112; Royal Society Part II, *supra* n. 1, p. xi.

[144] US Army Medical Command, 'Medical Management of Army Personnel Exposed to Depleted Uranium (DU)', OTSG/MEDCOM Policy Memo 05-003, 4 March 2005, p. 8 (Enclosure 1).

[145] R.H. Gwiazda, et al., 'Detection of depleted uranium in urine of veterans from the 1991 Gulf War', 86(1) *Health Physics* (2004) p. 12.

[146] M. McDiarmid, et al., 'Biologic monitoring for urinary uranium in Gulf War I veterans', 87(1) *Health Physics* (2004) pp. 51-56.

[147] E.A. Ough, et al., 'An Examination of Uranium Levels in Canadian Forces Personnel who Served in the Gulf War and Kosovo' (Canadian Military College, 17 July 2001) p. 17.

sight Board' had tested 437 veterans of both the 1991 Gulf War and Balkans conflicts;[148] only one veteran's urine contained small amounts of DU.[149]

4.1.2   *Balkans conflicts*

A Canadian study did not find evidence of DU in the urine or hair of veterans who served in the Balkans, or in a bone sample from a deceased veteran.[150] A study of 122 German soldiers deployed to the Balkans found no evidence that any of the soldiers had incorporated DU into their bodies.[151] Other assessments conducted by NATO countries with troops in the Balkans did not find evidence of either widespread DU exposure or DU-related health effects.[152] One study found traces of DU in the urine of civilians from Bosnia-Herzegovina and Kosovo, but at levels less than the amount of natural uranium normally found in the human body.[153] A second study found DU in three people from Serbia.[154] These last two findings suggest that civilians living near DU impact sites may have been exposed to DU dust, while soldiers who deployed into battlefield areas after the cessation of hostilities were largely spared exposure.

4.1.3   *Iraq*

Between 2003 and 2006 the US government tested more than 2,100 American Iraq war veterans for DU exposure and the UK government tested approximately 350 British veterans. There have reportedly been few positive test results but these results obscure problems with reporting, selection processes and testing methods. For example, between April 2004 and February 2005 the number of US troops that DoD reported had been tested inexplicably dipped by 25 percent (from 1,000 to

---

[148] UK Ministry of Defence, 'DUOB Testing Programme: Interim Summary of Results', <http://www.duob.org.uk/interim_summary.htm>; cf., UK Ministry of Defence, Veterans Policy Unit, 'The 1990/1991 Gulf Conflict: Health and Personnel Related Lessons Identified', undated (late 2004).

[149] Brian G. Spratt, 'Health hazards of depleted uranium munitions: estimates of exposures and risks in the Gulf War, the Balkans, and Iraq', forthcoming in *Critical Reviews in Chemistry* (2006).

[150] Ough, et al., *supra* n. 147, p. 17.

[151] P. Roth, E. Werner and H.G. Paretzke, 'A Study of Uranium Excreted in Urine' (GSF – National Research Center for Environment and Health, Institute for Radiation Protection, Neuherberg, Germany, January 2001) p. 32.

[152] See Office of the Special Assistant to the Deputy Secretary of Defense for Gulf War Illnesses, Medical Readiness, and Military Deployments (OSAGWI), 'Information Paper: Depleted Uranium Environmental and Health Surveillance in the Balkans' (Washington, D.C., U.S. Department of Defense 2001) <http://www.deploymentlink.osd.mil/du_balkans/>.

[153] N.D. Priest and M. Thirlwall, 'Early results of studies on the levels of depleted uranium excreted by Balkan residents', 9(4) *Archive of Oncology* (2001) p. 240; Royal Society Part II, *supra* n. 1, p. 123.

[154] S. Milacic, et al., 'Examination of health status of populations from depleted uranium-contaminated regions', 95 *Environmental Research* (2004) p. 5.

766), and then more than doubled (to 1,607),[155] suggesting political manipulation of test results, bureaucratic confusion or both.[156]

The DoD and the MoD have similar processes for selecting servicemembers for DU testing, but once identified, the MoD uses a more sensitive testing method than that employed by the DoD or VA.[157] The DoD selects servicemen and women to receive testing based on the results of the questionnaire 'Post Deployment Health Assessment' (DD 2796).[158] There is an assumption inherent in this self-identification process that servicemembers were fully aware of all the times and places they may have been exposed to DU, but even for those who believe they were exposed, testing has been incomplete. In September 2004 investigators from the US Government Accountability Office (GAO) reported that a review of several units' questionnaires showed that out of 32 veterans who indicated they were 'sometimes' or 'often' exposed to DU during service in Iraq, only three were provided with a DU test.[159] In addition to the questionnaires, military officials have identified three entire units for DU testing.[160]

The MoD also provides DU testing based on self-reported exposures and official identification of servicemembers possibly exposed to DU during and after two friendly fire incidents.[161] The MoD reportedly issued a card to its servicemen and women in Iraq that states: 'You may have been exposed to dust containing DU during your deployment…You are eligible for a urine test to measure uranium.'[162]

---

[155] D. Funk, 'Returning soldiers do not have dangerous DU radiation levels officials say', *The Army Times* (19 April 2004); W. Winkenwerder, Deputy Secretary of Defense for Health Affairs, 'Subject: Operation Iraqi Freedom Depleted Uranium Bioassay Results and Semi-Annual Data Submission', Memorandum for Assistant Secretaries of the US Army, Navy and Air Force, 10 September 2004; E. Embrey, M. Kilpatrick, and Lt. Col. M. Melanson (US Department of Defense), 'Depleted Uranium,' Briefing to the Honorable Bob Filner (D-CA), 3 February 2005 (in author's files); cf., W. Winkenwerder, Deputy Secretary of Defense for Health Affairs, 'Subject: Operation Iraqi Freedom Depleted Uranium Bioassay Results and Semi-Annual Data Submission', Memorandum for Assistant Secretaries of the US Army, Navy and Air Force, 14 February 2005.

[156] See D. Fahey, 'Summary of Government Data on the Testing of Veterans for Depleted Uranium Exposure During Service in Iraq', 10 February 2005, <www.danfahey.com>.

[157] DoD/VA use inductively coupled plasma mass spectrometry (ICP-MS); MoD uses multicollector (MC)-ICP-MS and sector field (SF)-ICP-MS. SF-ICP-MS is similar in quality to ICP-MS, but MC-ICP-MS is superior to both methods and is the primary method used by MoD. See R. Parrish, et al., 'Determination of 238U/235U, 236U/238U and uranium concentration in urine using SF-ICP-MS and MC-ICP-MS: An interlaboratory comparison', *Health Physics* 90(2) (February 2006) pp. 127-138; Spratt, *supra* n. 149.

[158] US Undersecretary of Defense for Personnel and Readiness, 'Post Deployment Health Assessment', DD 2796, April 2003. A positive response to questions 14, 17 or 18 triggers an evaluation and possible bioassay, according to the US Department of Defense Deployment Health Clinical Center (DHCC), 'Depleted Uranium Provider Reference Pocket Cards,' Post Deployment Health Clinical Practice Guideline, Version 1.0 (December 2003) Card 2.

[159] US Government Accountability Office (GAO), 'Preliminary Medical Screening Data: Operation Iraqi Freedom (OIF) Servicemembers Indicating Suspected Exposure to DU on Their Post-Deployment Health Assessment Forms', Briefing for The Honorable Ciro Rodriquez (D-TX) and the Honorable Bob Filner (D-CA), 30 September 2004 (in author's files).

[160] See D. Fahey, 'Summary of Government Data on Testing of Veterans for Depleted Uranium Exposure During Service in Iraq,' 10 February 2005, <www.danfahey.com>.

[161] Spratt, *supra* n. 149.

[162] N. Mackay and A. Wilson, 'MoD 'lied' over depleted uranium,' *The Sunday Herald* (UK) (29 February 2004).

The essence of the difference between the US and UK's test methods is that the DoD/Veterans Administration (VA)'s test must be administered within 180 days of exposure in order to reliably identify the presence of DU in a veteran's urine,[163] while the MoD's tests can identify DU in urine more than a decade after exposure.[164] The DoD/VA's test screens for total uranium levels in urine, but since everyone excretes small amounts of uranium, this method is not capable of detecting small amounts of DU in urine containing normal levels of uranium.[165] The DoD/VA's test is useful for identifying veterans retaining DU fragments from wounds, who typically excrete high levels of DU for years, but after 180-days it may not detect DU in the urine of veterans who had inhalation, ingestion or wound contamination exposures. By contrast, the MoD's test is capable of detecting small amounts of DU in urine containing normal levels of uranium. The 180-day window for the DoD/VA test to be considered effective is particularly troubling given that many servicemen and women serve one-year tours in Iraq.

It is encouraging to note that the DoD has tested more than 2,100 veterans and reportedly found only nine positive results (Table 3), but as the DoD has provided no detailed information about when these veterans were tested (i.e., before or after the 180-day window), it is difficult to evaluate the significance of these results. All of the veterans who tested positive were wounded by DU fragments; no veterans who had only inhalation, ingestion or wound contamination exposures have tested positive. Given the limitations of the DoD testing method, it is possible that some veterans who had inhalation, ingestion or wound contamination exposures but were tested more than 180 days after exposure may have received false negative test results. Although the vast majority of those tested served in Iraq, DoD has also tested 'a few' veterans who served at Karshi Khanabad (K-2) in Uzbekistan,[166] where DU from Soviet weaponry was found in the soil.[167]

---

[163] J.G. Webb, US Army Medical Command, 'Subject: Medical Management of Army Personnel Exposed to Depleted Uranium (DU)', OTSG/MEDCOM Policy Memo 05-003, 4 March 2005; W. Winkenwerder, Deputy Secretary of Defense for Health Affairs, 'Subject: Operation Iraqi Freedom Depleted Uranium Medical Management', 9 April 2004.

[164] Spratt, *supra* n. 149.

[165] Ibid., cf., Parrish, *supra* n. 157.

[166] The testing of veterans who served at K-2 was disclosed by DoD officials during a meeting with the Honorable Bob Filner on 4 February 2005, attended by the author. For more information on K-2, see GlobalSecurity.org's information page, 'Khanabad, Uzbekistan', <http://www.globalsecurity.org/military/facility/khanabad.htm>.

[167] US Deployment Health Clinical Center (DoD), 'Environmental Conditions at Karshi Khanabad (K-2)', 9 September 2002; D. Fahey, 'Unresolved Issues Regarding Depleted Uranium and Veterans of Operation Iraqi Freedom and Operation Enduring Freedom', 24 March 2004, <www.danfahey.com>.

Table 3. Reported Results of Urine Testing of American Iraq War Veterans

Time Period	Number Tested	Elevated Uranium	Confirmed DU Exposure
30 June 2003 – 31 March 2004[168]	766	14	5
1 April 2004 – 30 September 2004[169]	841	97	1
1 October 2004 – 31 March 2005[170]	363	25	1
1 April 2005 – 30 September 2005[171]	152	1	1
1 October 2005 – 31 March 2006[172]	57	0	1
1 April 2006 – 30 September 2006[173]	41	0	2
Totals[174] (*see footnote for explanation*) 30 June 2003 – 30 September 2006	2161	186	9

Table compiled by Dan Fahey

The MoD has tested 341 veterans of the Iraq war for DU exposure, including 14 with suspected Level I exposures (in or on vehicle at time of impact).[175] The MoD has been vague about the test results, saying initially that 'fewer than ten' tested positive for DU exposure,[176] although a 2007 report claimed there were zero posi-

---

[168] W. Winkenwerder, Deputy Secretary of Defense for Health Affairs, 'Subject: Operation Iraqi Freedom Depleted Uranium Bioassay Results and Semi-Annual Data Submission', Memorandum for Assistant Secretaries of the US Army, Navy and Air Force, 10 September 2004.

[169] Ibid., 14 February 2005.

[170] Ibid., 28 July 2005.

[171] Ibid., 27 February 2006.

[172] Ibid., 20 June 2006.

[173] Ibid., 5 February 2007.

[174] The total for number tested and confirmed exposures are not the sums of the columns, but rather the overall totals provided by the DoD in the 5 February 2007 test results report (Winkenwerder, ibid.). The figure for total elevated uranium reflects the DoD's 27 February 2006 test results. In its 27 February 2006 test results, the DoD noted that eight individuals had tested positive for DU; however the 20 June 2006 results stated that seven individuals had tested positive. I have not been able to account for the discrepancy in numbers. The Department of Veterans Affairs is responsible for testing Air Force and Navy/Marine servicemembers for DU exposure; the totals reported by DoD may therefore include results of testing by both DoD and VA. See A.J. Principi (US Department of Veterans Affairs), letter to The Honorable Bob Filner (D-CA), 18 January 2005 (in author's files).

[175] King's College London, 'King's Centre for Military Health Research: A Ten Year Report; What has been achieved by a decade of research into the health of the UK armed forces and UK?' (London, University of London, September 2006); cf., R. Brown, 'The UK Approach to the Assessment of DU Intakes from the Operational Use of DU Munitions', Presentation at the US Force Health Protection Conference, 11 August 2005; UK Ministry of Defence, Veterans Policy Unit, 'The 1990/1991 Gulf Conflict: Health and Personnel Related Lessons Identified', undated (late 2004) p. 20; Spratt, *supra* n. 149.

[176] I. Bruce, 'Fewer than 10 Gulf war troops had uranium poisoning,' *The Herald* (UK) (5 February 2004). See also UK Depleted Uranium Oversight Board, 'Interim Summary of Results', <www.duob.org.uk/interim_summary.htm>. Of those tested, there were '199 soldiers directly involved in fighting, 96 soldiers involved in other duties, 22 medical staff, and 24 people responsible for cleaning up or repairing contaminated vehicles'; see Anonymous, 'UK Army Personnel Involved in Iraqi Invasion Not at Risk from Depleted Uranium', Press Release, 2 July 2007, <http://www.eurekalert.org/pub_releases/2007-07/bsj-uap070207.php>.

tive test results.[177] Those who tested positive were wounded in March 2003 in one of two friendly fire incidents involving DU ammunition.[178]

5.   HEALTH EFFECTS

The health risks of DU derive from both its chemically toxicity and radioactivity.

The chemical toxicity of DU to humans varies depending on the nature of the compound, its solubility and the route of entry into the body. The chemical properties of DU may be detrimental to the biochemical processes in the body, potentially resulting in health effects such as cancer or neurological disorders similar to those caused by lead. Based on research conducted to date, the chemical toxicity properties of DU appear to constitute a greater hazard to human health than its radioactivity,[179] although recent research suggests that DU's radiation may be more harmful than previously suspected.[180] Additionally, the presence of other toxic metals in DU dust may contribute to adverse health effects.[181]

The radiation resulting from the decay of DU is predominantly alpha particles, although the decay chain includes beta and gamma radiation emitters.[182] Alpha particles pose little external risk due to their short range and limited ability to penetrate skin.[183] They can be hazardous inside the body, however, where the alpha radiation passes through cells causing cell damage or death and genetic damage that 'very rarely can lead to a sequence of events that ultimately results in cancer'.[184]

The chemical and radiological properties of DU may act independently or cooperatively to cause tissue damage,[185] leading to the development of disease. Research on human and animal populations exposed to uranium and DU has identified several potential health outcomes, including cancer, immune system damage,

---

[177] See Anonymous, ibid.

[178] According to Spratt, 'The first incident involved severe damage to a Challenger 2 tank loaded with DU munitions, and the second the suspected impact of small calibre DU rounds on two lightly armored vehicles known as Scimitars'. *Supra* n. 149; R. McCarthy, 'Friendly fire kills two UK tank crew,' *The Guardian* (UK) (26 March 2003); A. Gillan, 'I never want to hear that sound again', *The Guardian* (UK) (31 March 2003).

[179] AEPI Report 1995, *supra* n. 18, p. 108; J.L. Domingo, 'Reproductive and developmental toxicity of natural and depleted uranium: a review', 15 *Reproductive Toxicology* (2001) p. 604.

[180] A. Miller, et al., 'Genomic instability in human osteoblast cells after exposure to depleted uranium: delayed lethality and micronuclei formation', 64 *Journal of Environmental Radioactivity* (2003) p. 257.

[181] Mitchel and Sunder, *supra* n. 7, p. 66.

[182] Royal Society Part II, *supra* n. 1, p. 8. As DU decays, it degrades into other elements that release beta and gamma radiation. Beta and gamma radiation have a greater ability to penetrate skin than alpha radiation, but they pose less of a hazard if their source is inside the body than alpha radiation.

[183] Royal Society Part II, *supra* n. 1, p. 84.

[184] Royal Society Part I, *supra* n. 38, p. 3.

[185] US Institute of Medicine (IOM), Gulf War and Health, Vol. 1, 'Depleted Uranium, Pyridostigmine Bromide, Sarin, Vaccines' (Washington, D.C., National Academy Press 2000) p. 94.

nervous system disease, kidney dysfunction, non-malignant respiratory disease and reproductive effects.[186] The applicability of studies on adults for exposed children and the extrapolation of findings from animal studies to human populations are uncertain, thus limiting efforts to assess the possible health effects of exposure to DU.

Most of the evidence on the human health effects of uranium comes from studies of workers in uranium processing mills and other facilities. Some studies have found increased cancer rates, but other studies found no such increases. As noted by the US Institute of Medicine, however, the applicability of these studies to DU is limited:

> 'The literature on uranium miners is largely not relevant to the study of uranium per se because the primary exposure of this population was to radon progeny, which are known lung carcinogens. Although the studies of uranium processing workers are useful for drawing conclusions, the study settings have inherent weaknesses ... First, even studies involving tens of thousands of workers are not large enough to identify small increases in the relative risk of uncommon cancers. Second, few studies had accurate information about individual exposure levels ...Third, in these industrial settings, the populations could have been exposed to other radioisotopes (i.e. radium ore, thorium) and to a number of industrial chemicals that may confound health outcomes.'[187]

Nonetheless, some US government scientists, including the director of the VA study of DU-exposed veterans, point to the lack of evidence firmly linking uranium to health effects in workers as proof that DU is unlikely to cause any health problems among exposed soldiers or civilians.[188]

The current state of knowledge suggests that under certain conditions, soldiers and civilians can incorporate into their bodies sufficient amounts of DU to cause adverse health effects. Unfortunately, few studies have investigated the health of people exposed to DU during the Gulf Wars or in the Balkans. The most significant study of people exposed to DU is the US study of 1991 Gulf War veterans, which last examined just 35 Gulf War and two Iraq war veterans in 2007.[189] A complete understanding of the health effects of the use of DU weapons is unlikely until exposed populations are studied over a period of decades.

It is interesting to note that six months prior to the 1991 Gulf War, a US Army report warned of the potential consequences of using DU weapons:

---

[186] D.E. McClain, 'Project Briefing: Health Effects of Depleted Uranium', US Armed Forces Radiobiology Research Institute (Bethesda, MD, 1999) (in author's files); IOM, *supra* n. 185, Chapter 4.

[187] IOM, *supra* n. 185, p. 159.

[188] M. McDiarmid, 'Depleted uranium and public health', 322 *British Medical Journal* (20 January 2001) p. 123; R. Kathren, 'Health Effects of Depleted Uranium', US Health Physics Society news archive, 14 January 2001.

[189] M.A. McDiarmid, 'Depleted Uranium (DU) Follow-Up Program Update,' presentation to the US Institute of Medicine, 28 June 2007, Washington, D.C. (in author's files); cf., M.A. McDiarmid et al., 'Biological monitoring and surveillance results of Gulf War I veterans exposed to depleted uranium', 79 *International Archives of Occupational and Environmental Health* (2006) pp. 11-21.

'[A]erosol DU exposures to soldiers on the battlefield could be significant with potential radiological and toxicological effects. These health impacts may be impossible to reliably quantify even with additional detailed studies. It is not our intention to overstate this issue given other combat risks, nor to imply that the health of soldiers will definitely be compromised. We are simply highlighting the potential for levels of exposure to military personnel during combat that would be unacceptable during peacetime operations.'[190]

Cancer and kidney damage were listed as two possible health effects from exposure to DU.[191] In addition, the Army report predicted: 'Following combat, the condition of the battlefield and the long-term health risks to natives and combat veterans may become issues in the acceptability of the continued use of DU kinetic energy penetrators for military applications.'[192]

5.1   Cancer

Laboratory studies have clearly demonstrated that DU is carcinogenic, but the link between DU and cancer in humans remains uncertain. Some of the uncertainties are related to the long latency period for development of cancers related to DU and the fact that few exposed humans have been studied. While the use of DU weapons appears unlikely to cause widespread cancers, sufficient evidence exists to support concerns that exposure to DU may lead to an elevated risk of cancer in heavily exposed populations.

5.1.1   Laboratory studies

A US Army-funded study conducted by Lovelace Respiratory Research Institute in Albuquerque, New Mexico found that DU fragments caused soft tissue sarcomas in the muscles of rats.[193] The scientists who conducted the study caution that the findings from these studies cannot be directly extrapolated to humans.[194] Consequently, the military provided funding in August 2001 for an ongoing follow-up study of the relevance to humans of the finding of cancers in DU-exposed rats.[195]

Research conducted by the US Armed Forces Radiobiology Research Institute (AFRRI) found that DU transformed human cells to a pre-cancerous phase;

---

[190] M.E. Danesi, 'Kinetic Energy Penetrator Long Term Strategy Study' (Picatinny Arsenal, N.J., US Army Armament, Munitions, and Chemical Command, 1990) Appendix D, Vol. 1, pp. 4-5.

[191] Ibid., Vol. 1, p. 2-2.

[192] Ibid., Vol. 2, p. 3-4.

[193] F.F. Hahn, R.A. Guilmette and M.D. Hoover, 'Implanted depleted uranium fragments cause soft tissue sarcomas in the muscles of rats,' 110 *Environmental Health Perspectives* (2002) p. 51.

[194] Ibid., p. 59; see also F.F. Hahn, 'Toxic and Radiologic Health Effects of Uranium – Animal Studies', presentation to the US Institute of Medicine, 28 June 2007, Washington, D.C. (in author's files).

[195] Lovelace Respiratory Research Institute, 'What's New at Lovelace!' undated, <http://www.lrri.org/whatsnew.html>.

these cells then produced tumours when they were injected into mice.[196] The transformed cells also induced genetic instability and reduced production of a key tumour-suppressor protein.[197]

Other AFRRI studies found that DU causes DNA damage that might initiate and promote the formation of tumours.[198] The damage to DNA appears to be caused by both alpha radiation and chemical effects,[199] with delayed chromosomal damage observed in cells not directly irradiated by DU (the so-called 'bystander effect').[200] 'Considering that conventional understanding of potential DU health effects assumes that chemical effects are of greatest concern, results demonstrating that both radiation and chemical effects are involved in DU-induced cellular damage could have a significant impact on DU risk assessments.'[201]

A joint study by scientists at AFRRI and the University of Paris determined 'that internal exposure to embedded DU [in mice] significantly increased leukemia induction, [although] the mechanisms are largely unknown.'[202]

5.1.2   Literature reviews

Recent investigations by the World Health Organization and the Royal Society acknowledge the potential for DU to cause cancer in exposed humans if sufficient amounts of DU are incorporated into the body. The World Health Organization notes:

> 'There is a possibility of lung tissue damage leading to a risk of lung cancer if a high enough radiation dose results from insoluble DU compounds remaining in the lungs over a prolonged period (many years).'[203]

The Royal Society estimates a slight risk of bone cancer and leukaemia for people with heavy exposures,[204] and a slight risk of lung cancer for people with moderate[205] to heavy[206] exposures. However, the Royal Society cautions that 'any in-

---

[196] A. Miller, et al., 'Transformation of human osteoblast cells to the tumorigenic phenotype by depleted uranium-uranyl chloride', 106 *Environmental Health Perspectives* (1998) p. 465.

[197] Ibid., p. 470.

[198] A.C. Miller, et al., 'Depleted uranium-catalyzed oxidative DNA damage: absence of significant alpha particle decay', 91 *Journal of Inorganic Biochemistry* (2002) pp. 246-252 at 251; Miller, et al., *supra* n. 180, p. 248.

[199] Miller, et al., *supra* n. 198, pp. 246, 251

[200] Miller, et al., *supra* n. 180, p. 257.

[201] Miller, et al., *supra* n. 198, p. 251.

[202] A.C. Miller, et al., 'Leukemic transformation of hematopoietic cells in mice internally exposed to DU', 279 *Molecular and Cellular Biochemistry* (2005) p. 102.

[203] WHO, *supra* n. 90, p. 144.

[204] Royal Society Part I, *supra* n. 38, p. 12. The worst case estimate for individuals inside vehicles struck by DU could cause an excess of five leukemias or osteosarcomas per hundred thousand exposed.

[205] 'For the central estimate, the risk of lung cancer is 2.5 per hundred thousand... For the worst-case, the risk of lung cancer is 2.4 per hundred.' Royal Society Part I, *supra* n. 38, p. 13.

[206] 'For the central estimate, the excess risk of fatal lung cancer is about 1 per thousand... For the worst-case, the risk of lung cancer is 6.5 per hundred.' Royal Society Part I, *supra* n. 38, p. 12.

creased risk of lung cancer due to exposure to DU will also be very difficult to establish and, even if substantial, may take several decades to become evident'.[207] Both the World Health Organization and the Royal Society state that the risk of cancer development from brief exposures to small amounts of DU is negligible.

In 2000 the US Institute of Medicine stated that there is insufficient evidence to determine whether an association exists between exposures to DU and lymphatic or bone cancer.[208] The Institute of Medicine recommended additional studies of Gulf War veterans and uranium processing workers to clarify levels of exposure to uranium compounds and frequencies of health effects.[209] In 2006 the Institute of Medicine created a new committee to examine scientific literature on DU; the final report is expected to be released in early 2008.

A study by Sandia National Laboratories concluded: 'DU-induced cancer risks for most Gulf War veterans are extremely small', with only a slightly elevated risk for veterans inside vehicles struck by DU penetrators.[210] The report also stated that children playing in or near DU-damaged vehicles incur only a slightly elevated risk for lung and colon cancer, with the risk of leukaemia and other cancers insignificant.[211]

5.1.3   Human studies

Depleted uranium can theoretically cause cancer, but evidence of the development of cancers in humans exposed to DU is lacking. Large numbers of soldiers may have been wounded by DU fragments during combat and many soldiers and civilians may have been exposed to DU dust after the shooting stopped, but the development of cancers among people with known or suspected exposures to DU is unknown. Given that a ten to 30 year lag may exist after a person's exposure to uranium dust and the development of cancer,[212] it is possible that increased rates of cancers may be observed in the future among exposed populations.

Although there has been no credible scientific study of the fate of persons exposed to DU in southern Iraq, the Iraqi government under Saddam Hussein blamed DU for increased rates of cancers and birth defects among its southern population.[213] Anecdotal reports from journalists and human rights workers in Iraq indi-

---

[207] Royal Society Part I, *supra* n. 38, p. 22.

[208] IOM, *supra* n. 185, pp. 159-160.

[209] Ibid., p. 326.

[210] Marshall, *supra* n. 125, p. 116.

[211] Ibid., pp. 117-118.

[212] N.D. Priest, 'Toxicity of depleted uranium', *The Lancet*, 27 January 2001, p. 357; Hong Xia, et al., 'Spatio-Temporal Models with Errors in Covariates: Mapping Ohio Lung Cancer Mortality', 17 *Statistics in Medicine* (1998) p. 2038.

[213] Republic of Iraq, Ministry of Higher Education and Scientific Research, 'Conference on the Effects of the Use of Depleted Uranium Weaponry on Human and Environment [sic] in Iraq', 26-27 March 2002, posted at the web site of the International Depleted Uranium Study Team, <http://www.idust.org/>. Read the Bush administration's statement about the Iraqi claims at 'Depleted Uranium Scare', <http://www.whitehouse.gov/ogc/apparatus/suffering.html>. See also Agence France Presse, 'Iraqi experts say cancer cases have multiplied in south' (22 February 2002); BBC Monitoring

cate that DU exposures may have taken place after the war when civilians entered battlefield areas in search of usable equipment and scrap metal,[214] but this equipment may have also been contaminated with asbestos, PCBs (polychlorinated biphenyls – a class of organic compounds) and other toxins.[215] In addition, many Iraqi soldiers may have been exposed to DU dust on the battlefield, and others may retain DU fragments from wounds. While inhabitants of southern Iraq may have been exposed to DU in soil, air and water, in the last two decades they have also suffered malnutrition and been exposed to a variety of other toxins, including chemical warfare agents used by Iraq in the Iran-Iraq war, which could also explain some of their illnesses.[216] In sum, there is no strong evidence linking DU to cancer in Iraq,[217] but it is important to stress that there has also not been an adequate study of cancer rates and types among populations with actual or likely exposures.

Other investigations have not found widespread cancers. In Bosnia the World Health Organization investigated reports of increases in cancer cases in areas where DU weapons were used, but found there was no reliable information on cancer rates or trends.[218] The US study of Gulf War veterans found one lymphatic cancer and one bone tumour among 50 veterans examined in 1999, but the significance of this finding is unclear.[219] The same study identified chromosomal damage among some veterans with high DU exposures, but the significance of this finding is also unclear.[220] A study of Danish veterans deployed to the Balkans did not find increased occurrences of cancers that could be related to DU.[221] These findings are reassuring, but also perhaps inadequate; until studies and monitoring of exposed veterans and civilian populations takes place, evidence of an association (or lack thereof) between DU and cancer will remain elusive.

5.1.3.1 Hodgkin's lymphoma

The one cancer that has repeatedly shown up in surveys of veterans is Hodgkin's lymphoma (also known as Hodgkin's disease). Hodgkin's lymphoma develops in

International Reports, 'Saddam says western bombs have brought cancer to Iraq' (17 January 2002); C. Kapp, 'WHO sends team to Iraq to investigate effects of depleted uranium', 35 *The Lancet* (1 September 2001) p. 737; H.A. Ammash, 'Toxic pollution, the Gulf War, and sanctions', in A. Arnove, ed., *Iraq Under Siege* (Cambridge, M.A., South End Press 2000).

[214] S. Peterson, 'A rare visit to Iraq's radioactive battlefield', *The Christian Science Monitor* (29 April 1999), <http://www.csmonitor.com/atcsmonitor/specials/uranium/index.html>.

[215] United Nations Environment Programme (UNEP), 'Assessment of Environmental 'Hot Spots' in Iraq', (Geneva, UNEP 2005) pp. 116-120.

[216] United Nations Environment Programme (UNEP), 'Desk Study on the Environment in Iraq', (Geneva, UNEP 2003) pp. 36-38, 52-56; R.F. Mould, 'Depleted uranium and radiation-induced lung cancer and leukemia, commentary', 74 *The British Journal of Radiology* (August 2001) p. 680.

[217] A. Patel, 'No strong link between depleted uranium and cancer', Letter, 333 *British Medical Journal* (2006) p. 971.

[218] UNEP Bosnia, *supra* n.18, p. 239.

[219] See the discussion of the US DU Program at section 6 of this chapter.

[220] K.S. Squibb and M. McDiarmid, 'Depleted uranium exposure and health effects in Gulf War veterans', 361 *Philosophical Transactions of the Royal Society* (2006) pp. 639-648.

[221] H.H. Storm, et al., 'Depleted uranium and cancer in Danish Balkan veterans deployed 1992-2001', 42 *European Journal of Cancer* (2006) pp. 2355-2358.

the lymph nodes, and it is a rare form of cancer (2.58 cases per 100,000 people in more developed countries; 0.94 cases per 100,000 in less developed countries[222]) with no known risk factor.[223] According to the US Institute of Medicine:

> 'The lymphatic system is an important potential target for uranium radiation because inhaled insoluble uranium oxides can remain up to several years in the hilar lymph nodes of the lung. Studying the effect of uranium exposure on lymphatic cancer is more difficult than studying lung cancer because lymphatic cancer is much less common.'[224]

In general, Hodgkin's lymphoma occurs more often among men and in people aged 15-34 and over 55.

In the United States, one out of 50 veterans examined in 1999 by the Department of Veterans Affairs' DU Program had Hodgkin's lymphoma.[225] It is worth noting that although this cancer was discussed during an October 1999 meeting between the doctor in charge of the study and several Pentagon officials, in January 2001 a Pentagon official publicly denied the existence of this or any cancer among US veterans in the DU study,[226] suggesting political manipulation of study findings.

In August 2002 the UK Ministry of Defence released a study showing that deaths due to lymphatic cancers were nearly twice as high among Gulf War veterans compared to a control group.[227] There is no publicly available information about the number of cases of Hodgkin's lymphoma versus the more-common Non-Hodgkin's lymphoma, but of the 3,172 Gulf veterans seen at the UK Gulf Veterans' Medical Assessment Programme as of 31 January 2003, 11 cases of lymphoma (including Hodgkin's and Non-Hodgkin's) had been reported.[228] The Ministry of Defence denies a link between these cancers and DU, but has initiated an additional study to clarify this finding.

---

[222] In 1999 the incidence of Hodgkin's lymphoma among US residents was 2.8 per 100,000 people (3.0 for men, 2.5 for women). For men and women aged 25-29, the incidence was 5.4 per 100,000; for ages 30-34 the incidence was 4.1 per 100,000. L.A.G. Ries, M.P. Eisner, C.L. Kosary, B.F. Hankey, B.A. Miller, L. Clegg, B.K. Edwards, eds., *SEER Cancer Statistics Review, 1973-1999* (National Cancer Institute, Bethesda, MD 2002), <http://seer.cancer.gov/csr/1973_1999>. Incidence rates in other countries with forces that served in the Gulf War or Balkans are similar: Italy – 3.62; the Netherlands – 2.32; United Kingdom – 2.26; Saudi Arabia – 2.69; Kuwait – 4.33; Iraq – 2.10. J. Ferlay, F. Bray, P. Pisani and D.M. Parkin, GLOBOCAN 2000: Cancer Incidence, Mortality and Prevalence Worldwide, Version 1.0, IARC CancerBase No. 5, Lyon, IARCPress, 2001. Limited version available from <http://www-dep.iarc.fr/globocan/globocan.htm>.

[223] US National Cancer Institute, 'Information about detection, symptoms, diagnosis, and treatment of Hodgkin's disease', NIH Publication No. 99-1555, 16 September 2002.

[224] IOM, *supra* n. 185, p. 142.

[225] M. McDiarmid, et al., 'Surveillance of depleted uranium exposed Gulf War veterans: health effects observed in an enlarged 'friendly fire' cohort', 43 *J Occup Environ Med* (2001) p. 998.

[226] See discussion of US DU Program in section 6 of this chapter.

[227] UK Ministry of Defence, 'UK Gulf Veterans' Mortality Figures', August 2002, <http://www.mod.uk/issues/gulfwar/info/gen_reports/mortfigs_jun02.htm>.

[228] Statement of Lewis Moonie, UK Defence Minister, in response to Dr Gibson, House of Commons Hansard Written Answers for 11 February 2003, <http://www.parliament.the-stationery-office.co.uk/pa/cm200203/cmhansrd/cm030211/text/30211w04.htm>.

Among Italian soldiers who served in Bosnia and/or Kosovo, 'there is a disproportionately high number, which is statistically significant, of cases of Hodgkin's Lymphoma.'[229] Although the Italian Defence Ministry could not identify the causes of this increase, it stated: 'The results of sample studies carried out on Italian soldiers on duty in Bosnia and Kosovo have not shown evidence of depleted uranium contamination.'[230] Overall, the Defence Ministry found a smaller-than-expected number of cancer cases among these soldiers.[231]

5.2   Nervous system diseases

The Armed Forces Radiobiology Research Institute (AFRRI) conducted the first studies on DU's neurological effects. Studies of rats implanted with DU pellets found that DU 'penetrated the blood/brain barrier, accumulated in regions throughout the brain, and induced changes in normal neurophysiological parameters in the hippocampus, a region of the brain involved in memory and learning'.[232] 'The neurotoxicity associated with exposure to other heavy metals (e.g., lead and mercury) raises concerns about the neurological and behavioral consequences of uranium accumulation in the brain with chronic exposure to DU fragments.'[233] The AFRRI studies prompted additional investigations of the neurological effects of DU.

Several studies have validated the finding that DU crosses the blood-brain barrier.[234] One study found that the concentration of uranium in the brains of rats significantly increased after one injection of a uranium solution into the bloodstream.[235] Another study on rats exposed to DU in drinking water determined that DU crossed the blood/brain barrier and produced behavioural effects that increased with the duration of exposure.[236] This study also found that male rats appeared to

---

[229] Original in Italian: '*Esiste un eccesso, statisticamente significativo, di casi di Linfoma di Hodgkin*' Istituita dal Ministro Della Difesa, 'Relazione Finale Della Commissione Istituita dal Ministro Della Difesa Sull'Incidenza di Neoplasie Maligne Tra I Militari Impiegati in Bosnia e Kosovo,' 11 June 2002, p. 21.

[230] Original in Italian: *'I risultati dell'indagine a campione svolta sui militari italiani impiegati in Bosnia e Kosovo non hanno evidenziato la presenza di contaminazione da uranio impoverito.'Istituita dal Ministro Della Difesa, 'Relazione Finale Della Commissione Istituita dal Ministro Della Difesa Sull'Incidenza di Neoplasie Maligne Tra I Militari Impiegati in Bosnia e Kosovo.'* Ibid, p. 21.

[231] Original in Italian: '*Per le neoplasie maligne (ematologiche e non), considerate globalmente, emerge un numero di casiinferiore a quello atteso.*' Ibid., p. 21. See also C. Nuccatelli, et al., 'Depleted uranium: possible health effects and experimental issues', 79 *Microchemical Journal* (2004) p. 332.

[232] D.E. McClain, et al., 'Biological effects of embedded depleted uranium (DU): summary of Armed Forces Radiobiology Research Institute research', 274(1-3) *The Science of the Total Environment* (2001) p. 117; T.C. Pellmar, et al., 'Electrophysiological changes in hippocampal slices isolated from rats embedded with depleted uranium fragments', 20(5) *Neurotoxicology* (1999) pp. 785-792.

[233] Pellmar, et al., *supra* n. 135, p. 36.

[234] G.C.T. Jiang and M. Aschner, 'Neurotoxicity of depleted uranium: reasons for increased concern', 110 *Biological Trace Element Research* (2006) pp. 1-17.

[235] V. Lemercier, et al., 'Study of uranium transfer across the blood-brain barrier', 105 *Radiation Protection Dosimetry* (2003) pp. 243-245.

[236] W. Briner and J. Murray, 'Effects of short-term and long-term depleted uranium exposure on open-field behavior and brain lipid oxidation in rats', 27(1) *Neurotoxicity and Teratology* (2005) pp. 135-144.

be more sensitive to the behavioural effects of DU, and concluded that DU may have neurobehavioural effects in humans. Additional studies have determined that DU accumulates regionally in the brain, but the effects of accumulation remain unclear.[237]

Several French studies have examined uranium's central nervous system effects through a variety of exposure pathways. One study on rats fed water contaminated with uranyl nitrate 'demonstrated that chronic uranium ingestion caused subtle and transient central nervous system modifications at a molecular level...after both short-term and long-term exposure in discrete brain areas.'[238] Another study of rats chronically exposed to water contaminated with enriched uranium found a 'marked increase in REM [rapid eye movement] sleep *in vivo* mainly during the light period after [uranium] exposure.'[239] A related study found that both enriched uranium and depleted uranium deposit in the brain following chronic consumption of contaminated water; accumulation correlated with an increase in deep sleep, reduced memory capacities and increased anxiety in rats exposed to enriched – but not depleted – uranium.[240] A study of rats injected one time with DU found that the rats had shorter deep sleep and decreased food intake three days after exposure, but 'no significant effect was observed on the sleep–wake cycle'.[241] Another French study of rats with chronic DU inhalation exposures had two significant findings:

> 'This study demonstrates firstly that, after exposure through repeated inhalations, depleted insoluble uranium can enter the brain and accumulates differently from one brain area to the next. The brain transport mechanism is not clarified, however. Secondly, depleted uranium can produce behavioural changes in animals, shown by an increase in the spontaneous locomotion activity 1 day post exposure and a less-efficient spatial working memory 6 days post exposure.'[242]

The ability of DU to cross the blood-brain barrier and cause neurobehavioural effects in rats has been established, but the effects in humans remain unclear.

The US DU Program has assessed the neurocognitive functioning of a small number of DU-exposed veterans. The results of 1997 testing suggested a statistical relationship between high urinary uranium values and poor performance on certain

---

[237] V.A. Fitzanakis, et al., 'Brain accumulation of depleted uranium in rats following 3- or 6-month treatment with implanted depleted uranium pellets', 111 *Biological Trace Element Research* (2006) pp. 185-197; A.W. Dobson, et al., 'Depleted uranium is not toxic to rat brain endothelial (RBE4) cells', 110 *Biological Trace Element Research* (2006) pp. 61-72; Jiang and Aschner, *supra* n. 234.

[238] C. Bussy, et al., 'Chronic ingestion of uranyl nitrate perturbs acetylcholinesterase activity and monoamine metabolism in male rat brain', 27 *NeuroToxicology* (2006) pp. 245-252.

[239] P. Lestaevel, et al., 'Changes in sleep-wake cycle after chronic exposure to uranium in rats', 27 *Neurotoxicology and Teratology* (2005) p. 835.

[240] P. Houpert, et al., 'Enriched but not depleted uranium affects central nervous system in long-term exposed rat', 26 *Neurotoxicology* (2005) pp. 1015-1020.

[241] P. Lestaevel, et al., 'The brain is a target organ after acute exposure to depleted uranium', 212 *Toxicology* (2005) p. 219.

[242] M. Monleau, et al., 'Bioaccumulation and behavioral effects of depleted uranium in rats exposed to repeated inhalations', 390 *Neuroscience Letters* (2005) pp. 31-36.

tests assessing performance efficiency and accuracy.[243] A small number of veterans tested in 1999 demonstrated decreased cognitive performance.[244] The program tested just 32 veterans in 2003, finding clear evidence of neurological effects only in a few veterans who continue to excrete extremely high amounts of DU in their urine.[245]

The World Health Organization and the Royal Society state that, given the currently available information, it is difficult to draw any firm conclusions about the effects of DU on central nervous system function.[246]

5.3   Immune system damage

An Armed Forces Radiobiology Research Institute study found uranium concentrations in the lymph nodes of DU-implanted rats significantly higher than those observed in tissues collected from control animals. Based on this finding, 'impaired immune function resulting from the chronic accumulation of uranium in the lymphatic system as well as in bones and spleen might be a possible consequence of prolonged exposure to DU fragments.'[247] One *in vitro* study found alteration of immune functions in immune cells exposed to DU.[248]

The US Institute of Medicine states there is inadequate evidence to determine whether an association exists between exposure to uranium and immune system damage.[249] The Royal Society concludes that 'inhalation of DU on the battlefield is very unlikely to result in significant effects on immune function that would increase susceptibility to infection',[250] although it notes that DU in combination with other toxic exposures may affect immune system function.[251]

5.4   Kidney dysfunction

The kidney is the organ that helps the body to remove wastes from the blood so that they can be excreted. The kidney is generally considered to be the target organ for the chemical toxicity of uranium.[252] Animal studies indicate that long-term exposure to uranium causes effects in the kidneys ranging from small lesions to tissue death.[253] Evidence of kidney damage has also been found in workers with chronic

---

[243] McDiarmid, et al., *supra* n. 135, p. 178.

[244] McDiarmid, et al., *supra* n. 225, p. 9.

[245] McDiarmid, et al., *supra* n. 189, pp. 16-17; Squibb and McDiarmid, *supra* n. 220, p. 646.

[246] WHO, *supra* n. 90, p. 145; Royal Society Part II, *supra* n. 1, p. 66.

[247] Pellmar, et al., *supra* n. 135, p. 34.

[248] B. Wan, et al., 'In vitro immune toxicity of depleted uranium: effects on **murine** macrophages, CD4+ T Cells, and Gene Expression Profiles', 114(1) *Environmental Health Perspectives* (2006) pp. 85-91.

[249] IOM, *supra* n. 185, p. 158.

[250] Royal Society Part II, *supra* n. 1, p. 31.

[251] Ibid., p. 67.

[252] Ibid., p. 2; cf., M. Monleau, et al., 'The effect of repeated inhalation on the distribution of uranium in rats', 69 *Journal of Toxicology and Environmental Health, Part A* (2006) pp. 1629-1649.

[253] WHO, *supra* n. 90, p. 144.

exposures to uranium and in members of the public chronically exposed to uranium in drinking water.[254]

Recent research on veterans and rats, however, has prompted a reevaluation of the kidney as the target organ for uranium's effects. Test results of the 1997 examination of 29 US veterans did not find any sign of kidney disease: 'It may be at these relatively lower exposure concentrations, the kidney is not the critical organ and that another, possibly the neurological or reproductive system, is the 'sentinel' organ system, the system first perturbed biochemically.'[255] This finding is supported by US military research showing that DU-implanted rats suffered physiological changes in the brain without exhibiting any signs of kidney toxicity.[256] Other research on rats found that 'DU dust produced kidney damage at lower lung burdens and lower urine uranium levels than [natural uranium] dust, suggesting that other toxic metals in DU dust may contribute to the damage.'[257]

The US Institute of Medicine reported that studies of exposed animals and humans do not provide adequate evidence to link clinically significant renal dysfunction with exposure to low doses of uranium.[258] The Royal Society noted:

> 'The kidney is a resilient organ and even individuals who have received…high intakes of uranium appear to recover kidney function, although some abnormalities may remain detectable for several years. The long-term effects of acute uranium poisoning in humans are not known but clearly could lead to an increased likelihood of kidney failure later in life.'[259]

The World Health Organization recommended further studies 'to clarify our understanding of the extent, reversibility and possible existence of thresholds for kidney damage in people exposed to DU.'[260]

5.5 Respiratory disease

The potential for respiratory problems depends upon the size, solubility and amount of DU dust inhaled. In animal studies, evidence of lung damage also depended on the species studied.[261] In acute exposures, respiratory disease may be limited to lung tissue damage, eventually leading to emphysema or pulmonary fibrosis.[262] Several epidemiological studies have reported respiratory diseases in uranium mine

---

[254] WHO, *supra* n. 90, p. 144; cf., Dan Fahey, meeting notes for 'Expert Meeting on 'Depleted Uranium in Kosovo: Radiation Protection, Public Health and Environmental Aspects', Bad Honnef, Germany, 19-22 June 2001 (in author's files).

[255] McDiarmid, et al., *supra* n. 135, p. 179.

[256] Pellmar, et al., *supra* n. 232, pp. 785-792.

[257] Mitchel and Sunder, *supra* n. 7, p. 66.

[258] IOM, *supra* n. 185, p. 159.

[259] Royal Society Part II, *supra* n. 1, p. 16.

[260] WHO, *supra* n. 90, p. 148.

[261] Ibid., p. 144.

[262] ATSDR Report, *supra* n. 111, p. 65.

and mill workers who were also exposed to significant amounts of dust and other lung irritants, but not in uranium processing workers, who were not exposed to these potential aggravants.[263]

The US Institute of Medicine reported that there is insufficient evidence to determine whether an association exists between exposure to DU and nonmalignant respiratory disease.[264] The Royal Society stated:

> 'Large inhalation intakes of DU particles may result in short-term respiratory effects, as would a large intake of any dust, but long-term respiratory effects are not expected, except perhaps for the most heavily exposed soldiers, under worst-case assumptions, where some fibrosis of the lung may occur from radiation effects, in addition to an increased risk of lung cancer.'[265]

5.6   Reproductive and developmental effects

5.6.1   *Laboratory research and animal studies*

Research suggests that internal exposure to DU could potentially result in reproductive and developmental effects. Preliminary results of an Armed Forces Radiobiology Research Institute study on female rats found:

> 'DU from implanted pellets penetrated the placenta of pregnant rats and accumulated in the fetus, although at very low levels. Preliminary results also indicated that the longer one waits to breed the female after DU pellet implantation, the greater the chance for decreased litter size.'[266]

Another AFRRI study found that DU pellets implanted in male rats break down, enter the bloodstream and accumulate in various organs, including the testicles, suggesting the potential for unanticipated consequences of DU exposure.[267]

Other animal studies suggest that chronic uranium exposure in male rats might affect reproduction, while uranium exposure in female rats induces effects including decreased weight gain and food consumption.[268] Maternal exposure to uranium compounds has resulted in foetal effects, including reduced foetal body weight and length and an increased incidence of developmental abnormalities; postnatal effects in the offspring of male and female rats with chronic exposures to uranium compounds include decreased growth and survival rates compared to controls.[269] Another study on male rats determined 'there was no evidence of a detri-

---

[263] Ibid., p. 66.
[264] IOM, *supra* n. 185, p. 159.
[265] Royal Society Part II, *supra* n. 1, p. x.
[266] McClain, et al., *supra* n. 232, p. 118.
[267] Pellmar, et al., *supra* n. 135, p. 38.
[268] Domingo, *supra* n. 179, p. 606.
[269] Ibid., pp. 607-608.

mental effect of DU implantation on mating success ... sperm concentration, or sperm velocity.'[270]

5.6.2   Human studies

No birth defects were reported among the offspring of 50 DU-exposed US veterans examined in 1999 by the DU Program,[271] although birth defects have been reported in the offspring of DU-exposed veterans who have not been included in this study.[272] Developmental or behavioural effects have apparently not been measured. In 1997 the DU Program found elevated levels of DU in the semen of five out of 17 DU-exposed veterans.[273] Subsequent test results from the DU Program have not reported on the presence or absence of DU in semen, and reproductive health measures were generally reported to be normal.[274] The 2003 test results showed some abnormal hormone levels, but the clinical significance of this finding is unclear.[275]

Aside from the US DU Program, there has been no study of reproductive effects among populations with confirmed DU exposures.[276] A 2006 article reviewing literature about overall reproductive outcomes for Gulf War veterans notes:

> 'For male veterans, we conclude that there is no strong or consistent evidence in the literature to date for an effect of service in the first Gulf War on the risk of major, clearly defined, birth defects or stillbirth in offspring conceived after deployment. Effects on specific rare defects cannot be excluded at this stage, since none of the studies had the statistical power to examine them. For miscarriage and infertility, the picture is less clear. There is some evidence of small increased risks associated with service, but the role of bias cannot be ruled out. For female veterans, there is insufficient information to make robust conclusions, although the weight of evidence to date does not indicate any major problem. None of the studies discussed here have been able to examine risk according to particular exposures and so we cannot exclude the possibility of undetected adverse effects for small groups of veterans with high exposures to specific agents.'[277]

---

[270] D.A. Arfsten, et al., 'Evaluation of the effect of depleted uranium on male reproductive success, sperm concentration, and sperm velocity', *Environmental Research* (2006).

[271] McDiarmid, et al., *supra* n. 225, p. 999.

[272] See, e.g., 'Interview of dismount squad leader of C-22', Lead Sheet #19455, 15 October 1998 in the Office of the Special Assistant to the Deputy Secretary of Defense for Gulf War Illnesses, Depleted Uranium in the Gulf (II) (Washington, D.C., 2000), <http://www.gulflink.osd.mil/du_ii/du_ii_refs/n52en656/8288_010_0000003.htm>.

[273] McDiarmid, et al., *supra* n. 135, p. 172.

[274] McDiarmid, et al., *supra* n. 225, pp. 995, 999; Squibb and McDiarmid, *supra* n. 220.

[275] McDiarmid, et al., *supra* n. 189, p. 19.

[276] Some activists and scientists of dubious credibility have asserted widespread and severe reproductive outcomes as a result of the use of DU weapons. One literature review heavily influenced by 'gray literature' of questionable validity is R. Hindin, D. Brugge, and B. Panikkar, 'Teratogenicity of depleted uranium aerosols: A review from an epidemiological perspective', 4(17) *Environmental Health: A Global Access Science Source* (2005).

[277] P. Doyle, N. Maconochie and M. Ryan, 'Reproductive health of Gulf War veterans', 361 *Philosophical Transactions of the British Royal Society* (2006) pp. 571-584.

Some studies of the children of US Gulf War veterans have found evidence of increases in several birth defects,[278] but these studies have not assessed any particular risk factor, including DU exposure.[279]

The US Institute of Medicine reported that there is inadequate evidence to determine whether an association exists between exposure to uranium and reproductive and developmental problems.[280]

6.   THE UNITED STATES DEPLETED URANIUM PROGRAM

The most comprehensive study of humans exposed to DU is the US Department of Veterans Affairs' Depleted Uranium Program (DU Program). For a variety of reasons, however, this study has severe limitations that limit its significance and the applicability of its findings to other exposed populations. Among these limitations are the dissemination by Pentagon officials of false information about the health of veterans in the DU Program, suggesting politicisation of study findings; inclusion (as of 2007) of only about four percent of the veterans who had moderate to heavy exposures during the 1991 Gulf War; and the exclusion of veterans exposed to DU who have reported health problems, including birth defects in their children. These shortcomings take on political significance because findings from the DU Program have had a strong influence upon US and international debates about the use of DU weapons.

The DU Program was created in response to a 1992 congressional investigation, which bluntly criticised the Army for neglecting to inform soldiers about the battlefield hazards of DU. The investigation also faulted Army leadership for failing to monitor the health of veterans wounded by or otherwise exposed to DU on the battlefield.[281] The Congressional investigators recommended 'the testing of all crew members inside vehicles penetrated by DU munitions', and the Army agreed to begin testing 'all crew members' in July 1993.[282] Yet, for reasons that have never been explained, someone within the US Army Surgeon General's Office decided that not 'all' surviving crewmembers would in fact be tested. Although a review of records developed in the weeks and months after the war makes clear that at least 100 soldiers were 'inside vehicles penetrated by DU munitions',[283] in 1992 the

---

[278] Tricuspid valve insufficiency, aortic valve stenosis, renal agenesis or hypoplasia, and hypospadias.

[279] M.R.G. Araneta, et al., 'Prevalence of birth defects among infants of Gulf War veterans in Arkansas, Arizona, California, Georgia, Hawaii, and Iowa, 1989-1993', 67 *Birth Defects Research* (2003) pp. 246-260; H. Kang, et al., 'Pregnancy Outcomes Among U.S. Gulf War Veterans: A Population-Based Survey of 30,000 Veterans', 11 *Annals of Epidemiology* (2001) pp. 504-511.

[280] IOM, *supra* n. 185, p. 160.

[281] US General Accounting Office (GAO), 'Army Not Adequately Prepared to Deal with Depleted Uranium Contamination', GAO/NSIAD-93-90 (Washington, D.C., January 1993). See also Col. Robert G. Claypool, US Army Medical Corps, letter to Headquarters, US Army Chemical School, 'Subject: Depleted Uranium (DU) Safety Training', 16 August 1993.

[282] GAO, *supra* n. 281, pp. 7, 37.

[283] See, e.g., AMCCOM-SWA, *supra* n. 26.

Surgeon General's Office told Congressional investigators that there were only 35 such veterans, 22 of whom may have been wounded by DU fragments.[284] When the DU Program was created in 1993 at the Baltimore, MD VA Medical Center, just 33 veterans were enrolled in the study.[285]

6.1   'We have seen no cancers'

Perhaps the most troubling aspect of the DU Program is the promotion of incomplete and false information about the health of veterans in the study by Pentagon officials seeking to downplay public concerns about DU. Pentagon officials most notably misrepresented the health of veterans during the 2001 European DU controversy and the 2003 war in Iraq.

In late 2000 and early 2001, news reports throughout Europe speculated that DU shot by US A-10 aircraft might be causing cancer, leukaemia and other health effects among NATO soldiers who served in Bosnia and Kosovo.[286] In an attempt to downplay concerns about DU-induced cancers, Pentagon spokesman Michael Kilpatrick made an unambiguous statement to the NATO press corps: 'We have seen no cancers or leukemia in this group [participants in the DU Program], which has been followed since 1993.'[287] This denial is also contained in a Power Point presentation given by Kilpatrick and Col. Erik Daxon to the ambassadors of the North Atlantic Council.[288] In June 2001, at a DU conference in Germany attended by the author, US Army Colonel Francis O'Donnell echoed Dr Kilpatrick's statement, telling scientists from a dozen European governments and several United Nations agencies that there have been no cancers among the 60 veterans examined by the DU Program.[289]

As mentioned above (section 5.1.3.1), these explicit and public denials are contradicted by the fact that at least one of the 50 veterans examined by the DU Program in 1999 had a lymphatic cancer: Hodgkin's disease.[290] The existence of

---

[284] GAO, *supra* n. 281, p. 5.

[285] F. Hooper, et al., 'Elevated urine uranium excretion by soldiers with retained uranium shrapnel', 77 *Health Physics* (November 1999) p. 513. Of note is that, between 1993 and 1998, the Department of Defense told five consecutive US federal investigations of Gulf War veterans' illnesses that only 35-36 veterans were exposed to DU in friendly fire incidents; in 1998 the Pentagon admitted that approximately 104 veterans were exposed to DU in friendly fire incidents. See Fahey, *supra* n. 55, pp. 4-6.

[286] See, e.g., articles by the BBC, <http://news.bbc.co.uk/1/hi/in_depth/europe/2001/depleted_uranium/default.stm> and Christian Science Monitor, <http://www.csmonitor.com/atcsmonitor/specials/uranium/index.html>.

[287] Dr M. Kilpatrick and Col E. Daxon, US Department of Defense, 10 January 2001, NATO Press Briefing, Brussels, Belgium, <http://www.nato.int/docu/speech/2001/s010110b.htm>. Dr Kilpatrick is Director of the Office of the Special Assistant to the Deputy Secretary of Defense for Gulf War Illnesses, Medical Readiness and Military Deployments (as of 2005).

[288] See Anonymous, 'Medical Surveillance', <http://www.nato.int/du/010110pc/frame.htm>.

[289] Col Frank O'Donnell, 'Expert Meeting on 'Depleted Uranium in Kosovo: Radiation Protection, Public Health and Environmental Aspects', Bad Honnef, Germany, 20 June 2001 (author's notes).

[290] The Office of the Special Assistant to the Deputy Secretary of Defense for Gulf War Illnesses; 'Meeting with Dr M. McDiarmid and her staff on 15 October 1999 to discuss the Baltimore DU Fol-

this cancer was discussed privately during an October 1999 meeting between officials from the Department of Defense and the Department of Veterans Affairs. Among the meeting participants were Dr Kilpatrick and Col. O'Donnell, the very men who, 15 and 20 months later, respectively, told public audiences at the height of the European DU controversy that no cancers had been found.

In 2003, on the eve of the US-led invasion of Iraq, Dr Kilpatrick participated in a press conference at the Pentagon during which he offered a more nuanced explanation of cancers in veterans: 'We've looked at them [veterans in the DU Program] for cancers. There has been no cancer of bone or lungs, where you would expect them, to see that. We have seen no leukemias.'[291] Dr Kilpatrick's failure to mention the existence of Hodgkin's lymphoma is a notable omission not easily overlooked in the context of political controversy over DU weapons.

In May 2003 a US Army doctor in Baghdad falsely claimed that there have been no tumours among veterans in the DU Program.[292] In fact, at least one veteran enrolled in the DU Program had a non-malignant bone tumour in 1999,[293] but this finding is inexplicably missing not only from the 2001 public statements made by Dr Kilpatrick, Col. Daxon and Col. O'Donnell but also from the DU Program's published report about the findings of its 1999 examinations.[294] This is particularly puzzling considering that recent laboratory research confirms that DU may contribute to cellular changes that result in the initiation and promotion of tumours.[295] According to the US Institute of Medicine: 'Like the lymphatic system, bone is an important potential target for the effects of uranium because uranium is distributed to the bone, replaces calcium in bone matrix, and may remain in bone for several years.'[296]

The selective and misleading presentation of information about cancers and tumours is not restricted to Pentagon spokesmen, however. On 20 January 2001 the director of the VA's DU study wrote in the *British Medical Journal* that 'none of these veterans [15 with DU fragments] has leukaemia, bone cancer, or lung cancer.'[297] The article noticeably does not mention the fact that of 50 veterans examined in 1999, one veteran had Hodgkin's lymphoma and one had a bone tumour. When I questioned Dr McDiarmid at the 28 June 2007 meeting of the US Institute of Medicine in Washington, D.C. about why she did not mention the Hodgkin's

low-Up Program and the Extended Follow-Up Program,' undated, <http://www.gulflink.osd.mil/du_ii/du_ii_refs/n52en651/0089_005_0000001.htm>.

[291] US Department of Defense (DoD), 'Briefing on Depleted Uranium', 14 March 2003, <http://www.defenselink.mil/news/Mar2003/t03142003_t314depu.html>.

[292] D. Gray, 'US military says depleted uranium shells in Iraq pose no health dangers', Associated Press, 6 May 2003.

[293] The VA's DU Program told the veteran that the tumour was benign, but the tumour is not formally documented in a publicly released document. The veteran discussed his bone tumor in an interview with Akira Toshiro from the Hiroshima, Japan newspaper *Chugoku Shimbun* (4 April 2000), <http://www.chugoku-np.co.jp/abom/uran/us_e/000404.html>.

[294] See McDiarmid, et al., *supra* n. 225, pp. 991-1000.

[295] Miller, et al., *supra* n. 198, p. 251.

[296] IOM, *supra* n. 185, p. 143.

[297] McDiarmid, *supra* n. 188, p. 123.

lymphoma and bone tumour in this 2001 article, she stated that she had a word limit and could not report all study findings, a claim that rings hollow in the context of the political controversy surrounding the DU issue at that time. Similarly, a 2006 article summarising the findings of the DU Program for the entire period 1993-2005 is notable for its failure to mention either the Hodgkin's lymphoma or the bone tumour,[298] another exclusion of pertinent information about veterans' health that is difficult to understand or excuse.

The public denials of the existence of cancer and a tumour among the few veterans in the US study raises the possibility that other cancers or health effects have been observed, but not publicly reported. The presentation of incomplete and false information about veterans' health also suggests that the DU Program functions more as a political tool than a scientific study.

6.2   Small study size

Perhaps the most significant shortcoming of the US study is its small size, which limits the significance and applicability of its findings. When the DU Program was created in 1993, study designers noted: 'The small size of the [enrolled] population [33 veterans]…[makes it] highly unlikely that definitive conclusions concerning cancer induction will be obtained from the study.'[299] The study size is indeed small, but this has not prevented Pentagon spokesmen from claiming that the absence of cancer among this group (itself a lie) suggests that other veterans are unlikely to develop cancer as a result of DU exposures.[300]

The US Department of Defense conservatively estimates that approximately 866 to 932 soldiers had moderate to heavy DU exposures (above regulatory limits) during and after the Gulf War, including 104 soldiers who were inside vehicles at the time they were struck by DU penetrators.[301] In addition, the Department of Defense estimates that *thousands* of other veterans may have inhaled or ingested DU during entry into contaminated vehicles.[302] Although large numbers of troops were potentially exposed to DU, since 1993 the US study of exposed veterans has assessed the health status of a total of 70 veterans.[303] The DU Program often claims

---

[298] Squibb and McDiarmid, *supra* n. 220.

[299] US Department of Veterans Affairs, Baltimore VAMC, Department of Veterans Affairs Program for the Follow-up and Monitoring of Gulf War Veterans with Imbedded Fragments of Depleted Uranium, Draft (23 September 1993) p. 11, <http://www.gwu.edu/~nsarchiv/radiation/dir/mstreet/commeet/meet3/brief3.grf/tab_h/br3h1a.txt>.

[300] DoD, *supra* n. 291; Kilpatrick and Daxon, *supra* n. 287.

[301] OSAGWI, *supra* n. 4, p. 7.

[302] The Office of the Special Assistant to the Deputy Secretary of Defense for Gulf War Illnesses, Annual Report, November 1996-November 1997 (Washington, D.C., January 1998) p. 30. See also D. Fahey, 'The story of depleted uranium', in T. Ensign, ed., *America's Military Today* (New York, New Press 2004).

[303] During the period prior to the 2003 Iraq War, Pentagon spokesman Dr Michael Kilpatrick claimed that 90 veterans have been examined and DU Program director Dr Melissa McDiarmid claimed that 70 veterans had been examined: S.H. Mather, MD, MPH, Chief Public Health and Environmental Hazards Officer, US Department of Veterans Affairs, letter to Mr Dan Fahey, 30 April 2003; DoD, *supra* n. 291.

it has 70 veterans enrolled in the program to make the program seem larger than it is; in fact, many of the veterans counted by the DU Program have not been examined in at least seven years. Even in 2001, when European governments assessed the health of tens of thousands of their soldiers who served in the Balkans,[304] the DU Program assessed the health of just 39 US Gulf War veterans.[305]

As long as the DU Program surveys the health of only a few dozen veterans, the presence or absence of DU-induced health problems among the hundreds or thousands of veterans with moderate to heavy exposures remains unknown. Despite the small study size and the 1993 statement about the limitations of study findings in such a small group, Dr McDiarmid has claimed that 'findings in the chronically exposed cohort [39 veterans examined in 2001] offer guidance for predicting future health effects in other potentially exposed populations…'[306]

6.3   Exclusion of veterans reporting health problems

A third shortcoming of the DU Program, linked to the small study size, is its exclusion of DU-exposed veterans who have reported health problems, including kidney problems and birth defects.[307] One excluded veteran, who was in a vehicle penetrated by a DU round, told Pentagon investigators that he had fathered two children since the war who have serious medical problems.[308] Another excluded veteran, who in July 1991 cleaned up the DU-contaminated compound at Doha, Kuwait, fathered a son after the war with a limb reduction defect.[309] A Marine who fought in Kuwait developed serious kidney problems immediately after the war, as did a soldier who worked for several weeks inside 27 US vehicles contaminated by DU without any respiratory or other protection,[310] but the DU Program has also excluded these veterans.

In light of the controversial nature of DU weapons, the dishonesty of Pentagon spokesmen and subsequent silence of officials in charge of the DU Program

---

[304] See OSAGWI, *supra* n. 152, pp. 10-13.

[305] J. Stolte, Depleted Uranium Program, e-mail to Dan Fahey, 17 August 2001 (in author's files).

[306] M.A. McDiarmid, et al., 'Health effects of depleted uranium on exposed Gulf War veterans: A 10-year follow-up', 67 *Journal of Occupational and Environmental Medicine* (2004) pp. 277-296.

[307] See D. Fahey, 'Case Narrative: Depleted Uranium (DU) Exposures', Swords to Plowshares, National Gulf War Resource Center, Military Toxics Project (20 September 1998) pp. 97-163.

[308] 'Interview of dismount squad leader of C-22', Lead Sheet #19455, 15 October 1998, in the Office of the Special Assistant to the Deputy Secretary of Defense for Gulf War Illnesses, Depleted Uranium in the Gulf (II) (Washington, D.C., 2000), <http://www.gulflink.osd.mil/du_ii/du_ii_refs/n52en656/8288_010_0000003.htm>. This document notes: 'One [child] born in 92 or 93 has unexplained nosebleeds for which doctors can offer no explanation, and the other, born in 1996, was born legally blind and also has an unexplained rash that won't go away (neither condition has been diagnosed by doctors).' McDiarmid notes: 'The 50 Gulf War veterans [examined in 1999] fathered 35 children since returning from the Gulf War, all without birth defects.' *Supra* n. 225, p. 999) The significance of this finding is diminished by the exclusion from the study of hundreds of other veterans with moderate to heavy DU exposures, as discussed above.

[309] Fahey, *supra* n. 307, p. 140.

[310] Ibid., pp. 103, 126, 131.

may be seen as part of an intentional effort to downplay public concerns about the health and environmental effects of DU ammunition. Their acts also have other purposes and effects. Decision-making about expanding the size of the study group, or funding new DU cancer research, may be based on flawed risk assessments. Similarly, assessments of the hazards of DU on battlefields, testing ranges and manufacturing sites may be based on inaccurate information about the health of soldiers who experienced high exposures to DU. Moreover, US veterans may be denied health care and disability benefits based on flawed government decision-making about the relationship between DU and cancer or other illnesses.

7.   CONCLUSION

The environmental and health effects of DU weapons have been the subject of considerable controversy, but recent investigations by the Royal Society, United Nations Environment Programme, World Health Organization and other institutions and scientists have significantly advanced the state of knowledge on this issue. Overall, it appears that DU is more harmful than the US government would like to admit, but less harmful than Saddam Hussein, Slobodan Milošević, Yasser Arafat and some hyperbolic activists have asserted. The use of DU weapons has the potential to cause adverse environmental and health effects, but there has been inadequate study of contamination in battle zones and diseases among exposed populations. The potential for effects is highly variable depending on the method and amount of release, local geological and weather conditions, and the extent of human activity in contaminated areas.

The environmental effects of DU weapons change over time. In the short-term, the release of DU through impacts or fires may result in localised areas of significant air and soil contamination. Beginning shortly after the release, airborne DU will settle on the ground, infrastructure, equipment, plant and tree surfaces or in water. Intact DU penetrators and fragments slowly corrode and gradually release DU particles into the environment. Human activity in a contaminated area may result in resuspension of DU dust and further dispersal of DU contamination. Over the course of years or decades, the mobilisation of DU in soil could contaminate drinking water supplies, plants (through root uptake) and animals (including humans) that consume contaminated soil, water or plants.

Recent investigations by militaries and scientific and medical organisations have clarified the human exposure scenarios of greatest concern. Soldiers who are inside vehicles struck by DU weapons may inhale large amounts of DU dust, suffer wounds from DU fragments and internalise DU through contamination of wounds. Other soldiers likely to be heavily exposed include rescue and medical troops and those who work on or enter multiple contaminated vehicles. The local population is unlikely to experience short-term exposures as high as soldiers on the battlefield, but over time they may experience significant intakes of DU by ingestion (hand-to-mouth, consumption of contaminated food or water) and inhalation of resuspended dust. Those particularly at risk include children playing in contaminated vehicles or soil, people who chronically consume contaminated drinking water, and children or adults who discover and handle corroding penetrators.

The potential for health effects depends on a number of variables, including the route of exposure, the amount that is taken into the body, whether the exposure is acute or chronic, and the age and health status of the exposed individual. To date, there has been greater study of the effects of DU on laboratory rats than on exposed soldiers and civilian populations. The laboratory findings suggest that DU may cause cancers and a range of other health effects, but the largest study of exposed humans – the US DU Program – has examined only 81 veterans since 1993 (and only 46 Gulf War veterans between 2001-2007), and the health of even these few veterans has not been accurately and completely reported by Pentagon spokesmen or the doctors in charge of the study.

There are many uncertainties that limit a comprehensive evaluation of the environmental and health effects of the use of DU weapons, but the available evidence suggests the need for further research and a precautionary approach. Future research addressing the uncertainties should include studies of soldiers with confirmed or likely exposures to DU (particularly from the United States, United Kingdom, Iraq and former Yugoslavia). There should also be studies of civilians living or working in areas where environmental assessments determine there is air, soil, water, equipment or biological contamination. Political, legal, scientific and medical organisations should continue to urge countries using DU weapons to provide timely and accurate information about DU expenditure so that appropriate assessments and measures to protect public health may be taken as soon as possible after combat or accidental releases of DU.

Part Two
THE INTERNATIONAL LEGAL REGULATION OF DEPLETED
URANIUM WEAPONS

Chapter 3
THE LAW OF ARMS CONTROL
AND DEPLETED URANIUM WEAPONS

Guido den Dekker

1.   INTRODUCTION

As a rule, public international law prohibits the use of armed force by states in their international relations. At the same time, the law acknowledges that there are exceptions to the prohibition in circumstances where the international community as a whole benefits more from (limited) warfare than from a 'peaceful' situation in which certain states would be allowed to pose a serious threat to international peace and security. In the current international system, the United Nations (UN) Security Council may authorise the use of armed force against a state in order to restore international peace and security. Both in customary international law and in treaty law, the right of self-defence, either individually or collectively, constitutes a further exception.[1]

While the use of force may thus be legal under certain circumstances, this does not mean that states enjoy an unabridged freedom of choice in building up their armed forces and preparing for their (defensive) use. The separate questions of the limitations of that freedom of choice and the lawfulness of the use of armed force entail that such limitations apply independent of the answer to the latter question. Therefore, the use of prohibited weapons in self-defence is still contrary to international law. Likewise, the UN Security Council cannot authorise the use of prohibited weapons.[2]

Both the humanitarian laws of war and the law of arms control regulate the freedom of behaviour of states with regard to their national armed forces.[3] In this

---

[1] See the Charter of the United Nations, Arts. 2(4), 39, 42 and 51. For a typical example of the Security Council authorising the use of force, see UN Doc. S/678(1990), 29 November 1990. Already under the 1928 Kellogg-Briand Pact (Treaty for the Renunciation of War as an Instrument of National Policy), the existence of the right of individual and collective self-defence was not disputed, as was made explicit in the 1934 Budapest Articles of Interpretation, see M. Hudson, 'The Budapest Resolutions of 1934 on the Briand-Kellogg Pact of Paris', 20 *AJIL* (1935) p. 92.

[2] Arts. 24 and 25 of the UN Charter, respectively, require the Council to act in accordance with the Purposes and Principles of the Charter and the Member States to carry out the decisions of the Council in accordance with the Charter.

[3] The collective security system of the United Nations (UN) Charter, the humanitarian laws of war, as well as the law of arms control all set rules regarding the employment of armed force by states.

McDonald / Kleffner / Toebes (eds.), Depleted Uranium Weapons and International Law
© 2008, T·M·C·ASSER PRESS, *The Hague, The Netherlands and the Authors*

chapter, the focus will be on the law of arms control. This branch of public international law consists of the rules and principles of international law related to the control of armaments – 'control' being understood as encompassing the whole range of prohibitions of armaments, quantitative and qualitative arms limitations as well as obligations to disarm. For the purposes of this chapter, the main question is whether the law of arms control prohibits the use of depleted uranium (DU) weapons. Related questions that will be addressed more briefly are concerned with prohibitions on transfer or trade of DU weapons. The generic term 'DU weapons' is used here to refer to DU ammunition and armour, ranging, *inter alia*, from penetrating tips of cruise missiles to DU rounds for the US A-10 Warthog planes, Apache helicopters, Harrier airplanes and the cannon shells used by the US M1A1 Abrams tanks.[4]

In the following sections, it will first be argued why the law of arms control deserves separate consideration as a field of law from which prohibitions on the behaviour of states with regard to certain types of armaments arise (section 2). This also requires that the relationship between the law of arms control and the humanitarian laws of war be addressed (section 3). Next, an attempt will be made to answer the central question of this chapter by confronting the use of DU weapons with the rules of substantive law that are found in the sources of the law of arms control (section 4). Finally, the pros and cons of creating a special DU Weapons Convention will be briefly reviewed (section 5). Some conclusions appear at the end of the chapter (section 6).

2. THE LAW OF ARMS CONTROL

Ever since their creation, sovereign states have been responsible for safeguarding their own (military) security.[5] The introduction of the UN collective security system after World War II did not fundamentally alter this starting point. This is mainly due to the fact that Cold War opposition and a general lack of political will to cooperate blocked consensus in the UN Security Council, the organ primarily responsible for the maintenance of international peace and security (see Art. 24 UN Charter). Furthermore, the collective security system has never been fully imple-

---

These interrelated branches of law can be considered as sub-areas of the 'international law of military security', see E. Myjer, 'The settlement of disputes under the Chemical Weapons Convention and the case of the confidentiality commission', in D. Bardonnet, ed., *The Convention on the Prohibition and Elimination of Chemical Weapons: A Breakthrough in Multilateral Disarmament* (Dordrecht, Martinus Nijhoff 1995) pp. 537 at 547.

[4] For a detailed survey, see D. Fahey, Chapter 1 of this book.

[5] The 1648 Treaty of Westphalia is generally taken as the starting point of the creation of the modern international system composed of sovereign states, see, for example, L. Gross, 'The peace of Westphalia, 1648-1948', 42 *AJIL* (1948) p. 20. Long before that, rulers experienced difficulties in safeguarding their security; witness the many pacts and alliances, arms control treaties and peace treaties that were concluded as early as 3100 BC. For a historical overview, see A. Nussbaum, *A Concise History of the Law of Nations* (New York, McMillan 1961); T. Dupuy and G. Hammerman, eds., *A Documentary History of Arms Control and Disarmament* (New York, Bowker 1973).

mented in practice, since the projected stand-by armed 'UN forces' (as envisaged in Art. 43 UN Charter) have never been made available to the UN.[6] In the absence of an effective supranational 'enforcement authority', the close connection that exists between a state's national security and the security of other states has introduced a security dilemma in the international system. This dilemma entails that efforts to increase one state's security are generally achieved at the cost of increasing another's insecurity. Even though states may temporarily achieve an increase in their security through the development of their military capabilities, they may ultimately be negatively affected by offsetting measures undertaken by other states and the resulting deterioration in international security.[7] The most dramatic manifestation of this kind of 'action-reaction cycle' has been the nuclear arms race between the United States and the Soviet Union, which necessitated the starting of serious bilateral, as well as multilateral, arms control negotiations.

During the Cold War period, arms control was given priority in the pursuit of strategic stability between the superpowers and their alliances, the North Atlantic Treaty Organization (NATO) and the Warsaw Pact. In the early years of the Cold War, focus was on the comprehensive programme of general and complete disarmament under strict and effective international control, but soon full attention was drawn to the containment of the spread, production and testing of weapons of mass destruction rather than disarmament. Conventional weapons have received far less attention from the militarily powerful states than weapons of mass destruction. From the perspective of international (strategic) security, conventional weapons are less of a threat than weapons of mass destruction; from an arms trade perspective, conventional weapons are the most lucrative merchandise; from the perspective of international control, conventional weapons are the most difficult category.

Since the end of the Cold War in 1991 arms control has increasingly focused on the prevention of the spread of sophisticated weaponry to 'states of concern'. First and foremost, the non-proliferation of weapons of mass destruction more than ever represents the core interest of the militarily powerful states. For example, in 1992 the UN Security Council (dominated as it is by nuclear weapons-possessing states) declared that the proliferation of all weapons of mass destruction *eo ipso* constitutes a threat to international peace and security, and the members of the Council declared themselves committed to working to prevent the spread of technology related to the research for or production of such weapons and to take appropriate action to that end.[8] In 1999 NATO likewise emphasised the dangers of proliferation of weapons of mass destruction and their means of delivery and

---

[6] See, for example, R. Smith, 'The legality of coercive arms control', 19 *Yale JIL* (1994) pp. 455 at 459-462.

[7] See 'Concepts of security', 14 *United Nations Disarmament Study Series* (New York, UN 1986) p. 46; M. Clark, 'Arms control is not enough', in 40 *Orbis* (1996) pp. 71 at 73.

[8] See 'The Responsibility of the Security Council in the Maintenance of International Peace and Security', held at the level of Heads of State and Government, Note by the President of the Security Council, 31 January 1992, UN Doc. S/23500.

declared that it would enhance its political efforts to reduce those dangers.[9] As yet it remains to be seen to what extent the terrorist attacks on the World Trade Centre and the Pentagon of September 11 2001 and the US decision to withdraw from the Anti-Ballistic Missile (ABM) Treaty may (further) affect US policy on arms control, and other states' policies with it. The foreign and security policies of the Bush administration have represented a serious setback for the development of arms control law. At best it can be hoped that it will turn out to be only temporary.

In international relations, policies directed at maintaining and developing military capabilities and policies directed at arms regulation and disarmament are being pursued by states as a means to promote national and international security. International law does not restrict this free choice, or combination, of policies. As has been established by the International Court of Justice (ICJ), in international law there are no rules other than such rules as may be accepted by the state concerned, by treaty or otherwise, whereby the level of armaments of a sovereign state can be limited.[10] General international law does not require states to enter into arms control agreements, but instead allows them to arm and prepare for the eventuality of war as long as there is no aggressive intent.[11] The global system of collective security as laid down in the UN Charter, which is not only directed at responding to armed aggression but also at preventing large-scale international armed conflict, favours at least some kind of regulation of national armaments in peacetime.[12] The Charter, however, does not disclose a bias against either a high or a low level of armaments present in states; there only is a tendency against the lack of regulation and the arms race, albeit seemingly more out of economic than political concerns.[13] Without going into the possible motives for states to enter into arms control arrangements, over time states have found common ground to establish a growing body of norms relating to the limitation and prohibition of weapons together with their ancillary systems.[14] It can be observed that a quite impressive collection of treaties nowadays constitutes the basis for this branch of international law aiming at the regulation of armaments.

The law of arms control can be described as the part of public international law that deals with the restrictions internationally placed on the freedom of behaviour of states with regard to their national armed force, and with the applicable supervi-

---

[9] See 'The Alliance's Strategic Concept, NATO Doc. NAC-S(99)65, especially para. 40 and also paras. 21, 22, 41, 53(h) and 56.

[10] See ICJ, *Military and Paramilitary Activities in and against Nicaragua* (*Nicaragua* v. *United States of America*), Merits, *ICJ Rep.* (1986) 14, at p. 135, para. 269 (1986).

[11] Cf., the 'High Command Case' (*USA* v. *Von Leeb* et al.), Nuremberg 1948, 11 *NMT* pp. 462 at 488. See also Art. 51 of the UN Charter, which guarantees a right of individual or collective self-defence without setting any limitations as to the types or numbers of arms (to be) employed in self-defence.

[12] Cf., Arts. 11 and 26 UN Charter, dealing with the responsibilities in the field of arms control of the UN General Assembly and the UN Security Council, respectively.

[13] See A. Martin, *Collective Security, A Progress Report* (Paris, UNESCO 1952) pp. 58-60.

[14] A still highly recommended collection of papers on the subject can be found in J. Dahlitz and D. Dicke, *The International Law of Arms Control and Disarmament* (New York, United Nations 1991).

sory mechanisms. The law of arms control not only concerns the substantive norms agreed upon but also the institutional law – viz. the supervisory mechanisms and related rules.[15] The close interrelationship between arms control and security requires states to somehow gain confidence that they will mutually act in compliance with the arms control agreements they have entered into. In practice, the conditions under which states have been prepared to trust each other in the sensitive field of arms control have been embodied in supervisory mechanisms.

The term 'supervisory mechanism' refers to a legal arrangement on the basis of which control of compliance by the states, or by persons on their behalf, with arms control obligations can take place. For the most part, arms control treaties consist of supervisory mechanisms and other provisions of institutional law which do not set prohibitions or other behavioural standards but arrange the implementation of institutional legal structures, including, where applicable, international organisations with supervisory functions.[16] Viewed in a broad evolutionary context, the process of concluding arms control agreements is still inspired by the ultimate aim of achieving 'general and complete disarmament under strict and effective international control'.[17]

Despite the vastness and comprehensiveness of some of the arms control treaties, few substantive prohibitive rules can be identified. Moreover, in the law of arms control there has been a strong tendency to use a problem-solving approach. The essence of this approach is the fixation upon a single issue, which is addressed by means of very precise, clear and succinct drafting that is limited to particular problem-solving and does not seek to transcend it by trying to resolve other, not directly related, issues at the same time.[18] The substantive prohibitions that do exist in arms control treaties are specific and express, although not necessarily absolute.[19] Due to this nature of the law of arms control, the possibility of applying

---

[15] Cf., E. Myjer, 'The law of arms control, military security and the issues: an introduction', in E. Myjer, ed., *Issues of Arms Control Law and the Chemical Weapons Convention, Obligations inter se and Supervisory Mechanisms* (Dordrecht, Martinus Nijhoff 2001) p. 8.

[16] E.g., the Organisation for the Prohibition of Chemical Weapons (OPCW) has been established as part of the supervisory mechanism of the Chemical Weapons Convention (CWC, entry into force 1997), and the Comprehensive Test Ban Treaty Organisation (CTBTO) is to be established pursuant to the Comprehensive nuclear Test Ban Treaty (CTBT, 1996, not yet in force). Much earlier, in 1957, the International Atomic Energy Agency (IAEA) was established and became involved in the supervision of compliance with the Treaty on the Non-Proliferation of Nuclear Weapons (NPT, entry into force 1970) and the Nuclear Weapon Free Zone Treaties of Latin America (Treaty of Tlatelolco, 1967), the South Pacific (Treaty of Rarotonga, 1985), Africa (Treaty of Pelindaba, 1996) and Southeast Asia (Treaty of Bangkok, 1995).

[17] See, e.g., the preambles to the Limited Test Ban Treaty (LTBT, 1963), the NPT (1968), the Treaty on Conventional armed Forces in Europe (CFE-Treaty, 1990), the CWC (1993) and the CTBT (1996). See further G. den Dekker, *The Law of Arms Control: International Supervision and Enforcement* (The Hague, Martinus Nijhoff 2001) pp. 40-42.

[18] See N. Singh and E. McWhinney, *Nuclear Weapons and Contemporary International Law*, 2nd edn. (Dordrecht, Nijhoff 1989) p. 219.

[19] E.g., even the comprehensive prohibitions on chemical weapons in the CWC leave room for the application of toxic chemicals and their precursors for 'purposes not prohibited' under the Convention, as long as the types and quantities are consistent with such purposes (Art. II(1) of the CWC).

existing substantive prohibitions by analogy to new weapons is practically precluded.

3.   THE LAW OF ARMS CONTROL AND THE LAWS OF WAR

The law of arms control, like other branches of international law, occupies a place
of its own in the system of international law and politics and bears certain special
characteristics, which appear first and foremost in the sources of the law. Arms
control law is in essence treaty law; other sources play only a subsidiary role.[20] The
close connection between arms control and national as well as international security – in that arms control law contains restrictions and prohibitions on the quantity
and/or quality of the state's means of defence – has necessitated treaty-made law.
The importance of concluding treaties, as well as their modification by written
amendment procedures, is generally accepted in arms control law. An additional
incentive for the creation of treaties derives from the relative clarity, reliability and
certainty of treaty language as compared to rules emanating from the other formal
sources of international law – customary international law and general principles of
law. The relatively specific rules of treaty law provide more clarity and stability as
to their application than the vaguer rules and principles of the latter two sources.

It should be noted that arms control treaties are almost without exception
law-making treaties and do not represent a codification of pre-existing customary
international law. Arms control treaties are highly diverse, ranging from bilateral
treaties (almost all concluded between the United States and Russia) to regional
treaties (mostly nuclear weapon free zone treaties) and universal treaties, and seek
to control different types of armaments, including nuclear, chemical and biological
weapons as well as certain conventional weapons, or secure the demilitarisation of
areas such as the ocean floor and outer space.[21]

There obviously is a relationship between the law of arms control and the
humanitarian laws of war, if only because both branches of law are concerned with
restrictions on the *use* of weapons. The abhorrence of certain weapons, in particular
weapons of mass destruction and other means of warfare that are liable to cause
great losses, has given an impetus to demands for arms control measures (also) in
peacetime. One of the basic principles of the laws of war, which prescribes that
parties to a conflict and members of their armed forces do not have an unlimited
choice of methods and means of warfare, is upheld by the law of arms control.[22]

---

[20] See G. Lysén, *The International Regulation of Armaments: The Law of Disarmament* (Uppsala,
Iustus Vörlag 1990) p. 221. See also Fifth Report of the Committee on Arms Control and Disarmament
Law – National and International Verification Measures in *International Law Association, Report of
the 70th Conference* (London 2000) p. 3.

[21] For an overview of the contents of arms control treaties see J. Goldblat, *Arms Control, A Guide
to Negotiations and Agreements* (London, Sage Publications 1994); H. Blix, 'International law relating
to disarmament and arms control', in F. Kalshoven, ed., *The Centennial of the First International
Peace Conference, Reports & Conclusions* (The Hague, Kluwer Law International 2000) pp. 61-96.

[22] See for this principle, Art. 22 of the Regulations Annexed to the 1907 Fourth Hague Convention
on the Laws of Land Warfare and Art. 35(1) of 1977 AP I. The origins of this principle can be traced

This principle, in combination with the Martens Clause, may serve as a general legal basis to negotiate specific prohibitions to employ (current and future) means and methods of warfare of a nature so as to cause superfluous injury and unnecessary suffering.[23] Thus, it can be said that the principle of limited warfare and the Martens Clause can be used as a general legal basis to start negotiations which may result in prohibitions of certain behaviour with regard to specific weapons whose use may be contrary to the rules of humanitarian law. However, neither the principle nor the clause can be regarded as laying down a separate legal principle for judging the legality of weapons under existing international law. These principles of the laws of war may restrict behaviour using certain types of weapons in a given situation, but do not confer illegality *in abstracto*, outside a specific context.[24]

As previously mentioned, the nature of the law of arms control is such that prohibitions have to be specific and express. Even the 1981 Certain Conventional Weapons (CCW) Convention, which is by far the most 'humanitarian'-inspired arms control treaty in force today, clearly testifies to this.[25] Certain specifically defined conventional weapons have been prohibited in the protocols to the CCW Convention, and as long as a particular type of weapon does not fit the definitions in those protocols, there is no prohibition of that type of weapon. Another illustration is that the major military powers threatened to withhold their signatures of the 1977 Additional Protocols (APs) to the Geneva Conventions if specific weapons were so much as mentioned in those Protocols.[26] Apart from express prohibitions in treaties, states have not been known to lightly decide unilaterally to discard certain types of weapons, once introduced in their arsenals, because they are considered to cause unnecessary suffering.[27]

Clear distinctions between both branches of law are apparent, too. First and foremost, contrary to the humanitarian laws of war, considerations of prevent-

---

back even further, to the 1874 Brussels Conference. See C. Greenwood, 'The law of weaponry at the start of the new millennium', in M. Schmitt and L. Green, eds., *The Law of Armed Conflict: Into the Next Millennium* (Newport, Rhode Island, Naval War College 1998) pp. 185-231.

[23] Cf., D. Vagts, 'The Hague Conventions and arms control', in 94 *AJIL* (2000) pp. 31 at 36, who refers to both principle and clause as only a *moral* basis for restricting the usage of new weapons. The Martens Clause, which was introduced in the 1899 Hague Convention and repeated in Art. 1(2) of 1977 AP I, declares that in cases not covered by the Protocol or other international agreements, civilians and combatants remain under the protection and authority of humanity and the dictates of public conscience.

[24] Even with regard to nuclear weapons, pre-eminently fit to cause terrible suffering with indiscriminate effects, the ICJ in its 1996 Advisory Opinion was unable to establish an intrinsic or absolute prohibition of use. See ICJ Advisory Opinion on the *Legality of the Threat or Use of Nuclear Weapons*, 6 July 1996, 35 *ILM* (1996) p. 809, para. 105 (hereinafter, *Legality of the Threat or Use of Nuclear Weapons*).

[25] See Convention on Prohibitions or Restrictions on the Use of Certain Conventional Weapons which may be Deemed to be Excessively Injurious or to have Indiscriminate Effects, 19 *ILM* (1981) p. 1523.

[26] See Stockholm International Peace Research Institute, *World Armaments and Disarmament, SIPRI Yearbook 1979* (London, Taylor & Francis Ltd. 1979) pp. 453 at 455.

[27] See F. Kalshoven and L. Zegveld, *Constraints on the Waging of War*, 3rd edn. (Geneva, ICRC 2001) pp. 41-42, 92.

ing personal suffering (weighed against military necessity) do not play the main role in the law of arms control. Arms control treaties almost without exception prohibit weapons technologies rather than the (harmful) results of weapons' applications. The principal thrust of comprehensive arms control treaties such as the Chemical Weapons Convention (CWC) and the nuclear Non-Proliferation Treaty (NPT) is not so much to establish prohibitions of use, but rather to create a regime consisting of peacetime restrictions on production and possession of and, if possible, obligations to destroy a particular class of weapons. Naturally, this feature in itself also serves to reduce the consequences of armed conflict, but it is subordinate to the purpose of regulating in peacetime the military power relations between the States Parties to the treaties. The law of arms control is furthermore not concerned with the health effects of the use of weapons between combatants in general. Instead, the security of states and of the international community at large are central to it. Arms control law is preoccupied with preserving strategic stability as well as maintaining a global and, where applicable, a regional or sub-regional balance of power, with the overall objective of preventing the outbreak of large-scale armed conflict. The question whether the use of a particular kind of weapon would be reconcilable or not with the law of arms control lies primarily in the potential or actual impact of that weapon on the maintenance of international security and stability. Since weapons of mass destruction (nuclear, chemical and biological) can have such a (potentially) destabilising impact, the law of arms control has been primarily occupied with their control. In short, the principal rationale behind the law of arms control as it has developed over the last century can be described as an attempt to prevent the threat or large-scale use of force between states by prohibiting the possession, production, stockpiling and also the use of particular weapons and by having existing stockpiles destroyed, as far as possible under international supervision.

History shows that sometimes the eventuality of the use of a type of weapon has first to be outlawed before further prohibitions can be achieved. For example, as early as the late 1920s, the Geneva Protocol outlawed the use in war of asphyxiating, poisonous, or other gases, and bacteriological methods of warfare.[28] This treaty was a prelude to the 1972 Biological Weapons Convention (BWC) and the 1993 CWC, both of which encompass – apart from (repeated) prohibitions of use – comprehensive prohibitions on acquisition and possession of biological and chemical weapons, respectively, and obligations to destroy existing stockpiles of those weapons.[29] Nuclear weapons, in contrast, have been treated differently, in that qualitative and quantitative restrictions on their development have been agreed without

---

[28] See the Protocol for the Prohibition of the Use in War of Asphyxiating, Poisonous or other Gases, and of Bacteriological Methods of Warfare (1925 Geneva Protocol), 94 *LNTS* (1929) p. 65. See the even earlier Declaration (IV, 2) on the prohibition of use of projectiles 'the sole objective of which is the diffusion of asphyxiating or deleterious gases', signed at The Hague (29 July 1899).

[29] Although Art. I of the BWC does not include a prohibition of the use of biological weapons, it has been made out that such is the case, see Final Document of the Fourth Review Conference of the States Parties to the BWC, Part II, BWC/CONF.IV/9, p. 15 (1996).

the illegality of their eventual use being established.[30] Furthermore, the 1968 NPT introduced a fundamental distinction between Nuclear Weapon States (NWS) and Non-Nuclear Weapon States (NNWS), the latter being prohibited from acquiring nuclear weapons and being obliged to have their non-nuclear status verified by the International Atomic Energy Agency (IAEA) by way of full-scope safeguards agreements.[31] As such, this treaty introduced a fundamental difference between the few states that are allowed to possess nuclear weapons for their security and the (very many) states that are not.

Arms control treaties relating to conventional weapons, such as anti-personnel mines and small arms, fit less well into this strategic approach. This is clearly reflected in the scope and comprehensiveness of the treaties meant to control conventional weapons, in particular those that contain qualitative restrictions.[32] Treaties on conventional arms control do not contain strong supervisory mechanisms and carry relatively few institutional obligations for the States Parties. Obligations arise mainly in the regular exchange of information; transparency and confidence-building are the leading objectives in conventional arms control law. Furthermore, no special international organisations have been established to carry out supervisory tasks in the field of conventional arms control law. In contrast, the detection, deterrence and correction of non-compliance with the law, supervised if possible by specialised international organisations, are leading objectives of the control of weapons of mass destruction.

With respect to prohibited behaviour, the law of arms control is not necessarily linked to the humanitarian laws of war. The use of a weapon that is not prohibited by arms control law may still be contrary to the humanitarian laws of war in a given situation. If a legitimate weapon causes superfluous injury or unnecessary suffering in a particular situation because other weapons could be employed with identical military utility but causing less suffering, the use of that weapon may be prohibited under the principle forbidding such injury and unnecessary suffering. Similarly, the use of a legitimate weapon in a particular situation without any distinction between military targets and civilians or civilian objects may constitute a violation of the principle of distinction. Conversely, the law of arms control sometimes carries absolute prohibitions on certain types of weapons for the States Par-

---

[30] True, the ICJ in its 1996 Advisory Opinion came a long way but eventually could not give a definitive pronouncement on the legality of the use of nuclear weapons 'in the event of an extreme circumstance of self-defence, in which the very survival of a state would be at stake'. *Legality of the Threat or Use of Nuclear Weapons*, *supra* n. 24, para. 105(F). Restrictions on nuclear weapons have primarily focussed on testing restraints.

[31] See the Treaty on the Non-Proliferation of Nuclear Weapons (NPT), 7 *ILM* (1968) p. 809, Arts. I-III.

[32] For example, the 1990 CFE Treaty and its later adaptations, which contain quantitative restrictions on five categories of 'heavy' conventional arms, bear stronger resemblance to the treaties concerned with weapons of mass destruction than the 1980 CCW Convention and its protocols, which mainly illustrate that arms control law also has a strong humanitarian side.

ties to the treaties concerned, whereas situations can be imagined in which the use of those weapons would not violate the laws of war.[33]

4.   IS THERE A PROHIBITION OF DEPLETED URANIUM WEAPONS IN THE LAW OF ARMS CONTROL?

4.1   Introduction: starting point of analysis

As the ICJ has observed, state practice shows that the illegality of the use of certain weapons as such does not result from an absence of authorisation but, on the contrary, is formulated in terms of prohibition.[34] This practice is reflected in the substantive rules of arms control law. The consequence is that the use of a certain type of weapons cannot be deemed prohibited in each and every situation unless such prohibition arises from an arms control treaty. Failing such prohibition, as soon as a single situation can be imagined in which the use of a certain type of weapon does not necessarily violate the rules and principles of international law, it cannot be established that this weapon is prohibited 'intrinsically' or *per se*. It has already been mentioned that no such 'intrinsic' prohibition can result from the application *in abstracto*, outside a given context, of the general principles of the laws of war.

It follows that in order to answer the question whether employment of DU weapons is contrary to the law of arms control, it is necessary to try and find specific prohibitions of DU weapons in arms control treaties. An important question, which presents itself in that respect, is whether DU weapons can be qualified as nuclear, chemical or biological weapons within the scope of treaties that contain prohibitions regarding those weapons of mass destruction. After that, the role of secondary sources of law will be examined. A fundamental remaining question is under what conditions the use of DU weapons violates international humanitarian law in times of war. This question has to be resolved by reference to (careful analysis of) the laws of war, and will not be further dealt with here but is taken up in the following chapters.[35]

4.2   Arms control treaty law and the use of depleted uranium weapons

A preliminary observation is that the use of DU in ammunition and armour does not generate strategic instability between states or regions. DU weapons are not strategic weapons and have no role in deterrence; their military significance lies in the fact that they can penetrate hard targets at a faster rate than alternative munitions and with greater precision, potentially causing greater damage to targets. While DU

---

[33] For example, the use of environmental modification weapons in uninhabited territory, such as a desert, to protect the civilian population in cities behind that territory from destruction and looting by enemy forces would probably not be contrary to the laws of war.

[34] See *Legality of the Threat or Use of Nuclear Weapons*, *supra* n. 24, para. 52.

[35] For analyses of the relevant rules and principles of international humanitarian law, see Chapters 4-8 of this book.

armour piercing sabots may confer some military advantages in theatre, DU-armoured weapons platforms, such as tanks, are not invincible (as witnessed in the 1991 Gulf War 1991). The use of DU *per se* therefore does not upset any local or regional balance of power. Hence, from an international security and stability perspective, it is not obvious why the law of arms control would have established a prohibition of the use of DU weapons *per se*.

The possible harmful health effects of the use of DU weapons do not alter this conclusion. Obviously, if it were to be established that an effect of the use of DU weapons is to increase soldiers' chances of developing diseases such as 'Gulf War syndrome' or leukaemia, states might become more hesitant in acquiring and making use of this type of weapon since their own troops might be negatively affected,[36] although this would be far less of a potential problem with regard to the use of DU-tipped missiles or aircraft ammunition which, due to their range, are unlikely to affect the health of troops employing them than it might be with regard to the use of tank-fired DU rounds. However, the strategic or tactical value of using such missiles or aircraft in combat lies in the fact that a state can make use of such means of warfare at all; the DU-tipping of their payload or ammunition adds only a marginal additional advantage. Even if a particular use of DU weapons, for example, against a legitimate military target but proximate to a civilian population, was to prove contrary to international law because it could be shown to have long-term negative impacts on human health, this would not of itself suffice to generate an overall prohibition of DU weapons. Therefore, it remains important to establish whether DU weapons fall into any of the categories of nuclear, chemical, biological or incendiary weapons and hence are subject to the corresponding arms control regimes.

4.2.1   Nuclear weapons

Given that the use of nuclear weapons is not prohibited *per se*, it is very difficult to establish a prohibition of DU weapons based on their radioactivity. More important, however, is that DU weapons are much lower in the violence-spectrum than nuclear weapons. A definition of what constitutes a 'nuclear weapon' can be found in Protocol III to the Modified Brussels Treaty of 1954.[37] According to this defini-

---

[36] For the moment, the health effects of exposure to DU among US Gulf War veterans are inconclusive, see D. Fahey, Chapter 2 of this book. There are some indications that the DU used in Iraq (and also in Kosovo) may have been contaminated with very small percentages (0.0028%, as found by the United Nations Environment Programme (UNEP) in Kosovo) of uranium-236, a waste-product of reprocessed uranium and more radioactive than 'clean' DU (U-238). See Press Release UNEP/81, 16 January 2001 – UN Environmental Programme confirms uranium 236 found in depleted uranium penetrators, and J. Lichfield, 'Pentagon knew NATO shells contained dangerous nuclear waste', *The Independent* (29 January 2001), <www.independent.co.uk/news/World/Europe/2001-01/pentagon 290101.shtml>.

[37] Protocol III on the Control of Armaments, 23 October 1954, Annex II, 211 *UNTS* 364. See D. Rauschning, 'Nuclear warfare and weapons', in R. Bernhardt, ed., Vol. III, *Encyclopedia of Public International Law* (1997) p. 730.

tion, a nuclear weapon is, by explosion or other uncontrolled nuclear transformation of nuclear fuel or by radioactivity of nuclear fuel or radioactive isotopes, capable of mass destruction, mass injury or mass poisoning. Another, more recent, treaty defines nuclear weapons as any explosive device capable of releasing nuclear energy in an uncontrolled manner, not including the means, transport or delivery of such device if separable from and not an indivisible part thereof.[38] It will be clear that DU weapons do not fit either of these definitions of nuclear weapons. Even though inhalation or ingestion of DU will result in an enhanced radiation dose internally, the general scientific and medical consensus is that DU is more of a chemical problem than a radiological one.[39]

4.2.2   Biological weapons

DU weapons furthermore do not fit the definition of biological weapons as laid down in the BWC, since this Convention is concerned with 'microbial or other biological agents or toxins'.[40] Although the treaty does not define 'agents' and thus leaves some ambiguity, this term usually refers to living organisms or infective material (or their synthetic equivalent) obtained from them, that multiply inside the person, animal or plant attacked.[41] Toxins – not defined by the BWC but covered by it when used for hostile purposes – whether of a microbial, animal or vegetable nature, including their synthetically produced analogues, are substances that act like chemical agents but ordinarily are produced by biological or microbial processes.[42]

4.2.3   Chemical weapons

DU weapons do not correspond to the definition of chemical weapons as laid down in the CWC, either. DU is not amongst the 'toxic chemicals' with which the CWC is concerned and hence DU weapons are not included in the 'munitions or devices specifically designed to cause death or other harm through the toxic properties of toxic chemicals and their precursors', to which the prohibitions of the CWC apply.[43]

---

[38] Definition in Art. I(c) of the Treaty on the Southeast Asia Nuclear Weapon Free Zone, 35 *ILM* (1995) p. 635.

[39] See M. Clark, 'Depleted uranium', in *Radiological Protection Bulletin* No. 218 (December 1999). Cited in B. Vitali, 'New crimes against humanity: the military use of depleted uranium weapons', para. 2, <www.web-light.nl/VISIE/DUREPORT/vitale_du.html>, and see further D. Fahey, Chapter 2 of this book.

[40] '...of types and quantities that have no justification for prophylactic, protective or other peaceful purposes', see Art. I(1) of the Convention on the Prohibition of the Development, Production and Stockpiling of Bacteriological (Biological) and Toxin Weapons and on their Destruction, 1015 *UNTS* (1972) p. 163.

[41] IDDs' *Arms Control Reporter* (Cambridge, Mass. February 2001) para. 701.A.1.

[42] Ibid., para. 701.A.2. The range of toxins to which the BWC applies has been identified in the Final Declaration of the Second Review Conference of the treaty.

[43] See Art. II of the Convention on the Prohibition of the Development, Production, Stockpiling and the Use of Chemical Weapons and on their Destruction, 32 *ILM* (1993) p. 800.

The same may not necessarily be true with regard to the prohibitions of the 1925 Geneva Protocol, which are less stringent or 'fixed'. The Geneva Protocol prohibits the use in war of asphyxiating, poisonous or other gases, and all analogous liquids, materials and devices. Although the radiological properties of a DU weapon do not make it a poisonous weapon,[44] things might be less clear with regard to the chemical properties of DU. However, the question whether DU weapons could amount to poisonous weapons could only be answered in the affirmative if the effect of the damaging gas is the main result and not simply a side effect of the weapon. Only insofar as the intentional design of a weapon is to inflict poisoning as a means of combat would the use of this weapon qualify as the use of a 'poisonous gas'.[45] Clearly, DU weapons do not have as their main, let alone sole, objective the diffusion of poisonous gases. That, rather, is a side effect of the use of DU weapons, which may not even be militarily advantageous.[46]

4.2.4   Incendiary weapons

The pyrophoric properties of DU do not imply a prohibition on DU weapons under the law of arms control, either. Although 'incendiary weapons' are prohibited by Protocol III to the 1981 Certain Conventional Weapons Convention, DU weapons cannot be considered to qualify as such: Article 1(1)(b)(ii) of Protocol III stipulates that incendiary weapons do not include munitions designed to combine penetration, blast or fragmentation effects with an additional incendiary effect, such as armour-piercing projectiles, fragmentation shells, explosive bombs and similar combined-effects munitions in which the incendiary effect is not specifically designed to cause burn injuries to persons, but to be used against military objectives, such as armoured vehicles, aircraft and installations or facilities.

As regards the harmful effects of the use of DU weapons on the environment, the only specific arms control treaty that may restrain states' freedom of behaviour is the 1977 Environmental Modification (ENMOD) Convention.[47] It can, however, be established that this Convention is not applicable unless DU weapons were to be used as strategic weapons in order to chemically and radioactively pollute large parts of territory for months or even years. The ENMOD Convention only prohibits widespread, long-lasting and severe effects on the environment through the military or otherwise hostile use of environmental modification devices.[48] As

---

[44] In its 1996 Advisory Opinion, the ICJ stated that nuclear weapons are as such not to be considered poisonous weapons (see paras. 54-56). The same applies even more convincingly to DU weapons given their much more limited radiological properties as compared to nuclear weapons.

[45] See S. Oeter, 'Methods and means of combat', in D. Fleck, ed., *The Handbook of Humanitarian Law in Armed Conflicts* (Oxford, Oxford University Press 1995) p. 434.

[46] For a more detailed discussion of the question whether DU can be considered as a poison weapon, see J.K. Kleffner in Chapter 7 of this book.

[47] See Convention on the Prohibition of Military or any other Hostile Use of Environmental Modification Techniques, 1108 *UNTS* (1977) p. 151.

[48] See Art. I of the ENMOD Convention. In an 'Understanding Relating to Art. I', the meaning of these terms has been worked out as follows: 'widespread' – encompassing an area of the scale of

the use of DU ammunition is partly motivated by its high density and hardness, it is very unlikely that DU ammunition would be used in order to spread toxic aerosols and radioactive dust across large parts of a state's territory. Still, if it were to be, the state concerned would be acting in breach of the ENMOD Convention provided that state is a party to the Convention and the state against which the environmental modification techniques were used as a means of destruction, damage or injury is also a party.[49] Moreover, a state could, of course, incur responsibility under the law of state responsibility for acts that caused environmental damage, apart from the issue whether its behaviour was prohibited by the law of arms control.[50]

In summary, even though DU weapons are (low) radioactive, toxic (when ingested) and pyrophoric (on impact), they do not fall in either category of nuclear, biological, chemical or incendiary weapons for the purposes of arms control treaty law. The law of arms control does not otherwise entail treaty-based prohibitions of use that might be applicable to DU weapons.

4.3   Prohibitions other than use in arms control treaty law

The inability to establish a prohibition of use of DU weapons in arms control treaty law does not rule out the possibility of other prohibitions originating from the law of arms control, primarily, prohibitions of transfer and trade. The extensive institutional law present in arms control treaties that are meant to control weapons of mass destruction is considered necessary, *inter alia*, to be able to deal with the problem of dual-use technologies. Whereas the application of chemical, biological and nuclear agents and materials for military purposes has been severely restricted, the peaceful applications of the same agents and materials have to be addressed within the same arms control regimes. When taking into account that DU is a by-product of the uranium enrichment process that is typically suitable to develop nuclear weapons grade fissile material, it is not surprising that transfer and trade of DU is subject to internationally coordinated controls.

The most developed regime is contained in the IAEA safeguards system. Safeguards apply to nuclear material in all peaceful nuclear activity that has reached a certain stage in the nuclear fuel cycle present in a NNWS.[51] 'Nuclear material' means any 'source material' or any 'special fissionable material' as defined in Article XX of the Statute of the IAEA. 'Special fissionable material' as defined in

---

several hundred square kilometres; 'long-lasting' – lasting for a period of months or approximately a season; 'severe' – involving serious or significant disruption or harm to human life, natural or economic resources or other assets. See Stockholm International Peace Research Institute, *World Armaments and Disarmament, SIPRI Yearbook 1978* (London, Taylor & Francis Ltd 1978) p. 377.

[49] For a discussion of potential restrictions on the use of DU weapons arising under international humanitarian law based on their effects on the environment, see E.V. Koppe in Chapter 8 of this book.

[50] For consideration of the circumstances in which state responsibility for the use of DU weapons might arise, see T. Gries and M. Mohr in Chapter 10 of this book.

[51] See 'The Structure and Content of Agreements between the Agency and States Required in Connection with the Treaty on the Non-Proliferation of Nuclear Weapons', IAEA Doc. INFCIRC/153 (Corr.), para. 34(c) (1972).

Article XX(1) encompasses uranium-233 and uranium enriched in the isotopes 235 or 233,[52] whereas 'source material' encompasses uranium containing the mixture of isotopes occurring in nature and uranium depleted in the isotope 235 (Art. XX(3) IAEA Statute). It can be concluded that DU, which includes uranium depleted in the isotope 235 below that occurring in nature, is part of the nuclear material that is subject to IAEA safeguards, and hence trade or transfer restrictions on DU originate from IAEA safeguards agreements. Besides carrying out supervisory tasks related to non-proliferation, the IAEA has organised research projects and training courses on DU, and has conducted measurements of DU isotopes to evaluate the potential effects of DU on environment and health, without dealing with the weapon-side of things.

Over the years, many informal arrangements have been established and related institutions have been created to supplement the legally binding regimes with regard to the non-proliferation of nuclear, chemical and biological weapons. These arrangements and institutions mainly encompass international export control regimes, such as the Australia Group,[53] the Missile Technology Control Regime,[54] the Nuclear Suppliers Group,[55] and the Zangger Committee.[56] Perhaps the most important regional approach is found within the European Union (EU). A (legally binding) community regime has been set up for the control of exports of dual-use items and technology.[57] Within this community regime, 'dual-use items' means

---

[52] This has been defined as 'uranium containing the isotopes 235 or 233 or both in an amount such that the abundance ratio of the sum of their isotopes to the isotope 238 is greater than the ratio of the isotope 235 to the isotope 238 occurring in nature' (Art. XX(2) IAEA Statute).

[53] The Australia Group is an informal network of states that consult and harmonise their national export licensing measures on chemical and biological weapons. Participants aim to prevent any inadvertent contribution to chemical or biological weapons programmes. Thirty-three states from Europe, Asia-Pacific and the Americas and the European Commission participate in the Australia Group. See e.g., Press Release AG/Jun02/Press/7, <http://projects.sipri.se/cbw/research/AG-press-Jun02.html>.

[54] The Missile Technology Control Regime (MTCR) was established in 1987 with the aim of controlling exports of missiles capable of delivering weapons of mass destruction. The 33 states of the MTCR deal with such missiles, as well as related equipment and technology. See <www.mtcr.info>.

[55] The Nuclear Suppliers Group (NSG) is a group of nuclear supplier states which seeks to contribute to the non-proliferation of nuclear weapons through the implementation of two sets of Guidelines for nuclear exports and nuclear-related exports (one set for exports of items that are exclusively for nuclear use and the other set for the export of nuclear related dual-use items and technologies) and through the exchange of information, notably on developments of nuclear proliferation concern. The group consists of 34 Member States. See INFCIRC/539, Attachment, <http://projects.sipri.se/expcon/infcirc_539_1.htm>.

[56] The Zangger Committee (also known as 'Nuclear Exports Committee'), established in 1971, finds its origins in Art. III(2) of the NPT. The 33 Member States have agreed to exchange information about actual exports, or issue of licenses for exports, of source and special fissionable material, as well as of equipment and non-nuclear material, to any non-nuclear weapon states not party to the NPT, through a system of Annual Returns which are circulated on a confidential basis amongst the Member States. See INFCIRC/290/Rev.1 Annex, <http://projects.sipri.se/expcon/zangger.htm>.

[57] See Council Regulation (EC) No. 1334/2000, 22 June 2000, OJ L 159/1 (30/6/2000), as amended by Council Regulation (EC) No. 2432/2001, 20 November 2001, OJ L 338/1 (20/12/2001); Council Regulation (EC) No. 880/2002, 27 May 2002, OJ L 139/7 (29/5/2002); Council Regulation (EC) No. 149/2003, 27 January 2003, OJ L 30/1 (5/2/03).

items, including software and technology, which can be used for both civil and military purposes, and shall include all goods which can be used for both non-explosive uses and which assist in any way in the manufacture of nuclear weapons or other nuclear explosive devices (Art. 2). Even though the community regime is primarily directed at supporting the prevention of the proliferation of nuclear weapons, DU in the form of metal, alloy, chemical compound or concentrate, unless specially fabricated for the civil non-nuclear applications of shielding, packaging, ballast (not greater than 100 kg) or counterweights (having a mass not greater than 100 kg), as well as software and technology for the development, production or use of such DU materials, has been included in the list of dual-use items and technology.[58] In accordance with Article 3 of the community regime, a special authorisation shall be required for the export of those DU materials from the European Community to destinations outside.[59] The community regime implements internationally agreed dual-use controls, including the Wassenaar Arrangement,[60] the Missile Technology Control Regime, the Nuclear Suppliers' Group, and the Australia Group. Interestingly, for EU Member States, the consensus-based regulations of those informal regimes have, as such, been incorporated in the legally binding community regime. Of course, the EU Member States may unilaterally maintain additional (non-discriminatory) national controls of non-community regime origin, albeit that the national legislation or practices of the EU Member States concerning the control of technical assistance related to certain military end uses is subject to legislation of the Council of the European Union.[61]

Finally, prohibitions on arms transfers and the like may also result from arms embargoes applied by the UN Security Council as part of 'sanctions not involving the use of force', based on Article 41 of the UN Charter. In those instances, arms control law is not consensual but 'dictated' in nature. The best example in this respect, which includes obligations to destroy complete stockpiles of weapons capable of mass destruction, is the comprehensive arms control regime that was imposed on Iraq after its defeat in the 1991 Gulf War.[62] Dictated arms control law, too,

---

[58] See Annex I to Council Regulation (EC) No. 1334/2000 and its amendments, pp. 33-45 (Category 0C, 0D, 0E).

[59] With the exception of Australia, Canada, the Czech Republic, Hungary, Japan, New Zealand, Norway, Poland, Switzerland and the USA pursuant to a Community general export authorisation, see Art. 6 and Annex II of Council Regulation (EC) No. 1334/2000 and its later amendments. The decision whether or not to grant the authorisation (and the granting of the authorisation) are made by the authorities of the Member State where the exporter is established (Arts. 6 and 8).

[60] The Wassenaar Arrangement on export controls for conventional armaments and dual-use goods and technologies was established in 1996 by 33 Participating States, with the purpose of contributing to regional and international security, *inter alia*, by promoting transparency to prevent destabilising accumulations of conventional arms and dual-use goods and technologies, and to enhance cooperation to prevent the acquisition of armaments and sensitive dual-use items for military end-uses in situations or regions that are of serious concern to the Participating States. See the ACDA Fact Sheet on the Wassenaar Arrangement, <http://projects.sipri.se/expcon/acdawass.htm>.

[61] See European Council Joint Action of 22 June 2000 concerning the control of technical assistance related to certain military end-uses, doc. 2000/401/CFSP.

[62] The core of the arms control regime imposed on Iraq was formed by S/Res/678 (1990), S/Res/687 (1991), S/Res/1382 (2001), S/Res/1441 (2002).

has been primarily motivated by considerations of strengthening international security and non-proliferation efforts.

It appears that transfer of and trade in DU is subject to control under the IAEA safeguards system, as supplemented by some informal regimes. The reason for this type of control on DU, however, is not to prohibit the spread of DU weapons (ammunition and armour), but to prevent the more general risk of proliferation that may result from the application of dual-use materials such as DU. Although the transfer of and trade in DU weapons is not prohibited by international law, it is subject to export controls and licenses.

4.4   Customary international law on arms control

Unlike the humanitarian laws of war, the law of arms control has allowed almost no room for the development of specific rules of customary international law. A standing 'body of customary arms control law' cannot be identified. As we know, international custom must be evidenced in state practice and *opinio juris*. As with any rule of customary international law, acceptance by states of limitations on their armaments other than by treaty would require a certain level of uniformity and consistency of state practice in order for it to become binding law.[63] In the field of arms control not covered by treaty law, state practice is diffuse, the customary status of many instruments remains unclear, and the reliance by states on their national security interests seems to be the 'magic formula' to prevent any outside interference, through international law or otherwise, regarding the level of armaments present in a state, up to the present day.

Practice shows that several states consider it to be perfectly legal to include DU weapons in their arsenals. Furthermore, states have used DU weapons on several occasions, first in Kuwait and Iraq (1991) and later in Bosnia and Herzegovina and the Federal Republic of Yugoslavia (1994-1995), as well as in Kosovo (1999) and Iraq (2003), and possibly also in Afghanistan (2001-).[64] It has never been asserted by states in any of those situations that the use of DU weapons necessarily violated rules or principles of arms control law. Even though the use of DU weapons may still be contrary to the laws of war, this does indicate that at present there is no constant and uniform state practice or a corresponding *opinio juris*, and hence no established rule of customary law, that would render the production, possession or use of DU weapons unlawful from an arms control law perspective. Moreover, within the branch of arms control law, states have attached much importance to the element of reciprocity, in the sense that they have not been inclined to extend agreed prohibitions on the use of particular weapons towards 'third' states that have not taken on similar obligations towards them. Consequently, it is still disputed whether

---

[63] Cf., ICJ *Asylum* case, *ICJ Rep.* (1950) p. 227.

[64] See, e.g., UNEP Press Release 21 March 2000: 'NATO confirms to the UN use of depleted uranium during the Kosovo conflict'; IAEA/1345, UNEP/82, 25 January 2001 – 'UNEP and IAEA exploring depleted uranium missions to Bosnia and Herzegovina, Yugoslavia and Iraq'. See further D. Fahey, Chapter 2 of this book.

the prohibitions of use contained in arms control treaties such as the CWC, BWC and the 1925 Geneva Protocol have an *erga omnes* character. As long as universal membership of fundamental arms control treaties such as the ones just mentioned has not been attained, the issue of reciprocity may present an obstacle in the process of the development of arms control treaty law into rules of general customary international law.[65]

Furthermore, only a few general principles of international law play a role in the field of arms control. For example, the principles relating to environmental protection which can be found in the treaties prohibiting nuclear testing have secondary meaning that lies far behind the applicable strategic considerations.[66] Perhaps one of the few general principles of arms control law proper that can be established is that a state's behaviour with regard to its national armed forces should not be of such a destabilising nature so as to invite an arms race. An arms race, especially in the nuclear field, not only upsets the strategic balance and threatens international peace and security but is also a waste of precious labour and material. The prevention of an arms race has been one of the principal motivations behind the conclusion of several arms control treaties, the (few) arms control paragraphs of the UN Charter (Arts. 11 and 26), and, indeed, has been a rationale behind the law of arms control as a whole.[67] Whatever the specific application of any such principle in other cases, it is clear that the development of DU weapons cannot invite an arms race. There is no need for a prohibition of DU weapons from an international security perspective. As mentioned, ordinary shells or ordinary tank ammunition may still destroy a DU-armoured tank and, conversely, not every DU missile or grenade hits its target. In short, DU is hardly a revolutionary new weapon which (almost) guarantees victory and which is desired by (all) the parties to potential or actual armed conflicts, thus implying important strategic consequences.[68] States that have more limited military budgets than the United States and most European states and that do not have DU weapons in their arsenals are not very likely to start producing them, simply because the military advantages of DU weapons are not sufficiently great when compared with other improvements in military material which may be achieved with similar spending.

---

[65] It may nevertheless be contended that some arms control treaty provisions have effectively developed into rules of general customary international law, such as the prohibition to carry out atmospheric nuclear weapon tests (contained in the 1963 Limited Test Ban Treaty) and the 'peaceful use' provisions of the 1959 Antarctic Treaty and the 1967 Outer Space Treaty.

[66] See the preambles to the Limited Test Ban Treaty and the Comprehensive nuclear Test Ban Treaty. The prohibition of testing 'any type of weapons' as contained in the 1959 Antarctic Treaty and the 1967 Outer Space Treaty was also environmentally motivated.

[67] See, in particular, the bilateral arms control treaties concluded between the United States and former Soviet Union/Russia, such as the (now defunct) ABM Treaty, the INF Treaty and the START-I Treaty. See also the Treaty on the Conventional Armed Forces in Europe (CFE Treaty, 1991) and the 1999 Agreement on Adaptation of the CFE Treaty, and the Wassenaar Arrangement's (non-binding) 'Elements for Objective Analysis and Advice Concerning Potentially Destabilising Accumulations of Conventional Weapons', 3 December 1998.

[68] As, for example, was effectively the case with the introduction of the machine gun by the American colonists against the Indian tribes in the United States in the late nineteenth century, and, more recently, with the development of (thermo-) nuclear weapons.

Apart from its relatively low cost advantage, consideration of the risks that the use of DU ammunition may entail for their own troops may dissuade states from acquiring DU weapons. It must be acknowledged that the threat or use of any particular type of weapon by one side in a conflict might initiate the (preventive) build-up of a deterrent capacity with similar weapons by the other side, but given the relatively low additional value in war of DU-tipped shells and missiles when compared to similar weaponry without DU-tipping, the dangers of DU weapons-proliferation should not be overestimated.[69] Besides, the applicable export control regimes that are in place today provide an additional burden to would-be DU- 'proliferators'. At least the further development and (potential) use of DU weapons certainly cannot be held to invite arms races that would upset a regional or sub-regional balance of power or otherwise invite inter-state aggression.

It can be concluded that DU weapons are not intrinsically illegal under customary international law. Without prejudice to the possibility of the laws and customs of war restricting the use of DU weapons in certain circumstances, a customary law prohibition of DU weapons in all circumstances cannot be established.

4.5   Soft law developments?

Arms control has always been a rewarding subject for all kinds of political deliberations not producing legally binding documents. In recent years, attempts have been made to give some weight to these legally non-binding documents, by calling them 'politically binding' or norms of 'soft' law. Even if a 'hard' prohibition in arms control law – based in either treaty or customary international law or general principles of law – on the use or possession of DU weapons cannot be established, there may be slow tendencies in soft law which indicate this as a possible direction for the development of the law. The multilateral negotiating fora which may be used to initiate the development of new rules of arms control law are (the First Committee of) the UN General Assembly and the Conference on Disarmament (CD) and, to a lesser extent, the Disarmament Commission (DC) and the Organization for Security and Cooperation in Europe (OSCE). The decisions and resolutions of those fora, insofar as they deal with issues beyond their internal organisation and functioning, are not legally binding.

The use of DU in weaponry is a relatively new phenomenon.[70] The potentially harmful consequences of the use of DU weapons in combat only became known about a decade ago. Within the UN General Assembly, which has a role in arms control and disarmament assigned to it by Article 11(1) of the UN Charter, the issue of DU will most probably remain in the shadows of the more encompassing problems of achieving nuclear, chemical and biological disarmament and dealing with the related threats of proliferation of those weapons (including to terrorist

---

[69] On the question of DU weapons proliferation see D. Fahey, Chapter 1 of this book.

[70] The American military industry was the first to make use of DU in ammunition and to harden artillery, tanks and aircraft. See D. Fahey, Chapter 1 of this book.

groups).[71] Only Iraq made some attempts to introduce the subject of the effects of the use of DU in armaments on the agenda of the General Assembly, albeit on the untenable basis of the alleged prohibition of 'new types of weapons of mass destruction'.[72] Within the other fora mentioned above, the issue of DU so far has not received serious attention. The information and reports of NATO's Ad Hoc Committee on Depleted Uranium, which was established on 10 January 2001 to serve as a forum for the exchange of information on the possible health risks associated with DU and which acts as a clearing-house on this issue, so far have not shown results that are likely to give rise to legal developments in this respect.

It is difficult to predict future developments in this regard, but it is safe to assume that a 'hard' prohibition of DU weapons is not likely to emerge within a short period of time. The historical record of meaningful developments in arms control law shows that the support of the militarily most powerful states – in the first place, the nuclear weapons states – is indispensable for successful negotiations. The NWS play a prominent role in arms control negotiations, and have always dominated the negotiations in the CD, the world's single largest forum for multilateral arms control negotiations, as in the CD's predecessors.[73] With France, the Russian Federation, the United Kingdom and the United States producing DU ammunition and with over a dozen states in possession of stockpiles of DU weapons, the chances of serious negotiations leading to the outlawry of the use or possession of DU weapons seem remote.

However, in international politics, perceptions may count more than facts. It is conceivable that a kind of taboo on the use of DU weapons could develop, as some have argued may be the case with regard to the use of weapons of mass destruction.[74] Otherwise, if it can eventually be proven beyond doubt that the use of DU weapons bears with it great risks for the troops of the user state, this might serve as a more practical deterrent than any legal prohibition.

5.   A DEPLETED URANIUM WEAPONS CONVENTION – FEASIBLE? DESIRABLE?

The foregoing does not rule out the possibility that at some point in the future sufficient political support might be found for the adoption of a 'DU Weapons Con-

---

[71] Pursuant to Art. 11(1) of the UN Charter, the General Assembly may consider the general principles of cooperation in the maintenance of international peace and security, including the principles governing disarmament and the regulation of armaments, and may make recommendations with regard to such principles to the Members or to the Security Council or to both.

[72] See 14th Meeting of the First Committee, GA/DIS/3209, 24 October 2001. In GA/Res/54/44, 23 December 1999, the General Assembly 'reaffirmed that effective measures should be taken to prevent the emergence of new types of weapons of mass destruction'.

[73] See A. Eide, 'Armaments, inequality and humanitarian concerns: prohibiting the use of certain conventional weapons', in R. Akkerman, P. van Krieken and C. Pannenborg, eds., *Declarations on Principles – A Quest for Universal Peace* (Leiden, Sijthoff 1977) pp. 217 at 229-232.

[74] See R. Price, 'A genealogy of the chemical weapons taboo', in 49 *International Organization* (1995) pp. 73-103 and see M. Koskenniemi, 'Faith, identity, and the killing of the innocent: international lawyers and nuclear weapons', in 10 *Leiden JIL* (1997) pp. 137-162.

vention'. Clearly, a 'law of arms control like-prohibition' of DU weapons – one that is valid both in peacetime and in armed conflict – would have to be laid down expressly, in a treaty to be duly ratified by the participating states. Perhaps a strategy could be employed similar to the one that led to the prohibition of chemical weapons, in that first a DU Weapons Convention could be concluded which (only) prohibits the use in armed conflict of DU weapons. Reciprocity may play an important part in that type of convention, in that States Parties may make the standard reservation to extend the obligation of non-use of DU weapons only to other States Parties that have taken on the same obligation towards them.

Obviously, the best guarantee against their use would be a prohibition of the acquisition and transfer of DU weapons and an accompanying obligation to destroy existing arsenals under international supervision. To that end, a DU Weapons Convention, possibly in the form of a new protocol to the CCW Convention, could be drawn up prohibiting the 'offensive' use of DU in weapons such as grenades, bullets and mines, and the 'defensive' use of DU in armour. However, as long as the use of DU in upgrading conventional weapons is considered militarily useful and advantageous, the states already using DU weapons are not likely to join such a DU Weapons Convention.

As is *mutatis mutandis* the case with any arms control treaty, if the states that have DU weapons in their arsenals were to decide not to ratify any DU Weapons Convention, it would diminish its impact and value to a considerable extent. A comparison can be made with the 1997 Landmines (Ottawa) Convention, the first global arms control treaty ever to have come into being without the support of the world's most militarily powerful states.[75] Without intending to underestimate the relevance of the Ottawa Convention, it should be acknowledged that the absence of the major anti-personnel mine using and exporting countries has certainly diminished its importance and global impact. The same is likely to be the case if a convention prohibiting DU weapons were created. Failing the cooperation of the DU users and producers, a treaty of marginal importance would emerge, possibly introducing yet another partition between the 'haves' and 'have-nots'.[76] It can be fairly questioned whether the formal introduction of a kind of 'non-proliferation regime' for DU weapons would – if at all feasible – be desirable. Such measures could further contribute to the established inequality of states in terms of military security that was introduced by the 1968 NPT.

---

[75] Convention on the Prohibition of the Use, Stockpiling, Production and Transfer of Anti-Personnel Mines and on their Destruction (APM Convention), see 36 *ILM* (1997) p. 1507, entry into force in 1999.

[76] This can be illustrated with reference to the UK Select Committee on Foreign Affairs' Fourth Report on the NATO intervention in Kosovo. Regarding DU munitions, the UK Minister of State and the Ministry of Defence remarked that depleted uranium ammunition is not proscribed by any of the international agreements to which the UK is a party, and that although UK forces did not use depleted uranium ammunition in Kosovo, the Ministry of Defence reserves the right to issue depleted uranium-based weapons if the safety of British troops requires a capability against modern armour, see House of Commons – Foreign Affairs – Fourth Report, para. 51, <www.parliament.the-stationary-off.../pa/cm199900/cmselect/cmfaff/28/2814.htm>.

A further complicating factor is that international supervision of compliance with a DU Weapons Convention would be very difficult, if not impossible. Even though the introduction of supervisory mechanisms in an arms control treaty is not a legal duty, there is general agreement that all consensual arms control measures, especially when they involve disarmament proper, should be implemented from beginning to end under strict and effective international supervision so as to provide firm assurance that all parties are honouring their obligations.[77] Whatever the actual importance of the presence of supervisory mechanisms in a treaty on DU weapons, the absence of such mechanisms would certainly provide the opponents of DU weapons control with a valid argument by which to frustrate negotiations and prevent the coming into being of a DU Weapons Convention. After all, problems of verification of compliance have haunted arms control negotiations for many decades.[78] All in all, it seems highly unlikely that the necessary political support for the successful negotiation of a DU Convention can be found, at present or in the near future. Even if further research reveals beyond doubt that the use of DU in weapons poses considerable health risks, it will still be up to each individual state to decide whether these risks warrant supporting a treaty outlawing DU weapons.

As a far more modest approach, a voluntary system aimed at providing transparency, similar to the UN register of conventional weapons, could be applied which included DU weapons as a special category.[79] This might at least diminish the uncertainty regarding whether or not DU weapons have been used in a particular conflict, or, alternatively, whether DU weapons are likely to be used in the event that a conflict in a certain region is imminent. This type of confidence-building measure, however, is merely a prelude and not a true alternative to a specific arms control treaty.[80] Measures of transparency and openness cannot be 'enforced' in principle because they do not set enforceable behavioural standards in the first place. Transparency regimes do not involve prohibitions of state behaviour, but entail requesting the participating states to make (voluntary) contributions to the exchange of information, (voluntary) reporting, and other measures to increase transparency

---

[77] See e.g., McCloy-Zorin Joint Statement of Agreed Principles for Disarmament Negotiations, UNGA Doc. A/4879 (1961), principle (6); Final Document of the United Nations General Assembly Tenth Special Session on Disarmament (SSOD-I), UN Doc. S-10/2 (1978), para. 31, and see the Disarmament Commissions' 'Principles of Verification affirmed by the Disarmament Commission', annexed to S. Sur, *A Legal Approach to Verification in Disarmament and Arms Limitation* (New York, United Nations 1988) p. 67.

[78] See, e.g., P. Trimble, 'Beyond verification: The next step in arms control', in 102 *Harvard Law Review* (1989) pp. 885 at 885-886.

[79] See on the UN Register on Conventional Armaments, UNGA Res. 46/36 L, 9 December 1991 (especially para. 7), and UNGA Res. 49/75 C, 9 January 1995. See also the Report of the UN Secretary-General on the UN Register on Conventional Armaments, A/47/342, 14 August 1992. The Register covers the following categories of major conventional arms: battle tanks, armoured vehicles, large calibre artillery systems, combat aircraft, attack helicopters, warships and missiles and missile launchers.

[80] See G. den Dekker, 'The law of arms control and sub-regional arms control in the Former Yugoslavia: 'hard' law in a 'soft' law context', in 55 *NILR* (1998) at p. 384.

and openness in arms production and arms transfers.[81] In a way, the IAEA, the UN Environment Programme and NATO's Ad Hoc Committee on Depleted Uranium are collectively fulfilling this task already with respect to DU weapons.

In the end, transparency regimes will neither constrain the proliferation of DU weapons nor restrict the transfer and trade of DU. More effective restraints could be expected from restrictive unilateral arms export policies based on domestic laws. At that point, however, one leaves the realm of public international law and enters the field of domestic (economic) policies, where short-term profits may easily prevail over long-term security risks and assessments.[82]

6.   CONCLUSIONS

Currently, the behaviour of states with regard to DU weapons (production, possession, trade and use) is not prohibited by the law of arms control, simply because there is no treaty containing such prohibitions and customary international law has not sufficiently developed in the field of arms control to support a different conclusion. This does not exclude that the transfer of and trade in DU may be subject to export controls and licenses because of its dual-use characteristics.

For various reasons, among which are the relatively low additional strategic value of DU weapons and the near impossibility of supervision of compliance, it can be questioned whether a convention banning the employment of DU weapons would be feasible or even desirable. Furthermore, even if a DU Weapons Convention could be negotiated, the key players in the field (i.e., the states that produce and/or use DU weapons) may decide to remain outside the regime, thus potentially reducing the importance of such a convention to mere symbolism.

Probably the best guarantee against the use of DU weapons lies in the damage that may be done to the troops employing those weapons, and in the possible humanitarian consequences of the use of DU weapons. The law of arms control, however, is not so much concerned with the protection of individuals (civilians and combatants) or the implementation of humanitarian principles, but rather with the (strategic) behaviour of states and the prevention of war by enhancing international security and supporting a stable balance of power, on different levels.

A system of transparency measures regarding DU weapons may be useful and feasible, although it should be borne in mind that this type of (voluntary) measure is not a substitute for arms control treaty law.

---

[81] See for an example of legally binding transparency measures the OAS Inter-American Convention on Transparency in Conventional Weapons Acquisitions, AG/RES. 1607 (XXIX-O/99), 7 June 1999.

[82] To illustrate this point, one may think of the vast arms supplies from European countries to Iraq and the military construction works performed by European (state-owned) companies in Iraq at the time of the Iraq-Iran war. Within years, the Allies had to fight against their own (old) military material used by Iraq in the 1991 Gulf War.

Chapter 4
A MILITARY VIEW ON DEPLETED URANIUM

Burrus M. Carnahan

1. WHAT IS MILITARY UTILITY?

A proper application of the fundamental principles of international humanitarian law to any weapon, including depleted uranium (DU) weapons, requires a balancing or weighing of both military and humanitarian considerations. For example, the fact that it may cause suffering or injury to combatants cannot, in principle, make a weapon unlawful. As observed by Ambassador George Aldrich: '[w]hether the suffering a weapon causes is 'unnecessary' in the sense required to make it unlawful requires a balancing of this suffering against the military necessity for its use.'[1] A commentary on Article 35 of Additional Protocol I (hereinafter, AP I) has similarly stated that

> '... the suffering or injury caused by a weapon must be judged in relation to the military utility of the weapon. The test is whether the suffering is needless, superfluous, or manifestly disproportionate to the military advantage reasonably expected from the use of the weapon.'[2]

'Military necessity', 'military utility', and 'military advantage reasonably expected' must, then, be considered in the process of determining the lawfulness of any weapon.

In principle, military necessity differs from military utility or military advantage. The latter are solely military concepts, while military necessity is both a legal standard and a military concept. On the other hand, assessing the existence of military necessity requires examination of military utility and advantage. Indeed, military necessity may be thought of as a certain level of military utility or advantage. This chapter will therefore seek to discover the legally relevant military utility of DU weapons by examining sources that assess their military necessity.

As a legal concept, 'military necessity' first appeared in the mid 19th century, initially as a limitation on the old custom that allowed confiscation of private

---

[1] Speech before the American Society of International Law, 13 April 1973, reprinted in *Digest of United States Practice in International Law* (1973) p. 504. At the time, Ambassador Aldrich was Deputy Legal Adviser to the US State Department.

[2] M. Bothe, K.J. Partsch and W. Solf, eds, *New Rules For Victims Of Armed Conflicts* (The Hague, Martinus Nijhoff 1982) p. 196.

McDonald / Kleffner / Toebes (eds.), Depleted Uranium Weapons and International Law
© 2008, T·M·C·ASSER PRESS, *The Hague, The Netherlands and the Authors*

property of enemy civilians.[3] The most influential definition and discussion of military necessity from this period appeared during the American Civil War in General Order #100,[4] issued by the United States Army in 1863 and otherwise known as the Lieber Code after Dr Francis Lieber, who chaired the committee that drafted it.[5]

General Order #100 was intended to restate, in a concise, accessible form, the international laws and customs of land warfare as recognised by the United States in the mid 19th century. While the Lincoln administration never recognised the Confederate government, or the legitimacy of secession, by 1863 it did recognise that the US government was in a state of war with the armies of the Confederacy, and with the individual rebels that composed those armies. Lincoln's legal advisers reasoned that if the government was at war with its rebellious citizens in the South, then he could use all means permitted by the international laws and customs of war to suppress that rebellion. Hence the need for General Order #100.

The scope and limits of military necessity are addressed as follows in Articles 14 to 16 of the Code:

'Art. 14. Military necessity, as understood by modern civilized nations, consists in the necessity of those measures which are indispensable for securing the ends of the war, and which are lawful according to the modern law and usages of war.

'Art. 15. Military necessity admits of all direct destruction of life or limb of armed enemies, and of other persons whose destruction is incidentally unavoidable in the armed contests of the war; it allows of the capturing of every armed enemy, and every enemy of importance to the hostile government, or of peculiar danger to the captor; it allows of all destruction of property, and obstruction of the ways and channels of traffic, travel, or communication, and of all withholding of sustenance or means of life from the enemy; of the appropriation of whatever an enemy's country affords necessary for the subsistence and safety of the army, and of such deception as does not involve the breaking of good faith either positively pledged, regarding agreements entered into during the war, or supposed by the modern law of war to exist. ....

'Art. 16. Military necessity does not admit of cruelty – that is, the infliction of suffering for the sake of suffering or for revenge, nor of maiming or wounding except in fight, nor of torture to extort confessions. It does not admit of the use of poison in any way, nor of the wanton devastation of a district. It admits of

---

[3] See generally B. Carnahan, 'Lincoln, Lieber and the laws of war: The origins and limits of the principle of military necessity', 92 *AJIL* (1998) p. 213.

[4] In American military parlance, a 'general order' is a written command intended to apply to all members of a military organisation. Commanding officers often use general orders to establish general policies, procedures or standards of conduct throughout their commands. Because it was issued by Headquarters, US Army, and signed by President Lincoln as commander-in-chief, General Order #100 applied to all members of the Union Army after April 1863.

[5] Instructions for the Government of the Armies of the United States in the Field, 24 April 1863, in D. Schindler and J. Toman, eds., *The Laws of Armed Conflicts*, 3rd edn. (Dordrecht, Martinus Nijhoff 1988) p. 3.

deception, but disclaims acts of perfidy; and, in general, military necessity does not include any act of hostility which makes the return to peace unnecessarily difficult.'

Several conclusions about military necessity, significant for the law of weapons, can be drawn from the Lieber Code. First, military necessity only applies to military measures otherwise 'lawful according to the modern law and usages of war'. That is, it does not affect positive rules of the law of war. In the context of weapons law, military necessity cannot justify the use of weapons, such as chemical or biological agents, prohibited by treaty.

Second, the exclusion of cruelty, or 'the infliction of suffering for the sake of suffering', from the calculus of military necessity means that weapons may not be designed with the intent to terrorise the enemy through their hideous effects. This does not exclude all modification of weapons for psychological impact – the use of whistles on bombs in World War II has never been considered unlawful – but it does prohibit modifying or designing weapons to cause greater injury or destruction solely for greater *in terrorum* effects. A recent application of this corollary to the principle of military necessity would be the Protocol on Blinding Laser Weapons adopted in 1995 by the parties to the 1980 Convention on Certain Conventional Weapons.[6] The Protocol prohibits the use in armed conflict of laser weapons designed to permanently blind enemy personnel. Because the legitimate military purpose of a blinding laser can be achieved by temporarily degrading the sight of an enemy soldier, to deliberately design the weapon to permanently blind could only be intended to create terror due to the horrible nature of the injury.

Third, while Article 14 may suggest that military necessity sets a considerably higher standard of behaviour than mere military utility or advantage, the specific examples in Article 15 imply that, in practice, this distinction is more illusory than real. The killing or wounding of all armed combatants is thus declared justified, without suggesting the need for a case-by-case assessment of whether it is really necessary to attack a particular combatant at a particular time (e.g., one fleeing the battle in panic). There is no suggestion in the rules on destruction and seizure of property or obstruction of transportation that any threshold of necessity need be established higher than ordinary military advantage.

The most serious recent effort to clarify the idea of military necessity tends to support the Lieber Code's broad interpretation of the concept. During the negotiation of the Second Protocol to the 1954 Hague Convention on Cultural Property in armed conflict, the parties sought to tighten the provisions in the 1954 Convention allowing waiver of protection for reasons of military necessity. Without discussing all the nuances of the Second Protocol's text, it is significant that the standards it adopts tend to fall back on the idea of military advantage – 'definite military advantage', 'similar military advantage', and 'concrete and direct military advan-

---

[6] The Protocol on Blinding Laser Weapons (Protocol IV to the 1980 Convention), 13 October 1995, <http://www.icrc.org/ihl.nsf/>.

tage'.[7] Reading the Second Hague Protocol together with the Lieber Code suggests that military necessity exists whenever the anticipated military advantage is definite, concrete and direct, rather than merely speculative.

Fourth, military necessity may justify injury, suffering and destruction either to damage the enemy or to assist friendly forces. A few months after signing General Order #100, President Abraham Lincoln succinctly restated this principle when he wrote that '[c]ivilized belligerents do all in their power to help themselves, or hurt the enemy, except a few things that are regarded as barbarous or cruel.'[8]

That military necessity applies to measures helping friendly forces, not just to efforts to hurt the enemy, is an aspect of the law of war that has often been overlooked in recent years. For example, high-altitude air bombardments carried out by NATO in the former Yugoslavia and by the United States in Afghanistan have been criticised by human right organisations, who have argued that the attacks should have been carried out at lower altitudes because fewer collateral civilian casualties would have resulted. High-altitude attacks reduce the risk to the attacking force from anti-aircraft fire and missiles, but this justification has been dismissed as a callous and inhumane attribution of greater value to the lives of the attacking aircrews than to those of civilians near the target.[9] This viewpoint, however, ignores the military value of protecting the attacking force.

Unlike most other human organisations, armed forces do not exist to serve their members. From a strictly military standpoint, both an armed force and its members exist solely to execute their assigned missions. Increasing the rate of combat losses will, in turn, increase burdens on an armed force's supply and manpower support systems. Exposing aircraft, crew and equipment to increased danger will, thus, gradually erode the ability of an air force to carry out its missions. It is for such reasons, which might be crystallised in the phrase 'military necessity', that air

---

[7] Arts. 1(f), 6(a) and (b), and 7(c) and (d), respectively, of the Second Protocol to the Hague Convention of 1954 for the Protection of Cultural Property in the Event of Armed Conflict, The Hague, 26 March 1999, available at the ICRC website, <http://www.icrc.org/ihl.nsf/>. These phrases were adapted from Arts. 52 and 57 of Protocol I Additional to the Geneva Conventions of 1949, in Schindler and Toman, *supra* n. 5. Protocol I avoids the term military necessity, and uses these phrases in the course of defining military objectives and acceptable collateral injury. Protocol I uses the military necessity standard only in dealing with tangential issues, such as the right of a state to adopt a 'scorched earth' policy in the event of invasion (Art. 54), or the right of parties to a conflict to limit the activities of civil defence or relief workers (Arts. 62, 67 and 71).

[8] Public Letter to James C. Conkling, 26 August 1863, in R.P. Basler, ed., *Collected Works of Abraham Lincoln*, Vol. VI (New Jersey, Rutgers University Press 1953) p. 406.

[9] E.g., 'Criticism of the NATO bombing campaign has included allegations of varying weight: a) that, as the resort to force was illegal, all NATO actions were illegal, and b) that the NATO forces ... deliberately or recklessly caused excessive civilian casualties in disregard of the rule of proportionality by trying to fight a "zero casualty" war for their own side. Allegations concerning the "zero casualty" war involve suggestions that, for example, NATO aircraft operated at heights which enabled them to avoid attack by Yugoslav defenses and, consequently, made it impossible for them to properly distinguish between military or civilian objects on the ground.' *Final Report to the Prosecutor by the Committee Established to Review the NATO Bombing Campaign Against the Federal Republic of Yugoslavia*, 8 June 2000, para. 2.

forces prefer to conduct high-altitude bombardments, not because their governments attribute greater inherent value to the lives of their own personnel.

The application of this aspect of military necessity to the law of weapons
means that the development and use of a particular weapon might be justified not
because it can hurt enemy forces more effectively, but rather because it is easier to
transport, or is made of more readily-available materials, or for other logistical
reasons. For example, the government of the United States has listed a variety of
military factors that may legitimately be considered in the process of balancing the
suffering a weapon causes against the military necessity for its use. These include,
inter alia, the effectiveness of the weapon against particular military targets; effects
on enemy morale, stamina and cohesion; the availability and effects of alternative
weapons; cost; the logistics of providing the weapon where, when and in the quantities needed; and the security of the troops utilising the weapon.[10]

This chapter examines how these military considerations apply to the legal
status of DU weapons under the laws of war. As a vehicle for the examination, this
chapter uses two legal reviews of DU weapons conducted by the US government,
which has taken the leading role in the development and deployment of these munitions.

2.   LEGAL REVIEW OF DEPLETED URANIUM BY THE UNITED STATES

2.1   The United States weapons review process

In the early 1970s several weapons used by US forces in Southeast Asia, ranging
from napalm and cluster bombs to the M-16 infantry rifle, were heavily criticised
by non-governmental organisations and some neutral governments as violating,
inter alia, the rule against unnecessary suffering and superfluous injury. In response
to this history, and in preparation for the international conference on humanitarian
law held in Geneva from 1974 to 1977, the US Department of Defense issued an
instruction on 16 October 1974,[11] requiring legal review of weapons before their
purchase by the Departments of the Army, Navy and Air Force. Specifically, the
Instruction required that:

> 'The Secretary of each Military Department will ensure that a legal review by
> his Judge Advocate General is conducted of all weapons intended to meet a mili-
> tary requirement of his Department in order to ensure that their intended use in
> armed conflict is consistent with the obligations assumed by the United States
> under all applicable international laws including treaties to which the United

---

[10] Statements to the Lucerne Conference by Ronald J. Bettauer, 25 September 1974 and Waldemar
Solf, 26 September 1974 in *Digest of United States Practice in International Law* (Washington, Dept.
of State Publication 8809, 1974) p. 707. See also Bothe, Partsch and Solf, *supra* n. 2, p. 197.

[11] DoD Instruction 5500.15, 16 October 1974, reprinted in *Digest of United States Practice in
International Law* (Washington, Dept. of State Publication 8809, 1974) p. 710.

States is a party and customary international law, in particular the laws of war.'[12]

In conducting these reviews, each Judge Advocate-General relies on technical data provided by the research and development organisation of his Military Department, as well as medical information from the service medical corps.

2.2   The 1975 Air Force review

One of the first weapons to be reviewed under this instruction was a DU projectile, the 30 mm armour-piercing incendiary (API) ammunition for the GAU 8 rapid fire cannon in the nose of the A-10 Thunderbolt II ground attack fighter. Ironically, this was the very ammunition used in the 1999 Kosovo conflict, which gave rise to so much controversy in Europe a year later.

Intervention in a high level internal armed conflict, through intense application of air power, was not the military use for which either the A-10 Thunderbolt or its DU weapons were initially developed. In the mid 1970s, at the height of the Cold War, Western Europe faced the threat of ground attack from superior Warsaw Pact tanks and other armoured forces. Under the defence doctrine then in place, NATO might ultimately resort to the first use of tactical nuclear weapons to blunt such an attack.[13] Obviously, however, it would be preferable to develop credible conventional means to blunt a massive armoured attack. These would both reinforce the deterrent effect of NATO's nuclear weapons and, if war did occur, offer at least a hope of successful conventional defence.

To meet this need, the US Air Force developed the A-10. Designed to carry a heavy load of munitions at low altitude and slow speed, the A-10 was intended to be a tank-killer on the plains of central Europe. The same considerations applied to the munitions developed for its only internal armament, the GAU 8 cannon. While not directly mentioned in the legal review, this historical context must always be kept in mind when considering the initial military justification for API DU ammunition.

On 14 March 1975 the Judge Advocate-General of the US Air Force signed a legal review of both API ammunition and high explosive incendiary (HEI) ammunition for the GAU 8 cannon.[14] The review noted that the API munitions were

---

[12] Ibid., section IV.A, p. 710. The Judge Advocate-General is the highest-ranking uniformed legal officer in each of the three Military Departments (Army, Navy and Air Force).

[13] See, e.g., para. 7 of the Final Communiqué of the Spring Ministerial Session of the North Atlantic Council Meeting in Athens, Greece, 4-6 May 1962: 'The purpose of NATO is defense, and it must be clear that in case of attack it will defend its members by all necessary means. The Council has reviewed the action that would be necessary on the part of member countries, collectively and individually, in the various circumstances in which the Alliance might be compelled to have recourse to its nuclear defenses.' Available on the NATO website <http://www.nato.int/docu/comm/49-95/c620504a.htm>.

[14] Department of the Air Force, Headquarters United States Air Force, JA letter to AF/RDP, Subject: Legal Review of 30 mm Ammunition, 14 March 1975, with attached Legal Memorandum.

designed to be 'used against hard targets (heavy tanks, armored personnel carriers, etc.)', while the HEI munitions were intended for use 'against soft targets (personnel, trucks, vans, etc.)'. Only the API munitions used DU.

In its analysis of the API round's consistency with the principle of unnecessary suffering and superfluous injury, the legal memorandum emphasised that this ammunition is designed for use against *matériel* targets, not personnel. Indeed, the opinion observes that 'anti-personnel use [of the API round] would not be cost effective' from a military standpoint. It implied that the legal conclusions might change if DU weapons were deliberately chosen for anti-personnel use where alternate weapons were available.

The military justifications for DU use in an anti-*matériel* weapon were summarised as follows: 'Significant advantages are described in terms of availability, low cost and superior armor penetration capability.' 'Availability' and 'low cost' may be consolidated under the category 'logistic considerations'.

A byproduct of producing low-enriched uranium nuclear reactor fuel,[15] DU is certainly widely available at low cost. By the spring of 1999 the US Department of Energy had such a 'vast store' of the material that it created a special program to seek new practical uses for DU, and some non-governmental experts proposed that it be disposed of as waste.[16] In the Cold War context in which the 30 mm API round was developed – planning and preparing for a massive armoured attack on the plains of central Europe – the availability of large quantities of cheap ammunition would certainly be of major military significance.

One other logistical consideration was invoked in the review. While the United States is not a party to the 1868 Saint Petersburg Declaration,[17] it regards the Declaration's ban on the use of small calibre incendiary or explosive bullets as an expression of the principle against unnecessary suffering, at least in some circumstances of land warfare.[18] The 30 mm HEI round was explosive, and both the HEI and API rounds had incendiary capabilities. The review addressed the St. Petersburg Declaration issue by asserting that 'there is no rule in aerial warfare prohibiting, *per se*, projectiles below 400 grams which are explosive or incendiary in nature, assuming appropriate military purposes are served by these characteristics.' This exception for weapons used in aerial warfare was held to apply 'even though they are directed solely at combatants' (i.e., personnel). Support for this exception was sought in 'military considerations precluding resort to available alternate weapons at the time different targets are attacked'. While expressed in an elliptical manner, the point being made is that the ammunition in fighter planes is loaded while the aircraft is being prepared for flight on the ground. Once in the air, the pilot

---

[15] See D. Fahey, Chapter 1 of this book. Highly enriched uranium may also be produced for use in nuclear weapons. The United States ceased enriching uranium for weapons purposes in 1964.

[16] M. Conley, 'DOE researchers continue quest for alternate uses of depleted U', *NuclearFuel* (27 November 2000) p. 8.

[17] Reprinted in Schindler and Toman, *supra* n. 5, p. 95.

[18] See, e.g., US Dept. of the Air Force, AFP 110-31, *International Law – The Law of Armed Conflict and Air Operations* (19 November 1976) p. 6-2.

cannot change it. A fighter might be loaded with incendiary ammunition and sent on an anti-*matériel* mission, but then be diverted in flight to an anti-personnel mission (e.g., close-air support to ground forces), or come across an anti-personnel target of opportunity (e.g., a marching column of enemy troops). Although loaded with incendiary munitions, it would not be unlawful to attack enemy personnel because of the military necessity resulting from the impossibility of changing munitions in flight. In support of this point, the opinion invokes an important type of evidence of military utility, the 'practice and experience in two World Wars, and subsequently', as well as 'the widespread use of such [incendiary] projectiles in current inventories'. Similar evidence would later be invoked in support of the military utility of DU weapons in the 1994 Army Review.

This reasoning – that military necessity, supported by widespread state practice, permits the use of small calibre incendiary munitions by warplanes – should apply equally to the use of DU API munitions against enemy personnel. Ammunition for the GAU 8 cannon on the A-10 is loaded on the ground, and the pilot cannot change it in flight. While API ammunition is not designed for anti-personnel use, the practical impossibility of changing munitions in flight dictates that it may be used to attack enemy personnel in particular cases.

Aside from logistical considerations, the other major military justification for use of DU was its anticipated effect against the 'hard' material targets for which it was designed. Based on research data, the opinion noted that DU was selected for use in the API round because of its 'superior armor penetration capability', which derived from the 'kinetic effects' of 'depleted uranium on enemy armor'. This capability was not described in detail.

Taking all these military factors into account, the Judge Advocate-General of the Air Force concluded that the development and intended use of the 30 mm API round was consistent with the legal obligations of the United States. In order to avoid future questions, the legal memorandum attached to this finding suggested that this ammunition not be used 'solely against personnel' where 'alternate weapons are available'. As noted above, in air warfare alternate munitions would never be available once an A-10 Thunderbolt left its airbase on a mission.

2.3   The 1994 Army review

In late 1994, almost 20 years after the Air Force review, the US Army carried out its own legal review of the M829A2 tank ammunition cartridge.[19] This cartridge used a 25 mm diameter rod of DU to penetrate the armour on enemy vehicles. It would be carried as ammunition for the 120 mm main gun of the Army's M1A1 tank.[20]

---

[19] Department of the Army, Office of The Judge Advocate-General, DAJA-IO Memorandum for US Army Armament Research, Development and Engineering Center, Subject: M829A2 Cartridge, 120 mm, APFSDS-T (Depleted Uranium Tank Round); Law of War Review, 27 December 1994.

[20] This was actually the second 120 mm depleted uranium round developed for the M1A1 tank. The earlier round, designated the M829A1, was used in combat in the 1991 Gulf War. This 1994 opinion was chosen for analysis because it incorporated that combat experience.

Several historical developments since 1975 altered the focus of this review. By this time, of course, the Cold War was over, and US forces no longer faced the threat of a massive armoured assault in Europe. Of equal significance, DU ammunition had been used in battle by the US Army and Air Force during the Persian Gulf War of 1991. The reviewer now had the benefit of studies based on actual, rather than simulated, combat experience with DU weapons.

As might be expected, the 1994 review ignored the logistic considerations that were so prominent in the earlier review. There was no longer a need to ensure that large quantities of munitions were available to meet a Warsaw Pact juggernaut, and a tank crew can change munitions to meet the requirements of different targets. The opinion found the principal military justification for the use of DU in its capability to penetrate enemy armour. Because it describes how a DU round functions, it is worthwhile quoting the opinion at length.

'The long rod penetrator contained in this cartridge is designed to perforate the heavy armor surrounding the crew and ammunition compartments of enemy Main Battle Tanks. The lethality mechanism is simple kinetic energy, using a dense material [depleted uranium] impacting the target in a concentrated region at a supersonic velocity. Upon impact, extremely high pressures are generated at the interface between the penetrator and the target. This pressure deforms and erodes the tip of the penetrator and deforms the target material, creating a cavity in the target. The depth of the cavity increases as additional penetrator material feeds into the interface.

As the penetrator exits the final layer of armor, a radially expanding, conical cloud of fragments is deposited into the compartment. This behind-armor debris is composed of spall, which is target material released during perforation of the armor, and residual penetrator material. The debris particles contain (sic) both a high temperature and velocity which provides them the capability to destroy or debilitate objects within the vehicle. For example, if the behind-armor debris strikes the targeted tank's ammunition, it is likely that it will ignite the ammunition, resulting in catastrophic destruction of the tank.
....
In military applications, when alloyed (as in the M829A2), DU [depleted uranium] is ideal for use in armor penetrators. The penetrator has the speed, mass, and physical properties to perform exceptionally well against armored targets. DU provides a substantial performance advantage, well above other competing materials such as tungsten steel. This enables DU penetrators to defeat an armored target at a significantly greater distance.'

Two aspects of the last paragraph above should be noted. First, the opinion acknowledges that materials other than DU might be used for armour-piercing munitions, but gives great weight to the fact that DU weapons have a 'substantial performance advantage'. In assessing the legality of a weapon under the principle of unnecessary suffering and superfluous injury, it is always helpful and often necessary to compare the weapon under review with the characteristics of alternative weapons that might be used to accomplish the same military mission.

Second, the advantages of DU tank ammunition allow forces using it to successfully engage enemy armour at a greater distance, perhaps beyond the range of the enemy's tank guns. Depleted uranium munitions may thus contribute to the security of the forces using them.

As did the Air Force opinion when considering the incendiary effect of A-10 ammunition, the Army opinion looked to the practice of other states for confirmation of the military utility of this type of munition. By 1994, the opinion noted, the armed forces of the United Kingdom, Russia, Turkey, Saudi Arabia, Pakistan, Thailand and France deployed some form of DU weapons. Furthermore, 'other nations are developing or already have DU-containing weapon systems in their inventories'.

The opinion, signed for the Judge Advocate-General of the Army on 27 December 1994, concluded that: 'The greater penetration capability of DU is the *military necessity*[21] for its use. DU is not employed to increase the suffering of individual enemy combatants, and there is no evidence to suggest that DU increases combatant suffering.'

3.   SUMMARY AND CONCLUSIONS

These reviews found that DU had military utility from several different standpoints. The reviews also identified these factors as legally significant, in the view of one government, for assessing the legitimacy of DU weapons under the principle of unnecessary suffering and superfluous injury.

The most important military justification for the use of DU ammunition, invoked in both reviews, was its suitability against the intended targets of these munitions – tanks, armoured personnel carriers and other 'hard' *matériel* targets. Depleted uranium was found to be a superior penetrating material when compared with alternatives, such as tungsten steel. Evidence of this superior armour penetrating capability was found in research and development tests, in combat experience, and in the adoption of similar munitions by other nations. The security of the attacking force was also a consideration in the use of these munitions by ground forces, since DU weapons allow engagement of enemy armoured forces at greater distances.

Logistical considerations, in terms of cost and availability, were an additional military justification, at least when the anticipated threat was a massive assault by superior enemy armoured formations. While that threat existed, the need to deploy large quantities of superior armour-piercing munitions was a major, though secondary, military justification for the use of DU.

Note should also be taken of what these opinions did not say. Neither review attempted to justify the use of DU in an anti-personnel mode. No effort was made to justify use of DU because of any supposed radiological or toxic effects on

---

[21] Emphasis in the original.

enemy personnel; in fact, both reviews denied that such effects, if they existed, had played any role in the choice of DU as a component of the munitions under consideration. The earlier review even cautioned against using such munitions against enemy personnel, if alternative weapons were available (although it implicitly conceded that in air warfare that choice would rarely exist). Finally, it should be emphasised that, while the United States has asserted that effects on enemy morale and cohesion may legitimately be taken into account in determining the legality of weapons,[22] neither opinion claimed that DU weapons could legitimately be used to terrorise enemy combatants by instilling a fear of long-term radiation effects.

That the weighing of military necessity is an inherently subjective process is as true in the 21st century as it was when the principle was first enunciated in the 19th. Military utility and human suffering are both inexact concepts, which are quite dissimilar from one another. No effort to quantify either has yet proven persuasive over the long-term. While there are a few well-recognised limits to the principle of military necessity, such as the prohibition on increasing suffering for the sake of suffering, usually the best that can be expected of a government is that it honestly considers all the known facts about a weapon, such as DU weapons, and in good faith weighs its expected logistic and other military virtues against the reasonably anticipated suffering of the enemy, both combatant and non-combatant.

In seeking to identify the appropriate military utility of DU weapons, this discussion has necessarily focused on issues surrounding unnecessary suffering and superfluous injury, because these issues usually lie at the core of weapons acquisition reviews. It is not intended to suggest that use of DU in combat would raise no other humanitarian law issues, such as the requirement to minimise collateral civilian injury or the need to ensure that collateral injury to civilians or damage to civilian property is not excessive in light of the expected military advantage.

Resolution of these issues will involve a weighing of military utility against humanitarian considerations or environmental risks, similar to the balancing process in determining whether a weapon causes unnecessary suffering or superfluous injury. The balancing process used to resolve these issues will examine the same military advantages identified in the course of determining whether DU weapons cause unnecessary suffering or superfluous injury. In 1975 and 1994, when the two US military reviews of DU were done, there was little evidence that these munitions posed significant collateral threats to civilian life or health or to the environment; military utility appeared to clearly outweigh any remaining marginal risks. These issues were revisited by the International Atomic Energy Agency, the World Health Organization, the UN Environment Programme, and several national governments following the 1999 media furore over DU in the former Yugoslavia. Little new evidence was found that might have affected the earlier military assessment of these munitions.

The verdicts of the US Air Force and Army that the military advantages of DU weapons clearly outweigh the minimal collateral risks of their use are thus

---

[22] See n. 5, *supra*.

likely to stand, absent the emergence of new scientific evidence of medical or environmental damage. Only the passage of time may reveal such evidence, as more definite knowledge is gained on the long-term effects of the military use of DU.

Chapter 5
THE USE OF DEPLETED URANIUM AND THE PROHIBITION OF WEAPONS OF A NATURE TO CAUSE SUPERFLUOUS INJURY OR UNNECESSARY SUFFERING

Marten Zwanenburg[1]

1. INTRODUCTION

This chapter discusses the question of the compatibility of the use of depleted uranium (DU) weapons with the principle of international humanitarian law prohibiting superfluous injury or unnecessary suffering.

This assumes as a working hypothesis that there is potential incompatibility. Such a working hypothesis is justified by expressions of public opinion and, to a lesser extent, official reports that link the use of DU weapons with serious health effects.[2] The health effects referred to concern civilian populations or the military personnel of the forces using DU weapons, whereas the principle prohibiting superfluous injury and unnecessary suffering is concerned with the health effects of the use of weapons on military personnel of the opposing forces. It is obvious that the effects of DU will be the same for all concerned, however, as long as they are exposed to it in the same way. This explains why the compatibility of DU with the principle prohibiting superfluous injury and unnecessary suffering has been questioned, including by certain legal experts.[3]

This chapter first discusses the origins of the principle prohibiting superfluous injury and unnecessary suffering. It then examines the interpretation of the principle. The chapter continues with a discussion of the relationship between the principle and the so-called Marten's clause. It concludes with an analysis of the compatibility of the use of DU with the principle prohibiting superfluous injury and unnecessary suffering.

---

[1] This chapter was written in a personal capacity. The views presented in this chapter do not necessarily represent the views of the Ministry of Defence of The Kingdom of the Netherlands or any other part of the government of The Kingdom of the Netherlands.

[2] See D. Fahey, Chapter 2.

[3] See, e.g., M. Byers, The laws of war, US-style', 25 *London Review of Books* (2003).

McDonald / Kleffner / Toebes (eds.), Depleted Uranium Weapons and International Law
© 2008, T·M·C·ASSER PRESS, *The Hague, The Netherlands and the Authors*

2.   THE PRINCIPLE OF THE PROHIBITION OF SUPERFLUOUS INJURY OR
        UNNECESSARY SUFFERING

2.1   Treaty provisions

The principle prohibiting the use of weapons, projectiles or *matériel*[4] of a nature to
cause superfluous injury or unnecessary suffering was first codified in the Preamble to the 1868 St. Petersburg Declaration.[5] This Declaration prohibited the use
in time of war of explosive or inflammable projectiles with a weight of less than
400 grams. The significance of the Declaration lay in the general principle underlying that specific prohibition, set out in the Preamble, the relevant part of which
states:

> 'That the only legitimate object which States should endeavour to accomplish
> during war is to weaken the military forces of the enemy;

> That for this purpose it is sufficient to disable the greatest possible number of
> men;

> That this object would be exceeded by the employment of arms which uselessly
> aggravate the sufferings of disabled men, or render their death inevitable.

> That the employment of such arms would, therefore, be contrary to the laws of
> humanity.'

A similar provision was included in the 1874 Brussels Project,[6] this time in the
operative part. The principle was included as Article 23(e) of the Hague Regulations of 1899[7] and the later version of 1907.[8] The authentic French text of the 1899
Hague Regulation reads: 'Il est [...] interdit [...] d'employer des armes, des projectiles ou des matières propres à causer des maux superflus.' The 1899 English translation in the relevant part referred to 'arms, projectiles and materials of a nature to
cause *superfluous injury*', but the 1907 translation was changed to refer to 'weapons [...] calculated to cause *unnecessary suffering*'.
    When the principle was included in 1977 Protocol I Additional to the 1949
Geneva Conventions (AP I)[9] in Article 35(2), both the terms 'superfluous injury' as

---

[4] Where the term 'weapons' is used in this article, it includes projectiles and *matériel*.

[5] Declaration Renouncing the Use, in Time of War, of Explosive Projectiles under 400 Grammes
Weight, 11 December 1868 (1907 Supp.) 1 *AJIL* (1907) p. 95.

[6] Project of an International Declaration Concerning the Laws and Customs of War, Brussels,
27 August 1874.

[7] Convention (II) with Respect to the Laws and Customs of War on Land and its annex, Regulations Concerning the Laws and Customs of War on Land (1899), 26 *Martens Nouveau Receuil* (ser. 2)
p. 949.

[8] Convention (IV) with Respect to the Laws and Customs of War on Land and its annex, Regulations Concerning the Laws and Customs of War on Land (1907), 3 *Martens Nouveau Receuil* (ser. 3)
p. 461.

[9] 1977 Geneva Protocol I Additional to the Geneva Conventions of 12 August 1949, and Relating
to the Protection of Victims of International Armed Conflicts, 16 *ILM* (1977) p. 1391.

well as 'unnecessary suffering' were included. Article 35(2) AP I provides: 'It is prohibited to employ weapons, projectiles and materials and methods of warfare of a nature to cause superfluous injury or unnecessary suffering.'

2.2   Customary law

As well as being codified in several legal instruments, the principle prohibiting superfluous injury and unnecessary suffering is a norm of international customary law. This is uniformly stated by different authors, who frequently reserve a special place for it within the category of norms of international customary law.[10] The Nuremberg Tribunal concluded that, by 1939, the Hague Regulations 'were recognized by all civilized nations and were regarded as being declaratory of the laws and customs of war'.[11] In his report on the establishment of the International Criminal Tribunal for the former Yugoslavia (ICTY), the UN Secretary-General noted that the 1907 Hague Regulations were one of the parts of conventional international humanitarian law that had beyond doubt become part of customary law.[12]

In 1996 the International Court of Justice (ICJ) was categorical on the customary nature of the principle in its Advisory Opinion on the Legality of the Threat or Use of Nuclear Weapons. The Court stated:

> 'the cardinal principles contained in the texts constituting the fabric of humanitarian law are the following. The first is aimed at the protection of the civilian population and civilian objects and establishes the distinction between combatants and non-combatants; States must never make civilians the object of attack and must consequently never use weapons that are incapable of distinguishing between civilian and military targets. According to the second principle, it is prohibited to cause unnecessary suffering to combatants: it is accordingly prohibited to use weapons causing them such harm or uselessly aggravating their suffering.'[13]

The Court added that: '[T]hese fundamental rules are to be observed by all States whether or not they have ratified the conventions that contain them, because they constitute intransgressible principles of international customary law.'[14]

In this way, the Court not only confirmed the customary nature of the principle but it also appears to have given it an elevated status above 'ordinary' customary law.

---

[10] Solf, for example, calls it one of two 'basic principles which underlie customary international humanitarian law', W. Solf, 'Basic rules', in M. Bothe, K.J. Partsch and W. Solf, eds., *New Rules for Victims of Armed Conflicts* (The Hague, Martinus Nijhoff 1982) pp. 192 at 193.

[11] The Nuremberg Judgment (International Military Tribunal 1945-1946), reprinted in L. Friedman, ed., 2 *The Law of War: A Documentary History* (New York, Random House 1972) pp. 922 at 961.

[12] Report of the Secretary-General Pursuant to Paragraph 2 of Security Council Resolution 808 (1993) of 3 May 1993, UN Doc. S/25704, para. 35.

[13] *Legality of the Threat or Use of Nuclear Weapons*, Advisory Opinion of 8 July 1996, 35 *ILM* (1996) pp. 809 and 1343 at 827, para. 78.

[14] Ibid., para. 79.

It has been asserted that the principle is equally applicable as a customary norm to international and non-international conflicts.[15] This assertion is supported by the finding of the Appeals Chamber of the ICTY in its decision on jurisdiction in the *Tadić* case. In this decision the Appeals Chamber referred to the principle whereby weapons or other *matériel* or methods prohibited in international armed conflicts must not be employed in any circumstances as an example of the gradual extension to internal armed conflict of rules and principles concerning international wars.[16] The extension of the scope of application of the Certain Conventional Weapons Convention to non-international armed conflicts in 2001 also suggests that, at the least, there is an emerging principle of superfluous injury and unnecessary suffering in the law of non-international armed conflict. According to the ICRC's Customary International Humanitarian Law study, 'State practice establishes this rule as a norm of customary international law applicable in both international and non-international armed conflict.'[17] At present this observation is not of great significance, because the conflicts in which the use of DU weapons has been reported have until now been limited to international armed conflicts, although this is unlikely to remain the case indefinitely as more states acquire these weapons.

3.   INTERPRETATION OF THE PRINCIPLE

3.1   **Scope of applicability**

It is clear from the passage of the *Nuclear Weapons* Advisory Opinion cited above that the principle applies only to combatants and not to non-combatants. The principle is not directly relevant to the protection of non-combatants because it is forbidden to make them the object of attack. Consequently, the question of which weapons may be used in attacking them does not arise. Other aspects of the principle are, however, subject to some controversy.

The first question is whether the principle constitutes a prohibition of particular categories of weapons in all cases, or whether it refers to the particular circumstances at hand. In the latter case the principle would be of the same kind as the principle of proportionality, which prohibits conduct in certain circumstances but not in others.[18] The fact that the principle calls for a balancing test suggests that the test can have a different outcome in different circumstances, and that therefore the principle refers to the particular circumstances at hand. This is confirmed by the ICRC Commentary on Article 35(2) of Additional Protocol I (hereinafter, AP I),

---

[15] H. Meyrowitz, 'The principle of superfluous injury or unnecessary suffering: from the Declaration of St. Petersburg of 1868 to Additional Protocol I of 1977', 34 *IRRC* (1994) pp. 98 at 121.

[16] *Prosecutor* v. *Duško Tadić*, Case No. IT-94-1-AR72, Decision on the Defense Motion for Interlocutory Appeal on Jurisdiction, A. Ch., 2 October 1995, para. 118.

[17] J.-M. Henckaerts and L. Doswald-Beck, *Customary International Humanitarian Law*, Vol. 1: Rules (Cambridge, Cambridge University Press 2005) p. 237.

[18] See Solf, *supra* n. 10, who maintains that the principle prohibiting superfluous injury and unnecessary suffering is 'another way of stating the rule of proportionality', at p. 195.

which gives the example of the protection of combatants from incendiary weapons such as flame-throwers or napalm. It states that while the principle prohibiting superfluous injury and unnecessary suffering does not protect them from such weapons, it is generally admitted that these weapons should not be used in such a way that they will cause unnecessary suffering, which means that, in particular, they should not be used against individuals without cover.[19] Such an interpretation does not preclude the principle from prohibiting the use of a weapon in all circumstances. This is the case if the weapon concerned is designed to cause excessive suffering, in which case it is designed to do so in all circumstances. In this case, the infliction of suffering is for its own sake.[20] This interpretation of the principle is laid down in the Fourth Protocol[21] to the Conventional Weapons Convention[22] prohibiting the use of blinding laser weapons. Article 1 of this Protocol states, in part, that 'it is prohibited to employ laser weapons specifically designed as their sole combat function or as one of their combat functions, to cause permanent blindness to unenhanced vision.'

It has been suggested that the principle prohibiting superfluous injury and unnecessary suffering can also prohibit a weapon as such, and not only its use in particular circumstances, if excessive suffering is caused in the majority of cases of the use of a weapon, even if it is not designed to have this effect.[23] According to this interpretation, the principle could serve as a prohibitory rule without having to take into account every conceivable circumstance in which a weapon could be used.[24] Judge Shahabuddeen takes this approach rather far in his Dissenting Opinion in the *Nuclear Weapons* Advisory Opinion. He states that: 'in judging of the admissibility of a particular means of warfare, it is necessary, in my opinion, to consider what the means can do in the ordinary course of warfare, even if it may not do it in all circumstances.'[25] There appears to be no state practice to support this interpretation, however, and it is unlikely to be accepted in practice.

---

[19] Y. Sandoz, C. Swinarski and B. Zimmerman, eds., *Commentary on the Additional Protocols of 8 June 1977 to the Geneva Conventions of 12 August 1949* (Geneva, International Committee of the Red Cross 1987) (hereinafter, ICRC Commentary) p. 406, para. 1424.

[20] See, e.g., United States Air Force Intelligence Targeting Guide, Air Force Pamphlet 14-210 Intelligence, 1 February 1998, Attachment 4 (Targeting and International Law).

[21] Additional Protocol to the Convention on Prohibitions or Restrictions on the Use of Certain Conventional Weapons Which May Be Deemed Excessively Injurious or to Have Indiscriminate Effects (Protocol IV on Blinding Laser Weapons), 13 October 1995, 35 *ILM* (1996) p. 1218.

[22] Convention on Prohibitions or Restrictions on the Use of Certain Conventional Weapons Which May Be Deemed Excessively Injurious or to Have Indiscriminate Effects, reprinted in 19 *ILM* (1980) p. 1523.

[23] L. Doswald-Beck, 'Lawfulness of the anti-personnel use of laser weapons', in L. Doswald-Beck, ed., *Blinding Weapons, Reports of the Meetings of Experts Convened by the International Committee of the Red Cross on Battlefield Laser Weapons 1989-1991* (Geneva, ICRC 1993) pp. 330 at 331. The ICRC Commentary on Art. 35(2) AP I is ambiguous in stating that the article: 'lays down a prohibition relating to the results produced, though not directly a prohibition of the means'. ICRC Commentary, *supra* n. 19, at p. 404.

[24] J.H. McCall, Jr., 'Blinded by the light: International law and the legality of anti-optic laser weapons', 30 *Cornell International Law Journal* (1997) pp. 1 at 26.

[25] Nuclear Weapons Advisory Opinion, *supra* n. 13, Dissenting Opinion of Judge Shahabuddeen.

It is necessary to make a distinction at this point between the question whether a weapon has been designed to cause excessive suffering or whether it causes such suffering in fact, and the question whether excessive suffering is caused in virtually all cases or simply in a majority of cases. With respect to the first question, the mental state of the designers of a weapon will generally be very difficult to establish. For this reason there was an understandable wish to lay down objective criteria to 'objectify' this subjective standard. The so-called 'SIrUS project' was an outcome of this wish. A group of experts, mainly from the health sector, united in this project to attempt to establish standards to determine whether a weapon is of a nature to cause superfluous injury or unnecessary suffering.[26] On the basis of medical data, the project established a series of 'baselines' relating to injury and suffering resulting from the effects of conventional weapons. These are the design-dependent, foreseeable health effects of weapons that do not constitute superfluous injury or unnecessary suffering. Any other foreseeable effects of weapons would therefore constitute superfluous injury or unnecessary suffering. In other words, the project attempted to provide standards to objectify a subjective determination. This must be distinguished from the question whether, according to these standards, a weapon causes excessive suffering in a majority of cases or in (virtually) all cases. In the former case, the result of the application of the principle is that the use of the weapon concerned would be legal in some cases and illegal in others. Only in the latter case could it validly be argued that the weapon is prohibited as such. The SIrUS project encountered much opposition from states.[27]

A second contentious matter is whether the principle applies to anti-personnel weapons but not to anti-*matériel* weapons. There is substantial evidence for that position. The examples of prohibited weapons given in military manuals are limited to anti-personnel weapons. The United States Field Manual, for example, states that usage has established the illegality of the use of lances with barbed heads, irregular-shaped bullets, and projectiles filled with glass, the use of any substance on bullets that would tend unnecessarily to inflame a wound inflicted by them, and the scoring of the surface or the filing off of the ends of the hard cases of bullets.[28] In addition, specific treaty prohibitions of weapons on the basis of the principle prohibiting weapons of a nature to cause superfluous injury or unnecessary suffering have hitherto been limited to anti-personnel weapons. There is also support for this view in the literature. Solf gives the following example:

'An artillery projectile or missile designed to destroy field fortifications or heavy material may be expected to cause injuries to personnel in the vicinity of

---

[26] See for further details R.M. Coupland and P. Herby, 'Review of the legality of weapons: A new approach, the SIrUS Project', 81 *IRRC* (1999) at p. 583.

[27] See, e.g., the report of a meeting discussing this and others topics organised by the ICRC in 2001, which states that 'the ICRC's proposals were not broadly accepted in the form presented in the SirUS project', ICRC Expert Meeting on Legal Reviews of Weapons and the SirUS Project, Jogny sur Vevey, 29-31 January 2001, 83 *IRRC* (2001) pp. 539 at 541.

[28] United States Field Manual 27-10, The Law of Land Warfare, Department of the Army, 18 July 1956, Chapter 2, Section III, para. 34(b).

the target which would be more severe than necessary to render these combatants hors de combat, but no authority has questioned the lawfulness of such projectiles despite the gravity of their incidental effect on personnel.'[29]

Limiting the scope of application of the principle to anti-personnel weapons is consistent with its restrictive interpretation as applicable only to weapons designed to cause superfluous injury or unnecessary suffering: anti-personnel weapons are designed with combatants and the amount of suffering and injury inflicted on them in mind while anti-*matériel* weapons are not (at least not primarily).

There are good arguments for applying the principle to anti-*matériel* as well as to anti-personnel weapons. This position is much more in tune with the nature of modern warfare insofar as it involves developed states. Operation Allied Force, Operation Enduring Freedom and Operation Iraqi Freedom demonstrated that modern inter-state warfare to a large extent involves bombing the adversary from a distance with the help of satellites and guided missiles. These missiles largely target military or combined military/civilian *matériel* and installations rather than individual combatants or formations of combatants. The statement in the Preamble of the 1868 St. Petersburg Declaration that to weaken the military forces of the enemy it is sufficient to disable the greatest possible number of combatants no longer gives the whole picture at a time when the destruction of military *matériel*, the restriction of military movement, the interdiction of military lines of communication and the weakening of the enemy's war-making resources are often of much more military importance than disabling men and women. The principles of international humanitarian law, including the principle prohibiting superfluous injury and unnecessary suffering, must be interpreted against the background of this reality. However, it is fair to say that these considerations do not represent current international law on the issue, as is confirmed by the state practice discussed above.[30]

A further controversial issue is whether the principle is autonomous, i.e., whether it can prohibit the use of a certain weapon by its own force, or whether specific treaty prohibitions are necessary to give effect to the principle. Sir David Hughes-Moran of the United Kingdom, in his legal analysis delivered at the 1974 Lucerne Conference – a conference of experts which discussed conventional weapons and which preceded the 1980 Conference on Certain Conventional Weapons – stated:

> '[A] prohibition so framed is of value in forbidding the use of any weapon which, while inflicting severe wounds, has no corresponding military significance. The difficult cases, in which the results of the proportionality equation are in dispute, can only be resolved by the agreement of states to prohibit the use of specific weapons.'[31]

---

[29] Solf, *supra*. n. 10, at pp. 196-197. But see Meyrowitz, *supra* n. 15 at pp. 109-110.

[30] For a contrary interpretation supporting the applicability of the principle to anti-personnel as well as to anti-*matériel* weapons, see A. McDonald, Chapter 12 of this book.

[31] Cited in F. Kalshoven, 'Conventional weaponry: The law from St. Petersburg to Lucerne and beyond', in M. Meyer, ed., *Armed Conflict and the New Law* (London, British Institute of International and Comparative Law 1989) pp. 251 at 260.

This point of view has been defended by referring to the *travaux préparatoires* of the St. Petersburg Declaration and the scarcity of occasions on which the principle has been invoked in state practice.[32] According to Cassese, the principle is an empty shell, which must be filled by specific prohibitions.[33] The ICRC Commentary to Article 35(2) appears to support this point of view.[34]

The better view, however, is that the principle prohibiting weapons of a nature to cause superfluous injury or unnecessary suffering is autonomous.[35] One reason is that every treaty rule must be presumed to have a meaning and not be superfluous, as expressed by the maxim *ut res magis valeat quam pereat*. Since the rule is expressed autonomously and without reference to specific weapons regulated in treaty law, it must be presumed to operate independently of such specific prohibitions. Another reason is that the history of the principle's formulation in humanitarian law instruments leads to the conclusion that it was separated from specific prohibitions.[36] Thirdly, accepting the point of view that the superfluous injury and unnecessary suffering principle is merely an empty vessel would lead to the absurd conclusion that if a new weapon was introduced that all states agreed caused superfluous injury or unnecessary suffering, its use would not be prohibited until a diplomatic conference could be convened. Finally, if the principle did not have an autonomous meaning it would deprive Article 36 of AP I of any meaning in relation to Article 35(2). The former provision, requiring parties to the Protocol to make sure that a new weapon is not prohibited by international law, only makes sense if the principle is given a generic meaning. In this context, it is interesting that some authors recommend weapons reviews on the basis of Article 36 as the best approach toward determining whether a specific weapon is in breach of the principle prohibiting superfluous injury and unnecessary suffering. They present this mechanism as an alternative to the conference mechanism.[37]

A last bone of contention is whether the principle codified in Article 35(2) AP I only applies to weapons developed before 1977 and therefore cannot function as a prohibitory rule. In discussions in the UN General Assembly in 1973, several states 'expressed the conviction that the principle [...] has no legal impact on new weapons causing unnecessary suffering.'[38] However, the ICJ firmly rejected this approach when it stated in the *Nuclear Weapons* Advisory Opinion that 'such a conclusion would be incompatible with the intrinsically humanitarian character of

---

[32] See, e.g., H. Meyrowitz, 'Problèmes juridiques relatifs à l'arme à neutrons', 27 *AFDI* (1981) p. 108.

[33] A. Cassese, 'Weapons causing unnecessary suffering: Are they prohibited?', 58 *Rivista di Diritto Internazionale* (1975) pp. 18 at 34.

[34] ICRC Commentary, *supra* n. 19, p. 402.

[35] For an extensive discussion supporting this conclusion see E. David, *Principes de Droit des Conflits Armés* (Brussels, Bruylant 1999) pp. 287-293.

[36] Ibid., pp. 287-288.

[37] See D.M. Verchio, 'Just say no! The SirUS Project: Well-intentioned, but unnecessary and superfluous', 51 *Air Force Law Review* (2001) p. 183.

[38] Cassese, *supra* n. 33 at p. 34.

the legal principles in question which [...] applies [...] to all kinds of weapons, those of the past, those of the present and those of the future.'[39]

3.2   A balancing test

The principle calls for a balancing of humanitarian concerns (the suffering and injury inflicted) and military necessity. The test is not whether there is a vast amount of injury or suffering but whether the suffering is needless, superfluous or manifestly disproportionate to the military advantage reasonably expected from the use of the weapon.[40] The comparison must to a certain extent be of a subjective character, because the values to be balanced are of a different nature and neither side of the equation is easy to quantify.

3.2.1   *The humanitarian side of the balancing test*

The relevant factors on the humanitarian side are the injury and suffering inflicted. The 1899 Hague Regulations only referred to 'superfluous injury' in the English text, while the 1907 version only referred to 'unnecessary suffering'. In Article 35(2) AP I both terms have been included to make clear that the principle expresses 'simultaneously the sense of moral and physical suffering'.[41]

It is not easy to find objective standards to measure injury and suffering. 'Suffering' poses most difficulties in this regard because it is more a psychological than an 'objective' medical term. For this reason, the trend has been to focus on 'injury'.[42] Even for 'injury' it is difficult to come up with practicable criteria. One criterion that is used is the incidence of permanent damage or disfigurement.[43] The more health effects are lasting, the less likely they will be justified by military necessity.

A second criterion is the feasibility of treatment under field conditions. This feasibility decreases with the necessity of sophisticated equipment and expertise for the treatment of injuries. It also decreases with the lack of information on the health effects of a particular weapon. Other criteria that are suggested are the painfulness or severity of the wound and mortality rates.[44]

As noted earlier, the SIrUS project attempted to establish standards to determine whether a weapon is of a nature to cause superfluous injury or unnecessary suffering. On the basis of medical data, the project established a series of 'baselines' relating to injury and suffering resulting from the effects of conventional weapons. These are the design-dependent, foreseeable health effects of weapons that do not constitute superfluous injury or unnecessary suffering. Any other foreseeable ef-

---

[39] Nuclear Weapons Advisory Opinion, *supra* n. 13, para. 86.
[40] Solf, *supra* n. 10, at 196. See further B. Carnahan, Chapter 4 of this book.
[41] ICRC Commentary, *supra* n. 19 at p. 407.
[42] Ibid., p. 408.
[43] Solf, *supra* n. 10 at p. 196.
[44] Ibid.

fects of weapons would therefore constitute superfluous injury or unnecessary suffering. One of the four criteria that was proposed by the project was that a weapon exert effects for which there is no well recognised and proven treatment. The experts held in this context that: '[t]he effects of other new weapons are not fully known and so treatment is unlikely to be successful.'[45] The ICRC proposed that the data on weapons injuries gathered by the project 'be taken into account in determining which weapons may cause superfluous injury or unnecessary suffering.'[46] The organisation stressed that the standards developed by the project were only a tool and did not give a definition of the notion of superfluous injury or unnecessary suffering. In other words, there would still be a need for a balancing test.

3.2.2   *The military necessity side of the balancing test*

Military necessity is the other part of the equation. The test, set out in the St. Petersburg Declaration quoted above, is whether the suffering or injury is needless, superfluous or manifestly disproportionate to the military advantage reasonably expected from the use of the weapon.[47] That this view of military necessity was also held by a number of the states that negotiated Article 35(2) AP I is clear from the report of the Rapporteur of the committee in question, Committee III, who stated that: 'several representatives wished to have it recorded that they understood the injuries covered by that phrase to be limited to those which were more severe than would be necessary to render an adversary hors de combat.'[48] The ICRC Commentary on AP I also adopts this view.[49] However, as submitted above, the Preamble of the St. Petersburg Declaration no longer adequately reflects the reality of modern warfare. Therefore, other factors must also be taken into account, such as a weapon's effectiveness in neutralising military *matériel*, interdicting military lines of communication and weakening the enemy's war-making resources.

A number of criteria are suggested to measure the military necessity of a weapon. One is the availability of alternative weapons systems. Other criteria are the logistics of providing the weapon at the place where it is to be used when needed and the security of the troops involved.[50]

4.   THE MARTENS CLAUSE

Closely linked to the prohibition of the use of weapons, projectiles and materials of a nature to cause superfluous injury or unnecessary suffering is the Martens Clause.

---

[45] R.M. Coupland, 'The SirUS Project: Towards a determination of which weapons cause 'superfluous injury or unnecessary suffering'', in H. Durham and T.L.H. McCormack, eds., *The Changing Face of Conflict and the Efficacy of International Humanitarian Law* (The Hague, Martinus Nijhoff 1999) pp. 99 at 114.

[46] Statement by the ICRC to the United Nations General Assembly, 54th session, First Committee, 20 October 1999.

[47] See *supra* n. 5.

[48] O.R. XV, at p. 267, CDDH/215/Rev.1, para. 21.

[49] ICRC Commentary, *supra* n. 19 at p. 409.

[50] Solf, *supra* n. 10 at p. 197. See further B. Carnahan, Chapter 4 of this book.

This clause first featured in the Preamble to the 1868 St. Petersburg Declaration. A modern version is to be found in Article 1(2) of AP I, which reads as follows:

> 'In cases not covered by this Protocol or by other international agreements, civilians and combatants remain under the protection and authority of the principles of international law derived from established custom, from the principles of humanity and from the dictates of public conscience.'

The clause was meant to address the rapid evolution of military technology.[51] If legal developments have not been able to keep pace with technological developments, the Martens Clause rules out the argument that just because there is no specific prohibition it follows that a means or method of warfare is legal.[52] The ICJ in the Nuclear Weapons Advisory Opinion specifically referred to the Martens Clause in relation to the unnecessary suffering of combatants.[53] But it did not specify what consequences the clause has for the principle. In his dissenting opinion, Judge Shahabuddeen did take this additional step. In his opinion, the Martens Clause can in and of itself prohibit what is not prohibited in conventional or customary law:

> 'Thus, the Martens Clause provided its own self-sufficient and conclusive authority for the proposition that there were already in existence principles of international law under which considerations of humanity could themselves exert legal force to govern military conduct in cases in which no relevant rule was provided by conventional law.'[54]

This resembles the interpretation given to the clause by the US Military Tribunal in the *Krupp* case, which said that the clause:

> 'is much more than a pious declaration. It is a general clause, making the usages established among civilized nations, the laws of humanity and the dictates of public conscience into the legal yardstick to be applied if and when the specific provisions of the Convention and the Regulations annexed to it do not cover specific cases occurring in warfare, or concomitant to warfare.'[55]

It is submitted, however, that 'principles of humanity' and 'dictates of the public conscience' provide no firm legal basis to prohibit the use of weapons by themselves.[56] This conclusion would seem to be supported by the fact that the majority of judges in the *Nuclear Weapons* Advisory Opinion did not consider the Martens

---

[51] Nuclear Weapons Advisory Opinion, *supra* n. 13, para. 78.

[52] See generally T. Meron, 'The Martens Clause, principles of humanity, and dictates of public conscience', 94 *AJIL* (2000) p. 78.

[53] Nuclear Weapons Advisory Opinion, *supra* n. 13 at p. 78.

[54] Ibid., Dissenting Opinion Judge Shahabuddeen, para. 408.

[55] *In re Krupp and others*, 15 *Ann. Dig.* (U.S. Mil. Trib. 1948) pp. 620, 622.

[56] C. Greenwood, 'Historical development and legal basis', in D. Fleck, ed., *Handbook of Humanitarian Law in Armed Conflicts* (Oxford, Oxford University Press 1995) p. 129.

Clause as an autonomous rule of international humanitarian law.[57] As presiding Judge Cassese of Trial Chamber II of the ICTY stated in the *Kupreskić* case:

> 'True, this Clause may not be taken to mean that the "principles of humanity" and the "dictates of public conscience" have been elevated to the rank of independent sources of international law, for this conclusion is belied by international practice. However, this Clause enjoins, as a minimum, reference to those principles and dictates any time a rule of international humanitarian law is not sufficiently rigorous or precise: in those circumstances the scope and purport of the rule must be defined with reference to those principles and dictates.'[58]

5.   CONCLUSION: IS THE USE OF DEPLETED URANIUM PROHIBITED BY THE PRINCIPLE?

It was submitted above that under current international law the principle prohibiting weapons of a nature to cause superfluous injury or unnecessary suffering applies to anti-personnel weapons but not to anti-*matériel* weapons. In consequence, the principle would only apply to the use of DU weapons in very limited circumstances, given that DU weapons are mostly used to penetrate armour.[59]

Whether the use of DU weapons against personnel is prohibited by the principle depends on the outcome of a balancing test between its adverse health effects on combatants and the military necessity of using this type of ammunition. It has been submitted above that its use is prohibited as such if the injury or suffering caused is disproportionate to the military necessity in (virtually) all cases. In this case, the weapon must have been designed to cause superfluous injury or unnecessary suffering, even if this subjective criterion cannot be established directly but only using objective standards. If the use of DU is not generally prohibited on this ground, it may be that it violates the superfluous injury and unnecessary suffering principle in specific circumstances. Whether this is the case must be determined on the basis of the balancing test discussed above and can only be determined on a case-by-case basis.

On the humanitarian side of the equation, it is difficult to quantify the injury and suffering caused by DU due to incomplete information. What can be said is that at least some of the potential effects of DU, in particular cancer and kidney disease, would in many cases lead to permanent damage.

It can be queried if the question raised by the SIrUS project as applied to DU, i.e., whether the effects of DU could be treated in the field, would have served any useful purpose. The potential effects of DU seem to be such that manifest themselves only after a certain period of time. In view of the many uncertainties regarding the actual health consequences of the use of DU weapons, the statement by the

---

[57] Nuclear Weapons Advisory Opinion, *supra* n. 13.

[58] *Prosecutor* v. *Kupreskić*, Case No. IT-95-16-T, Judgment, Tr. Ch. II, 14 January 2000, para. 525.

[59] See D. Fahey in Chapter 1.

SIrUS project that '[t]he effects of other new weapons are not fully known and so treatment is unlikely to be successful'[60] should be noted.

On the military necessity side of the equation, it has been noted above that factors other than the effectiveness of rendering combatants *hors de combat* should be taken into account. The attractiveness of DU is partly based on its ability to penetrate armour.[61] There seems to be no doubt that DU is highly effective in this regard. The United Kingdom government, for example, justifies the use of DU with the argument that: 'no satisfactory alternative material to DU exists to achieve the level of penetration necessary to defeat Main Battle Tanks'.[62] Another aspect of DU's attractiveness lies in its availability in large quantities and its cost effectiveness. These were some of the reasons given by the US Army for switching from tungsten to DU in a 1974 report.[63]

A factor that is partly determinative of DU's military necessity is the availability – or lack of – of alternatives. Tungsten is the main competitor with DU for use in armour-piercing penetrators. The density of available tungsten alloys is roughly equal to that of DU. There are conflicting statements as to the comparative advantages of tungsten and DU. The Pentagon states that 'depleted uranium will penetrate more armor of a given character and type at a given range than tungsten will'.[64] On the other hand, the US Navy in 1989 switched back to the use of tungsten primarily because of the 'overwhelming performance advantages' of tungsten, according to a Navy spokesman.[65] In view of these conflicting statements, it is difficult to draw definite conclusions concerning the relative penetrating capacities of tungsten and DU.

Another criterion is the security of the troops involved. The need to take precautionary measures to protect one's own personnel from the possible effects of DU detracts from its military necessity. This need has arisen in respect of DU. The United Kingdom, for example, issues detailed safety instructions to all service personnel and to all units serving in operational theatres where there may be a risk of exposure. The Dutch Minister of Defence has stated that Dutch service personnel are informed of possible health effects of ionising radiation, and that special attention is paid to precautionary measures.[66]

---

[60] *Supra* n. 45.

[61] For a discussion of aspects of military necessity of DU weaponry, see B. Carnahan, Chapter 4.

[62] Government Response to the House of Commons Select Committee's Seventh Report – Gulf Veteran's Illnesses, Annex A, <http://www.mod.uk/index.php3?page=1537>.

[63] Department of the Army, Office of The Judge Advocate-General, DAJA-IO Memorandum for US Army Armament Research, Development and Engineering Center, Subject: M829A2 Cartridge, 120 mm, APFSDS-T (Depleted Uranium Tank Round); Law of War Review, 27 December 1994. For analysis of this report see B. Carnahan, Chapter 4, and A. McDonald, Chapter 12 of this book.

[64] United States Department of Defense, Briefing on Depleted Uranium, 14 March 2003, <http://www.defenselink.mil/news/Mar2003/t03142003_t314depu.html>.

[65] Cited in S. Peterson, 'Tungsten: One alternative to a risky 'favorite round?' *Christian Science Monitor* (30 April 1999).

[66] Handelingen TK 1999-2000, 1089 (answers of the Netherlands Minister of Defence to questions by Members of Parliament).

It is fair to conclude that the military effectiveness of DU is recognised but that evidence of significant health effects resulting from exposure to DU is currently lacking. It must be underlined that the latter is at least in part due to the relative dearth of scientific and medical studies on exposed populations and contaminated battlefields. It is important that further research be conducted into the health effects of DU but, at this stage, it cannot be said that there is evidence that the use of DU weapons causes superfluous injury and unnecessary suffering.[67] It would be commendable if the states that possess DU would err on the side of caution and abstain from its use until more research is done. However, this is not a very realistic assumption. Armies cannot be expected to ban a weapons system on the basis of the possibility of some unspecified future complications. Of course, if ongoing or future research finds evidence of significant health effects resulting from the use of DU weapons, their legality should be reconsidered. Until then, it must be concluded that their use does not violate the principle prohibiting weapons of a nature to cause superfluous injury or unnecessary suffering.

---

[67] But see M. Mohr, 'Uranwaffeneinsatz: eine humanitär-völkerrechtliche Standortbestimmung', 11 *Humanitäres Völkerrecht – Informationsschriften* (2001) p. 27.

Chapter 6
THE USE OF DEPLETED URANIUM AND THE PRINCIPLES OF DISTINCTION, PROPORTIONALITY AND PRECAUTION

Jann K. Kleffner and Théo Boutruche

1.   INTRODUCTION

While undeniable military advantages of the use of depleted uranium (DU) exist,[1] certain evidence suggests that DU ammunition and armour, when shot or breached, corroding or consumed by fire, have several health impacts on the civilian population, which range from cancer to reproductive and developmental effects.[2] Conclusive evidence to that effect may only emerge after a considerable time has lapsed or, indeed, not at all. Further research is clearly required and, pending any definite result, states using DU weaponry cannot be expected to know the unknown. However, neither can civilians who might potentially be affected be left without any (legal) protection in the light of current research on the subject, which limits the 'unknown' at least to some extent. The dichotomy of these two considerations is in part reflected in the diametrically opposed positions on DU of those who claim that the use of DU weaponry is permissible as long as no conclusive evidence to the contrary is available, on the one hand,[3] and those declaring such use to be generally prohibited on the basis of the evidence that is currently available, on the other.[4] The precautionary approach to the assessment of the legality of the use of DU weaponry

---

[1] See B. Carnahan, Chapter 4 of this book.

[2] See D. Fahey, Chapter 2 of this book.

[3] In the words of Ms Aileen Carroll, Parliamentary Secretary to the Minister of Foreign Affairs of Canada, '[t]he use of depleted uranium munitions is not prohibited or restricted under the 1980 UN Convention on prohibitions or restrictions on the use of certain conventional weapons and related protocols, nor is it otherwise prohibited by international humanitarian law. This is because it is not deemed to be excessively injurious or to have indiscriminate effects.' Canada, 37th Parliament, 2nd Session, edited Hansard Number 114, 9 June 2003, <www.parl.gc.ca/37/2/parlbus/chambus/house/debates/114_2003-06-09/han114_1845-E.htm>.

[4] See, for instance, the Application of the Government of the Federal Republic of Yugoslavia in Legality of Use of Force (*Serbia and Montenegro* v. *Belgium et al.*) requesting the International Court of Justice to adjudge and declare, *inter alia*, that 'by taking part in the use of weapons containing depleted uranium, [the respondent States have] acted against the Federal Republic of Yugoslavia in breach of [their] obligation not to use *prohibited weapons* and not to cause far-reaching health and environmental damage'. (Emphasis added.) International Court of Justice, Application instituting proceedings filed in the Registry of the Court on 29 April 1999, General List No. 105.

McDonald / Kleffner / Toebes (eds.), Depleted Uranium Weapons and International Law
© 2008, T·M·C·ASSER PRESS, *The Hague, The Netherlands and the Authors*

pursued in the present book[5] seeks to move beyond this dichotomy by analysing those rules of international humanitarian law that could be at play if the present concerns were to materialise.

It is the aim of the present chapter to adopt such a precautionary approach in assessing DU weaponry in the light of one of the core principles of international humanitarian law, namely the principle of distinction between combatants and military objectives, on the one hand, and civilians and civilian objects, on the other. To that end, the principle of distinction and its regulation in contemporary international humanitarian law will briefly be introduced in order to set the groundwork for our analysis (section 2). We will then turn to a more detailed analysis of the question whether and to what extent the use of DU potentially violates specific rules that derive from the principle of distinction. Amongst these rules, Article 51 of Additional Protocol I (hereinafter, AP I) on the protection of the civilian population and the corresponding rule of customary international humanitarian law stands central (section 3). Additional rules of pertinence to the legal assessment of DU weaponry are the prohibition to render useless objects indispensable to the survival of the civilian population (section 4) and the obligation to take precautionary measures when using DU (section 5).

2.   THE PRINCIPLE OF DISTINCTION

The protection of civilians and civilian objects from attacks and dangers arising from military operations during armed conflict is one of the most fundamental objectives of international humanitarian law. Achieving this objective requires that parties to a conflict distinguish between combatants and the civilian population as well as between military objectives and civilian objects and direct their operations only against combatants and other military objectives.[6] The principle of distinction, which is the most fundamental expression of the principle of limited warfare, encapsulates this aim. The principle finds its legal basis in both conventional[7] and customary[8] international law and translates into a number of more specific rules.

---

[5] See Introduction.

[6] As to who qualifies as 'combatant', see Arts. 4(A)(1)-(3) and (6) Geneva Convention (GC) III and Arts. 43-44 Additional Protocol (AP) I. 'Civilians' are all those who do not qualify as combatants thus defined, cf., Art. 50 AP I. 'Military objectives' are those 'which by their nature, location, purpose or use make an effective contribution to military action and whose total or partial destruction, capture or neutralization, in the circumstances ruling at the time, offers a definite military advantage', cf., Art. 52(2) AP I. Although the definition is limited to 'objects', it is generally recognised that members of the armed forces are military objectives, cf., Y. Sandoz, C. Swinarski and B. Zimmermann, eds., *Commentary on the Additional Protocols of 8 June 1977 to the Geneva Conventions of 12 August 1949* (Dordrecht, Martinus Nijhoff Publishers/Geneva, ICRC 1987) p. 635, para. 2017 (hereinafter, Commentary); Y. Dinstein, *The Conduct of Hostilities under the Law of International Armed Conflict* (Cambridge, Cambridge University Press 2004) pp. 84-85. On the customary status of the aforementioned rules, see J.M. Henckaerts and L. Doswald-Beck, eds., *Customary International Humanitarian Law* (Cambridge, Cambridge University Press 2005) Vol. I: Rules, Rules 3-9.

[7] Cf., Art. 48 AP I.

[8] Henckaerts and Doswald-Beck, *supra* n. 6, Rules 1 and 7. Legality of the Threat or Use of Nuclear Weapons, Advisory Opinion of 8 July 1996, 35 *ILM* (1996) pp. 809, 827, paras. 78-79, where the Court

These rules include those on the general protection of the civilian population and individual civilians, including the prohibition of indiscriminate attacks, and on the general protection of civilian objects.[9] Further rules specifically protect cultural objects and places of worship[10] as well as objects indispensable to the survival of the civilian population,[11] the natural environment[12] and works and installations containing dangerous forces.[13] Additionally, the principle of distinction gives rise to rules that oblige parties to armed conflicts and their individual members to take precautions in attack and against the effects of attacks.[14]

Issues raised by the use of DU ammunition and armour with regard to the protection of the natural environment are addressed in more detail in another contribution to this book,[15] while the rules protecting civilian or cultural objects and places of worship and those protecting works and installations containing dangerous forces do not raise any questions peculiar to the use of DU as it is immaterial whether such objects are attacked by DU or other weapons – such attacks are prohibited regardless of the specific weapon used. In the following, these rules will therefore be discarded from our analysis, which is confined to the idiosyncratic issues that the use of DU weaponry raises in relation to three expressions of the principle of distinction. These are, first, the rules of international humanitarian law on the protection of the civilian population, including the prohibition of indiscrimi-

---

included the principle of distinction in the 'cardinal principles contained in the texts constituting the fabric of humanitarian law' and as one of the fundamental rules that 'are to be observed by all States whether or not they have ratified the conventions that contain them, because they constitute *intransgressible* principles of international customary law', emphasis added; S. Oeter, 'Methods and means of combat', in D. Fleck, ed., *Handbook of Humanitarian Law in Armed Conflict* (Oxford, Oxford University Press 1995) p. 120, margin 404, at paras. 1 and 6. As regards non-international armed conflict, the applicability of the principle of distinction has been affirmed, *inter alia*, in UN GA Resolution 2675 (XXV) of 9 December 1970 'Basic principles for the protection of civilian populations in armed conflicts'; ICTY Appeals Chamber, *Prosecutor* v. *Duško Tadić*, Case No. IT-94-1-AR72, Decision on the Defense Motion for Interlocutory Appeal on Jurisdiction, 2 October 1995, para. 127; Inter-American Commission on Human Rights, *Abella* v. *Argentina (La Tablada)*, Case 11.137, judgment of 18 November 1997, OEA/Ser.L/V/II.98, para. 177.

[9] Arts. 51 and 52 AP I; Henckaerts and Doswald-Beck, *supra* n. 6, Rules 1, 7, 11-12.

[10] Art. 53 AP I; 1954 Hague Convention and Protocols; Henckaerts and Doswald-Beck, *supra* n. 6, Rules 38-40.

[11] Art. 54 AP I; Henckaerts and Doswald-Beck, *supra* n. 6, Rule 54.

[12] Art. 55 AP I; Henckaerts and Doswald-Beck, *supra* n. 6, Rules 43-45.

[13] Art. 56 AP I; Henckaerts and Doswald-Beck, *supra* n. 6, Rule 42.

[14] Arts. 57-58 AP I. As to the customary status of these rules, see C. Greenwood, 'Customary law status of the 1977 Geneva Protocols', in A.J.M. Delissen and G.J. Tanja, *Humanitarian Law of Armed Conflict – Challenges Ahead, Essays in Honour of Frits Kalshoven* (Dordrecht/Boston/London, Martinus Nijhoff 1991) pp. 93-114, p. 111, noting that these provisions 'are based on customary international law but go beyond it in certain respects'. Likewise, David notes that Art. 57 'codifies the principle and develops it substantially', but asserts that, with respect to the Kosovo War, 'NATO authorities have never disputed the applicability of the principle as embodied in Article 57', notwithstanding the fact that some NATO members, such as the United States and France, were not parties to AP I at the time of the war, E. David, 'Respect for the principle of distinction in the Kosovo war', 3 *YIHL* (2000) pp. 81 at 85. See generally on the customary status of the rule of precaution in attack and against the effects of attacks, Henckaerts and Doswald-Beck, *supra* n. 6, Rules 15-24.

[15] See E. Koppe, Chapter 8 of this book.

nate attacks; second, the prohibition to render useless objects indispensable to the survival of the civilian population; and third, the obligation to take precautionary measures. The ensuing analysis proceeds on the assumption that these three expressions of the principle of distinction, to be found in Articles 51, 54, 57 and 58 of AP I, largely reflect customary law, notwithstanding some differences in the exact wording of the rules of customary international humanitarian law if compared to the rules of AP I.

3.   DEPLETED URANIUM AND THE PROTECTION OF THE CIVILIAN
      POPULATION

The most pertinent treaty regulating the protection of the civilian population against the effects of hostilities is the First AP. Article 51 of AP I sets forth a number of rules on how this protection is to be achieved. The first introductory paragraph of the provision contains the general rule that 'the civilian population and individual civilians shall enjoy general protection against dangers arising from military operations'[16] and clarifies that the stipulated rules 'are additional to other applicable rules of international law'. This paragraph is followed by more detailed rules of application which aim at reducing dangers to the civilian population to a minimum.[17] These more meticulous rules fall into two different categories. A first category concerns attacks which are prohibited regardless of any expected military advantage gained by such an attack.[18] A second category concerns disproportionate attacks, i.e., those which 'may be expected to cause incidental loss of civilian life, injury to civilians, damage to civilian objects, or a combination thereof, which would be excessive in relation to the concrete and direct military advantage anticipated.'[19]

An assessment of DU weaponry in the light of these rules requires as a preliminary matter a determination of the – actual or potential – dangers for the civilian population and individual civilians arising from military operations involving its use (section 3.1). In the light of that determination, the following specific prohibitions in Article 51 AP I need to be analysed: first, the prohibition of making

---

[16] The term 'military operations' refers to all the movements and activities carried out by armed forces related to hostilities, Commentary, *supra* n. 6, p. 617, para. 1936.

[17] Commentary, *supra* n. 6, p. 617, para. 1935.

[18] Cf., Art. 51(2), (4), (5)(a), (6) and (7).

[19] Cf., Art. 51(5)(b). See on the distinction between 'absolute prohibitions' and the proportionality principle, H. Fischer, 'Methoden und mittel der kriegsführung', in H. Schöttler and B. Hoffmann, *Die Genfer Zusatzprotokolle – Kommentare und Analysen* (Bonn, Osang 1993) pp. 160 at 167. For criticism of this merging of the notion of indiscriminate attacks and the proportionality principle in Art. 51 of AP I, see J. Gardam, *Necessity, Proportionality and the Use of Force by States* (Cambridge, Cambridge University Press 2004) pp. 94-96 (noting the conceptual differences between indiscriminateness and proportionality and observing on p. 95 that 'the tests of discrimination and proportionality operate cumulatively with the result that, if the attack is indiscriminate under the Protocol, it is irrelevant whether in fact it meets the proportionality requirement or not. The reverse is also the case, namely, that, even if the attack is discriminate, it will be prohibited if it fails to meet the proportionality requirement.').

civilians as such the object of attack (section 3.2); and second, the prohibition of indiscriminate attacks (section 3.3).

3.1 Dangers for the civilian population arising from military operations involving the use of depleted uranium

The actual or potential dangers arising from the use of DU weaponry are discussed comprehensively elsewhere in this book.[20] However, a number of characteristics of DU weaponry need to be recalled briefly because of their relevance when considering the specific dangers for civilians.

The two concerns related to the health effects of DU in military applications are heavy metal toxicity and toxicity from radiation.[21] When DU ammunition and armour is shot or breached, corroding or consumed by fire, the release of DU may contaminate soil, water, air, plant and animal life. Once such contamination has occurred, civilian populations may be exposed to DU through a number of routes, especially inhalation of DU dust; ingestion of DU directly or in contaminated food, soil and water; and dermal absorption.[22] The level of this exposure and the ensuing health risks depend on a number of factors, such as the type of ammunition used; the type of target and soil; the oxygen content of the casings of DU rounds or fragments; the presence and salinity of water; the intensity and duration of fires; the force of explosions; wind; and time.[23] In addition, other factors including age, sex, diet, family history, health status, and lifestyle of those exposed may affect the overall health consequences of exposure. These factors need to be taken into account when assessing the dangers for civilians that the use of DU weapons entails. However, any such danger does not result from their primary effects, i.e., their 'primary injuring mechanism',[24] or, more generally, the main purpose for which they are used.[25] Rather, the dangers for civilians emanate from the *secondary* ef-

---

[20] See D. Fahey, Chapter 2 of this book.

[21] Ibid., section 5 at pp. 53 et seq.

[22] Ibid., section 4 at pp. 46-48. Fahey also mentions the two other exposure routes of injection of fragments through wounds and wound contamination by DU dust. While these may also be of relevance in the context of exposure of civilians, they are primarily exposure routes relevant for combatants.

[23] Ibid., sections 2 and 3.

[24] See Henckaerts and Doswald-Beck, *supra* n. 6, p. 277.

[25] Cf., ibid., p. 272. Many rules banning or limiting the use of certain weapons refer to the 'primary effect' (Protocol on Non-Detectable Fragments (Protocol I) of the 1980 United Nations Convention on Prohibitions or Restrictions on the Use of Certain Conventional Weapons Which May be Deemed to be Excessively Injurious or to Have Indiscriminate Effects (CCW Convention)) or 'munition which is primarily designed to' (Protocol on Prohibitions or Restrictions on the Use of Incendiary Weapons (Protocol III) of the CCW Convention). Consequently, concerning the prohibition contained in Protocol I, the new military manual from New Zealand specifies, for example, that '[t]he prohibition does not extend to non-detectable fragments which are incidental to a weapon's principal design, e.g., plastic casings of anti-vehicle mines'. See New Zealand, Military Manual, forthcoming new edition, para. 7.13.2 (on file with the authors). In that latter example, the plastic casing has a specific military utility, which is making the weapon's detection more difficult. Cf., Dinstein, *supra* n. 6, p. 65. The report of the working group on that matter (Report of the Working Group, Committee IV, Diplomatic Confer-

fects of the use of DU ammunition in as much as these effects are only indirectly linked to the principal utility or design of the weapon. DU weaponry is *not specifically designed* with a view to deny the civilian population the general protection to which it is entitled under international humanitarian law, and the health risks stemming from DU for civilians will often be time delayed. The question arises whether and to what extent such secondary effects fall within the ambit of those rules of international humanitarian law which regulate means and methods of warfare.

3.2   Depleted uranium and the prohibition to make the civilian population and individual civilians the object of attack

A first set of rules of international humanitarian law, which might appear of *prima facie* relevance, is embodied in Article 51(2). This provision prohibits making 'the civilian population as such, as well as individual civilians the object of attack[26] and 'acts or threats of violence the primary purpose of which is to spread terror among the civilian population.' The latter prohibition can be discarded at the outset, as it does not bear specific relevance for the use of DU weaponry: acts or threats of terror are prohibited regardless of the weapon used.

With regard to the first prohibition, however, the question arises whether the use of DU weaponry would make the civilian population as such and/or individual civilians the object of attack.[27]

It would appear that the use of DU weaponry does not *per se* fall foul of the prohibition. A first argument in support of this assertion flows from the terms 'as such' and 'object of attack'. This wording implies that the prohibition only covers attacks that are directed against the civilian population and/or individual civilians, suggesting an intentional element. It would thus appear to refer to premeditated attacks that are aimed at hitting the civilian population or individual civilians and whose rationale is to do so.[28] Accordingly, the prohibition is commonly understood

---

ence on Humanitarian Law, CDDH/IV/224/Rev.l) refers to 'weapons which were designed to injure by such fragments'. Cf., J. Roach, 'Certain Conventional Weapons Convention: Arms control or humanitarian law?', 105 *Military Law Review* (1984) p. 70. See also H. Parks, Memorandum of Law – Review of Weapons in the Advanced Combat Rifle Program, 21 May 1990, reproduced in *The Army Lawyer* (July 1990) pp. 18 at 19. In light of those remarks, the technical expression 'primary effect' could therefore be interpreted as meaning the main purpose justifying the use of a weapon.

[26] Compare the corresponding prohibition in Art. 52(1), dealing with civilian objects. The analysis in this section applies to civilian objects *mutatis mutandis*. For the identification of attacks which are not directed at a specific military objective as indiscriminate attacks, see *infra* section 3.3.1.

[27] 'Attacks' are defined in Art. 49(1) as 'acts of violence against the adversary, whether in offence or in defence'. The provision further clarifies that '[t]he provisions of this Protocol with respect to attacks apply to all attacks in whatever territory conducted including the national territory belonging to a Party to the conflict but under the control of the adverse Party,' (para. 2).

[28] The commentary of Bothe, Partsch and Solf takes the same approach, noting that '[t]he term "as such" implies that there can be no assurance that attacks against combatants and other military objectives will not result in civilian casualties with respect to civilians in or near such military objectives', and then makes reference to the principle of proportionality. M. Bothe, K.J. Partsch and W.A. Solf, *New Rules for Victims of Armed Conflicts – Commentary on the Two 1977 Protocols Additional to the*

to relate to attacks in which the civilian population or individual civilians are 'a target or [...] a tactical objective'.[29] A second argument is a contextual interpretation of the provision. The prohibition to make the civilian population and individual civilians the object of attack in the first sentence of Article 51(2) is directly followed by the prohibition of acts or threats of violence the primary *purpose* of which is to spread terror among the civilian population contained in the second sentence. A contextual interpretation thus indicates that the entire second paragraph of Article 51 addresses attacks which have the civilian population as their primary, if not exclusive, target. The broader context of the first sentence of Article 51(2) also supports this view. Were one to understand the provision to not only cover the premeditated attack of civilians but equally extending to the *effects* of attacks that made legitimate targets the object of attack, the distinction with other prohibitions in Article 51 would become blurred and bear the potential of making some of them superfluous,[30] in contravention of the rule of treaty interpretation of *ut res magis valeat quam pereat*.[31]

Naturally, the prohibition to make the civilian population or individual civilians the *object* of the attack *can* be violated by using DU weaponry, for instance if a tank targets a civilian home with DU-tipped shells. However, such violation stems from the fact that, in this instance, the attack is directed against a civilian object, rather than the use of any particular weapon. It is the way in which the weapon is used (i.e., method) and not the properties of that weapon which triggers Article 51(2). Conversely, attacks –including attacks with DU-ammunition – would not constitute a violation of this particular rule as long as the civilian population or individual civilians are not used as a target or as a tactical objective.

3.3 Depleted uranium and the prohibition of indiscriminate attacks

Paragraphs 4 and 5 of Article 51 prohibit and define indiscriminate attacks.[32] The general prohibition in the opening sentence of Article 51(4) is followed by three specific categories of indiscriminate attacks[33] and two examples of such attacks in

---

Geneva Conventions of 1949 (The Hague/Boston/London, Martinus Nijhoff Publishers 1982) p. 300, at 2.3 and fn. 4. In other words, they are of the view that Art. 51(2) only refers to *intentional* attacks on civilians.

[29] Sandoz, et al., *supra* n. 6, p. 618, para. 1938. See also ICTY Trial Chamber, *Prosecutor* v. *Galić*, Judgment of 5 December 2003, para. 53, where the Trial Chamber held that 'the prohibited conduct set out in the first part of Article 51(2) is *to direct an attack* (as defined in Article 49 of Additional Protocol I) against the civilian population and against individual civilians not taking part in hostilities'. Emphasis added.

[30] Cf., especially Art. 51(4)(b), (c) and 51(5)(b), *infra* sections 3.3.2-3.3.5.

[31] According to this rule, an interpretation is not admissible which would make a provision meaningless or ineffective because parties are assumed to intend the provisions of a treaty to have a certain effect, and not to be meaningless. See R. Jennings and A. Watts, *Oppenheim's International Law*, 9th edn. (London, Longman 1996) Vol. 1, p. 1280.

[32] Some aspects of the prohibition are continued in paras. 7 and 8 and other provisions of the Protocol, namely Arts. 55-60; cf., Bothe, et al., *supra* n. 28, p. 304, at 2.5. For a definition of the term 'attacks', cf., Art. 49(1) AP I, *supra* n. 27.

[33] Cf., Art. 51(4)(a) to (c).

Article 51(5).[34] For the purpose of our analysis, the following types of indiscriminate attacks need to be clarified: attacks which are of a nature to strike military objectives and civilians or civilian objects without distinction because (1) they are not directed at a specific military objective (section 3.3.1); (2) they employ a method or means of combat which cannot be directed at a specific military objective (section 3.3.2); (3) they employ a method or means of combat the effects of which cannot be limited as required by AP I (section 3.3.3); and (4) disproportionate attacks (section 3.3.4).[35] While these standards of indiscriminateness may be clear in theory, their practical application is far from obvious. Suffice it to say at the outset, however, that, as will be shown, the two last mentioned are especially relevant.

3.3.1   *Attacks which are not directed at a specific military objective*

Subparagraph (a) of Article 51(4) identifies as indiscriminate attacks those which strike military objectives and civilians and civilian objects without distinction because they 'are not directed at a specific military objective'.[36] The central question with regard to this provision is what is meant by a 'specific military objective'. Clearly, an area may contain several military objectives[37] or a certain area may, indeed, constitute a military objective in itself.[38] Article 51(4)(a) does not prohibit belligerents from attacking such military objectives, either with DU weaponry or other means of warfare. Rather, it obliges belligerents to identify these objectives[39] and direct their attack against *specific and separable* targets.[40] In other words, Article 51(4)(a) is concerned with the way in which a weapon is used.[41] A clear

---

[34] Art. 51(5)(a) and (b).

[35] The fifth category, defined in Art. 51(5)(a) and prohibiting attacks by bombardment by any methods or means which treat as a single military objective a number of clearly separated and distinct military objectives located in a city, town, village or other area containing a similar concentration of civilians or civilian objects, does not raise any peculiar questions in relation to the use of DU weaponry. It will therefore not be analysed in the following.

[36] For a definition of what constitutes a military objective, see Art. 52(2) AP I, *supra* n. 6.

[37] An example of the former would be several tanks that are widely scattered.

[38] Several States Parties to the Protocol made declarations or statements of understating to the effect that a specific area of land may be a military objective if the requisite elements of Art. 52(2) are met, i.e., if, because of its nature, location, purpose or use make an effective contribution to military action and their total or partial destruction, capture or neutralisation, in the circumstances at the time, offers a definite military advantage. See declarations H of Canada, 7 of the FR Germany, F of Italy, 7 of the Netherlands, (j) of the United Kingdom (all reproduced in A. Roberts and R. Guelff, *Documents on the Laws of War*, 3rd edn. (Oxford, Oxford University Press 2000) pp. 502, 505, 507, 508, 511). The ICRC Commentary notes that, although these interpretations were not discussed during the Diplomatic Conference 'they appear to be reasonable', Sandoz, et al., *supra* n. 6, p. 637, para. 2026. Oeter notes that 'the notion of 'combat objectives' [...] indicates objects or topographically exposed ground positions which in the course of a specific military operation suddenly become important for both sides, and the possession or destruction of which accordingly becomes decisive in achieving operative goals (e.g., church or other towers, hills, steep slopes, exposed farms, etc., the control of which both sides try to achieve).' *Supra* n. 8, p. 160, margin 443 at para. 5.

[39] In that connection, note the obligation to take precautions in attack, Art. 57 and *infra* section 5.

[40] Oeter, *supra* n. 8, p. 119, para. 404.

[41] M. Schmitt, 'Precision attack and international humanitarian law', 87 (859) *IRRC* (2005) pp. 445-466 at 454.

example of an attack that would violate the prohibition would be a missile attack directed at an area the size of a town[42] or area bombing attacks such as those of World War II.[43] Article 51(4)(a) does not regulate a situation, however, in which an attack affects civilians or civilian objects as an incidental and unavoidable consequence of directing an attack at a specific military objective.[44] Nor does it address the inherent properties of weapons whose use may lead to the conclusion that a given means of warfare *cannot* be directed at a specific military objective.[45]

Thus understood, the use of DU weaponry does not raise specific issues in relation to the prohibition of attacks which are not directed at a specific military objective. Such attacks are prohibited regardless of the weapon used.

3.3.2 *Attacks which employ a method or means of combat which cannot be directed at a specific military objective*

A second form of indiscriminate attacks, regulated in Article 51(4)(b), are attacks which employ a method or means of combat which cannot be directed at a specific military objective and therefore strike military objectives and civilians or civilian objects without distinction. In contrast to the preceding sub-paragraph, it does not address the targeting of unspecified military objectives, but the (technological) properties and characteristics of weapons, tactics and strategies employed to weaken the adversary, which together are referred to as 'methods and means of combat'.[46]

It appears clear that DU weapons do not contravene the prohibition *per se* if one understands the term 'cannot be directed at a specific military objective' narrowly in the sense that it refers only to the capability of accurately targeting a specific military objective. DU ammunition can very well be used with a high degree of accuracy. An alternative approach would be, however, to construe the term broadly as equally prohibiting the employment of a method or means of combat which, while capable of accurately targeting a specific military objective, has effects which cannot be confined to a specific military objective. However, this latter interpretation meets a number of objections. First, the very expression 'cannot be directed at' would seem to indicate that the provision addresses the degree to which a weapon can be targeted.[47] Secondly, an expansive interpretation that equally cov-

---

[42] See A.P.V. Rogers, *Law on the Battlefield*, 2nd edn (Manchester, Manchester University Press 2004) p. 24.

[43] L.C. Green, *The Contemporary Law of Armed Conflict* (Manchester, Manchester University Press 1993) p. 152.

[44] Such side effects have to be assessed in the light of the principle of proportionality. See *infra* section 3.3.4

[45] This is, however, regulated in Art. 51(4)(b), and will be discussed in the following section.

[46] The words 'methods and means' include weapons in the widest sense ('means'), as well as the way in which they are used, i.e., tactics or strategies ('methods'), see Sandoz, et al., *supra* n. 6, pp. 398, at para. 1402.

[47] According to the *Concise Oxford English Dictionary*, the verb 'direct' means, *inter alia*, to 'point, aim or cause (a blow or missile) to move in a certain direction', *The Concise Oxford English Dictionary*, 9th edn. (Oxford, Clarendon Press 1995) p. 382.

ers the consequences of a given method or means of warfare would make the following sub-paragraph of Article 51(4) redundant, thus contravening accepted rules of treaty interpretation.[48] Thirdly, a narrow interpretation is also confirmed by the drafting history of the provision and subsequent academic writing. Both indicate that Article 51(4)(b) is concerned with '"blind" weapons, which cannot, with any reasonable assurance, be directed against a military objective'.[49] Often cited examples of a weapon that would contravene the prohibition are the V2 rockets used in World War II and the Scud missiles launched by Iraq against Israel and Saudi Arabia during the First Gulf War.[50]

While the use of DU weaponry by itself does thus not amount to a violation of Article 51(4)(b), such a violation could occur, however, if the weapons system that includes DU components is itself 'blind'. Thus, the use of an imprecise weapons system that carries DU-tipped warheads, for instance, could amount to a violation of Article 51(4)(b). Again, such a violation does not depend on the use of DU in shells that such weapons systems may deliver, but on the inaccurate properties of the weapons system as a whole.

3.3.3 *Attacks the effects of which cannot be limited as required*

The last form of indiscriminate attacks that Article 51(4) addresses are those which are of a nature to strike military objectives and civilians or civilian objects without distinction because they 'employ a method or means of combat the effects of which cannot be limited as required by [... Additional] Protocol [I]'[51] or, as far as the corresponding rule of customary international law is concerned, 'as required by international humanitarian law'.[52]

The reference to 'effects' in Article 51(4)(c) makes clear that it regulates the result or consequences of a given method or means of warfare for civilians and civilian objects.[53] In contrast to the two preceding subparagraphs, it is neither concerned with a purposive and intentional attack against unspecified objectives nor with the capability of a method or means of combat to be accurately targeted. In other words, even if a given method or means could be and actually were directed at

---

[48] See *supra* n. 31 and text.

[49] Bothe, et al., *supra* n. 28, p. 305, at 2.5.2.2. See also H. Fischer, *Der Einsatz von Nuklearwaffen nach Art. 51 des I. Zusatzprotokolls zu den Genfer Konventionen von 1949 – Völkerrecht zwischen humanitärem Anspruch und militärpolitischer Notwendigkeit* (Berlin, Duncker & Humblot 1985) p. 193 ['Das "cannot be directed" in Abs. 4 lit. b) bezieht sich *nur auf Fragen der Waffensteuerung und nicht auf die mit der Waffenwirkung zusammenhängenden Probleme*.' Emphasis added, footnote omitted]. Fischer further refers to the writings of Blix and Carnahan, who use the terms 'imprecise' and 'properly directed' when interpreting the provision, see Fischer, ibid., fn. 71-72 and text.

[50] See amongst others Sandoz, et al., *supra* n. 6, p. 621, para. 1958; Rogers, *supra* n. 42, p. 24; Green, *supra* n. 43, p. 152; Fischer, *supra* n. 19, p. 167; Schmitt, *supra* n. 41, p. 456.

[51] Art. 51(4)(c).

[52] Henckaerts and Doswald-Beck, *supra* n. 6, Rule 12(c).

[53] *The Concise Oxford Dictionary*, 9th edn. (Oxford, Oxford University Press 1995) defines 'effect' as 'the result or consequence of an action'.

a specific military objective, Article 51(4)(c) further limits the permissible consequences for civilians and civilian objects of such an attack.

However, beyond this general notion that the provision does limit the effects of methods and means of combat, Article 51(4)(c) could best be described as a provision *in nubibus*, having been referred to as the vaguest and most problematic subparagraph in Article 51(4).[54]

The most pertinent element of the provision that needs to be clarified is the phrase 'limited as required by [the] Protocol' or 'as required by international humanitarian law'. A considerable range of opinions have been expressed as to what these limitations are. The United States, for example, seems to understand the sole such limitation to be the principle of proportionality. The US Air Force Pamphlet puts it in the following words:

> 'Some weapons, though capable of being directed only at military objectives, may have otherwise uncontrollable effects so as to cause *disproportionate* civilian injuries or damage. [...] Uncontrollable refers to effects which escape in time or space from the control of the user as to necessarily create risks to civilian persons or objects *excessive in relation to the military advantage anticipated*. International law does not require that a weapon's effects be strictly confined to the military objectives against which it is directed, but it does restrict weapons whose foreseeable effects result in *unlawful disproportionate* injury to civilians or damage to civilian objects.'[55]

However, the US Air Force Pamphlet is an isolated incident of state practice shedding some light on what 'limitations required by international humanitarian law' are for the purpose of understanding the prohibition of the type of indiscriminate attacks under consideration here. Instead, the majority of states' military manuals which incorporate the prohibition simply reiterate the generic requirement that the effects of attacks must be 'limited as required by international humanitarian law'.[56] It would therefore appear farfetched to derive from state practice any agreement regarding the interpretation of the prohibition to the effect that it is neither more nor less than an implicit reverberation of the prohibition of disproportionate attacks. To adopt the latter position would be objectionable when applying the generally accepted rules on treaty interpretation to the prohibition in Article 51(4)(c). More specifically, a systematic interpretation of that provision would suggest that the limitations referred to do not exhaust themselves in the principle of proportionality. For Article 51(4) is followed by subparagraph (5), which makes it quite clear that disproportionate attacks are *an example* of indiscriminate attacks.[57] Such an exem-

---

[54] Bothe, et al., *supra* n. 28, p. 305, at section 2.5.2.3., referring to writings of Rauch and Kalshoven (fn. 20 and 21). Rogers shares this view, *supra* n. 42, p. 26 ['unfortunately vague'].

[55] US Air Force Pamphlet (1976), § 6-3(c), reproduced in J.M. Henckaerts and L. Doswald-Beck, eds., *Customary International Humanitarian Law* (Cambridge, Cambridge University Press 2005), Vol. II (Practice) Part 1, p. 286.

[56] See for examples Henckaerts and Doswald-Beck, supra n. 55, pp. 285-286, paras. 256-257.

[57] Cf., Art. 51(5)(b).

plary mentioning of the principle of proportionality militates against its being understood as the *only* limitation under the Protocol which is relevant in the context of Article 51(4)(c).

Several commentators also include restrictions relating to the protection of the natural environment[58] and of works and installations containing dangerous forces,[59] in addition to the principle of proportionality.[60] This interpretation finds support in the fact that the mentioned provisions do indeed require that the *effects* of attacks be limited. However, it is unclear what further limitations, if any, on the effects of methods and means of warfare can be derived from AP I or, as far as the customary rule is concerned, general international humanitarian law.[61]

Nothing in Article 51(4)(c) and the corresponding rule of customary international law indicate that the prohibition is restricted to any specific limitation(s). Instead, the reference to limitations 'required by this Protocol' or 'by international humanitarian law' suggests that the provision covers all limitations under the Protocol that have a bearing on the effects of methods and means of warfare, provided that these effects have the consequence of striking 'military objectives and civilians or civilian objects without distinction'.[62] This would not only support the view that the prohibition to attack, destroy, remove or render useless objects indispensable to the survival of the civilian population is a further such limitation,[63] but also suggests that the general limitations – notably the principle of distinction contained in Article 48 AP I and customary Rules 1 and 7, and the general protection against dangers arising from military operations to which civilians are entitled in accordance with Article 51(1) – are of equal relevance.[64] Accordingly, attacks which

---

[58] Arts. 35(3) and 55 AP I; Henckaerts and Doswald-Beck, *supra* n. 6, Rules 43-45. For a detailed discussion, see in this book E. Koppe, Chapter 8, pp. 161 et seq.

[59] Art. 56 AP I; Henckaerts and Doswald-Beck, ibid., Rule 42.

[60] Arts. 51(5)(b) and 57(2)(a)(ii) and (iii); Henckaerts and Doswald-Beck, ibid., Rule 14. See for instance Fischer, *supra* n. 49, p. 194; Bothe, et al., *supra* n. 28, pp. 305-306, at 2.5.2.3.; Oeter, *supra* n. 8, p. 176, margin 455 at 4.

[61] Green, for instance, takes a somewhat broader approach, referring to limitations required by the Protocol 'concerning protection of the wounded and sick, civilians and civilian objects, as well as prohibitions relating to protected places and excessive non-military damage, and which are of a nature likely to strike military objectives and civilians or civilian objects without distinction'. *Supra* n. 43, pp. 151-152. On this lack of clarity see also Rogers, who notes that '[t]here is no provision of Protocol I that specifically limits the effects of methods and means. It may be a reference to the rule of proportionality in Art. 57 (precautions in attack). If so, it is superfluous, because Art. 57 applies anyway. If it is a reference to Protocol I as a whole, it lacks the precision necessary for a provision the breach of which may result in a person's being charged with a war crime [under Article 85 (3)(b) of AP I].' *Supra* n. 42, p. 26.

[62] Cf., the closing part of Art. 51(4) and of Customary Rule 12, Henckaerts and Doswald-Beck, *supra* n. 6, p. 40.

[63] Cf., Art. 54. For more detailed analysis, see *infra* section 4.

[64] Such reasoning seems to underlie David's assertion that a high risk of harmful consequences for people's health from handling DU fragments and contaminated remnants would justify regarding DU as producing 'indiscriminate effects, and in this respect, their use would violate the principle of distinction'. Rather than referring to any of the specific limitations contained in the Protocol, he seems to base his assertion on the only other provision regulating indiscriminate *effects,* i.e., Art. 51(4)(c), which he regards as a customary rule (cf., *supra* n. 9), and to derive such limitation from the general principle of

employ a method or means of combat the effects of which cannot be limited to combatants and military objectives would equally violate Article 51(4)(c) if these effects have the consequence of striking military objectives and civilians or civilian objects without distinction.[65] The same would apply to methods and means of combat which entail effects that can be qualified as 'dangers arising from military operations' but against which the civilian population and individual civilians cannot be protected.

This is not meant to suggest that *any effect* would violate the aforementioned general limitation, which emanates from the principle of distinction. Hardly any method or means of warfare leaves civilians or civilian objects completely unaffected. For example, military aircraft may fly over civilians to attack a military objective at an isolated spot far removed from civilians or any civilian object, but they nevertheless affect civilians because they cause loud blasts when reaching supersonic speed. Assuming that the use of military aircraft qualifies as a method or means of warfare,[66] it would clearly verge at the brink of absurdity to argue that such effects of aerial bombardments would render the attack unlawful because the effects of the attack (noise) cannot be limited to the military objective or that it runs counter to the protection against dangers arising from military operations to which civilians are entitled.[67] Such an argument would disregard one of the fundamental precepts underlying international humanitarian law, which is to strike a balance between considerations of humanity and military necessity, rather than prohibiting armed forces from the conduct of military operations. In order to take account of this precept, it would therefore be reasonable for a violation of Article 51(4)(c) to require that the effects of a given method or means of combat the effects of which cannot be limited to military objectives, or one whose effects create dangers for civilians, reach a minimum threshold of intensity and severity.

---

distinction, i.e., Art. 48 of AP I. See David, *supra* n. 14, p. 96. See also L. Doswald-Beck, 'International humanitarian law and the Advisory Opinion of the International Court of Justice on the legality of the threat or use of nuclear weapons', 316 *IRRC* (1997) pp. 35-55, p. 40: 'presumably means, especially in the light of the paragraph's final phrase, that the effects do not otherwise violate the principle of distinction'].

[65] Note that methods or means of combat *the effects of which cannot be limited to military objectives* must be distinguished from those *which cannot be directed at a specific military objective,* discussed *supra* part 3.3.2. The former are concerned with the *effects* of a given method or means, the latter with the capability of being directed at a specific military objective. The capability of a weapon to be accurately targeted does not say anything about whether or not the effects of the use of such a weapon can be limited to military objectives. An example would be a high-precision weapons system that employs components which, upon impact of the target, emit poisonous gases (notwithstanding that such a weapons system may also violate the prohibition of poison; for a discussion of the latter prohibition, see J. Kleffner, Chapter 7 of this book).

[66] For the notions of 'methods' and 'means', see *supra* n. 46.

[67] See in that regard the judgment of the Amsterdam District Court, Fourth Chamber (Civil) (*Gerechtshof Amsterdam, Vierde meervoudige burgerlijke kamer*) of 6 July 2000 in the case of *Dedović v. Kok et al.*, rejecting the plaintiffs' claim that stress and tensions that were consequences of the NATO air strikes against Yugoslavia amounted to indiscriminate attacks in violation of AP I, Judgment at para. 5.3.23.

The ICRC Commentary seems to equate such a threshold with 'significant losses among the civilian population and extensive damage to civilian objects'.[68] Others have instead referred to a 'high risk of harmful consequences'.[69] One can also infer a hint – albeit a slight one – to a certain level of intensity and severity from the language of Article 51(4), which requires that the effects of a given method or means of combat must be 'of a nature to strike military objectives and civilians or civilian objects without distinction'.[70] Although this wording may not require full parity between the effects for military objectives and civilians or civilian objects, the words 'to strike' may be understood to indicate that not *any* effect of a method or means of combat contravenes the prohibition under Article 51(4)(c).

It follows from the preceding analysis that Article 51(4)(c) and the corresponding rule of customary international law equally prohibit those attacks as indiscriminate which employ a method or means of combat the effects of which cannot be limited to military objectives or expose civilians to dangers, provided such effects reach a certain threshold. The prohibition imposes limitations on belligerent parties, which amount to a standard of control and controllability over the effects of methods and means of warfare, so as to ensure that the level of intensity which would contravene Article 51(4)(c) is not reached.[71] It is worth noting that the United States, as one of the main users of DU weaponry, while not being a party to AP I but instead bound by the corresponding rule of customary international law,[72] subscribes to such a standard of controllability, albeit with the aforementioned difference that the United States understands this rule as being violated only if the effects cause disproportionate civilian injuries or damage.[73] In contrast to the US position, however, our interpretation suggests that the assessment whether a certain method or means of combat complies with the prohibition of this form of indiscriminate

---

[68] Sandoz, et al., *supra* n. 6, p. 623, para. 1963, emphasis added '[Some means of warfare] can [...] be completely out of the control of those using them, causing significant losses among the civilian population and extensive damage to civilian objects.'

[69] David, *supra* n. 14, p. 96

[70] Cf., the closing part of Art. 51(4) and of Customary Rule 12 cited *supra* n. 62 and text.

[71] Sandoz, et al., *supra* n. 6, p. 623, para. 1963 [asserting that a violation of Art. 51(4)(c) occurs if a means of warfare, such as fire or water, is used, which, while being able to have restricted effects in certain circumstances, is 'completely out of the control of those using them'.]. See also Oeter, *supra* n. 8, p. 119, para. 404 [use of force 'must also employ specific weapons and methods of warfare that are limited (or at least limitable) in their results']; UK Ministry of Defence, *The Manual of the Law of Armed Conflict* (Oxford, Oxford University Press 2004) p. 68, fn. 93, stating that the prohibition in Art. 51(4)(c) 'covers two situations: where the attacker is unable to control the effects of the attack, such as dangerous forces released by it, or where the incidental effects are too great.'; Institut de Droit International, Resolution adopted during the Session of Edinburgh – 1969, 'The Distinction Between Military Objectives and Non-Military Objects in General and Particularly the Problems Associated with Weapons of Mass Destruction', para. 7. ['Existing international law prohibits the use of all weapons which, by their nature, affect indiscriminately both military objectives and non-military objects, or both armed forces and civilian populations. In particular, it prohibits the use of weapons the destructive effect of which is so great that it cannot be limited to specific military objectives or is otherwise *uncontrollable* (self-generating weapons), as well as of "blind" weapons.' Emphasis added].

[72] On the customary status of Art. 51(4), see *supra* n. 9.

[73] Henckaerts and Doswald-Beck, *supra* n. 6, p. 43.

attacks involves a two-fold test. First, it needs to be assessed whether an attack employs a method or means the effects of which cannot be limited to military objectives and expose civilians to significant dangers. If such an assessment suggests an answer in the affirmative, the attack is prohibited. If the answer is in the negative, the principle of proportionality enters the equation, with the consequence that an attack which may not otherwise be prohibited would be illegal if it entails effects which may be expected to cause incidental loss of civilian life, injury to civilians, damage to civilian objects, or a combination thereof, which would be excessive in relation to the concrete and direct military advantage anticipated.[74]

To deduce from Article 51(4)(c) and the corresponding rule of customary international law such a standard of control and controllability is not the end of the matter when analysing the legality of the use of DU weapons. For the concerns relating to DU weaponry are not limited to the immediate consequences of an attack, such as those for civilians in the vicinity of a tank while the latter is hit by a DU-tipped anti-tank round or shortly thereafter while the tank is still burning. Rather, these concerns include more distant effects, for instance those produced by DU-remnants which enter ground- or freshwater resources or are dispersed by winds. DU's long-term mobility and its deposition in the soil, and the fact that it can be resuspended by wind, water or the movement of people and vehicles, means that the affected area can be wide-ranging. Any chemo-toxic and radiation effects that DU weaponry may have, especially in its after-use aerosol consistency, therefore cannot be contained in either space or time and can subsequently produce effects in areas removed from where the initial attack occurred after considerable time has lapsed. These attributes raise the question whether the prohibition of indiscriminate attacks extends to such consequential effects or whether it is confined to effects that arise directly and immediately from the use of a certain weapon. Again, the wording of the provision in AP I and customary international law are inconclusive in this regard.

The few instances when pronouncements have been made with respect to the temporal and geographical reach of the prohibition would seem to suggest that it is not limited to immediate effects. Thus, when the International Court of Justice applied the principle of distinction to the question whether the use of nuclear weapons could be permissible,[75] it took into account the fact that the effects of nuclear weapons 'cannot be contained in either space or time'[76] in coming to the conclusion that 'the use of such weapons seems scarcely reconcilable with respect for [the principle of distinction and the prohibition of superfluous injury and unnecessary suffering to combatants]', notwithstanding the fact that the Court was unable to conclude with certainty that the use of nuclear weapons would necessarily be at variance with the principles and rules of law applicable in armed conflict in any circumstance.[77] Naturally, this is not to equate DU weapons, which are conven-

---

[74] For further discussion, see *infra* section 3.3.4.

[75] Legality of the Threat or Use of Nuclear Weapons, Advisory Opinion, *supra* n. 8, paras. 90-95.

[76] Ibid., para. 35.

[77] Ibid., para. 95.

tional weapons, with nuclear weapons, which are weapons of mass destruction. However, the fact that the ICJ applied the general rules and principles of international humanitarian law to nuclear weapons[78] makes its findings equally relevant for other weapons, including conventional ones such as DU ammunition. The Court's view supports taking into consideration not only the immediate effects of a weapon when assessing its use in the light of the prohibition of indiscriminate attacks under Article 51(4)(c) and customary international humanitarian law but also more distant effects. Again, the United States would seem to share this view, when the US Air Force Pamphlet defines as uncontrollable effects 'which escape in time or space from the control of the user as to necessarily create risks to civilian persons or objects',[79] a standard that was applied in the 1975 US review of DU weaponry.[80] Similarly, the New Zealand military manual defines weapons which are indiscriminate in their effects, *inter alia*, as 'those which, because of their physical characteristics, [...] cannot be limited in their destructive effects. This may be [...] because other destructive forces such as fire, poison or radiation will be released that cannot be contained.'[81] It has also been argued in academic literature that 'indiscriminate weapons' are meant to cover cases where the weapon, even when targeted accurately and functioning correctly, is likely to take on 'a life of its own' and randomly hit combatants or civilians to a significant degree.[82] Admittedly, it may prove difficult or indeed impossible to establish a causal connection between the initial use of DU weaponry and those effects which are far removed from it in terms of space and time. However, if such a causal link could be established, nothing justifies depriving civilians and civilian objects from the protection which flows from the prohibition of indiscriminate attacks for the reason alone that the effects only materialise after a certain period of time has lapsed or at places removed from where an attack occurred.

Does, then, the use of DU weaponry violate Article 51(4)(c) and the corresponding rule on customary international law, assuming that the concerns raised by

---

[78] Cf., ibid., para. 86.

[79] Cf., *supra* n. 55, emphasis added.

[80] Department of the Air Force, HQUSAF (JA) Legal Memorandum concerning review by Harold R. Vague, Major General, USAF, The Judge Advocate-General United States Air Force, of the High Explosive Incendiary and Armour Piercing Incendiary munitions, 14 March 1975. For a detailed analysis of this, see in this book, B. Carnahan, Chapter 4, pp. 99 et seq.; A. McDonald, Chapter 12, pp. 281 et seq.

[81] New Zealand, Military Manual, *supra* n. 25, § 7.2.4 (b) and § 7.2.8 (on file with the authors).

[82] Doswald-Beck, *supra* n. 64, p. 41: 'The second hypothesis, which this author prefers, is not to try to find the answer in other parts of Article 51 of the Protocol, but rather to decide on the basis of the essential meaning of the principle of distinction. This principle presupposes the choice of targets and weapons in order to achieve a particular objective that is lawful under humanitarian law and which respects the difference between civilian persons and objects on the one hand, and combatants and military targets on the other. This requires both planning and a sufficient degree of foreseeability of the effects of attacks. Indeed, the principle of proportionality itself requires expected outcomes to be evaluated before the attack. None of this is possible if the weapon in question has effects which are totally unforeseeable, because, for example, they depend on the effect of the weather. It is submitted that the second test of "indiscriminate weapons" is meant to cover cases such as these, where the weapon, even when targeted accurately and functioning correctly, is likely to take on "a life of its own" and randomly hit combatants or civilians to a significant degree'.

its use were to materialise? Even if one accepts, as we do, that inherently indiscriminate weapons exist[83] – a matter that is not entirely beyond dispute[84] – it is impossible for us to contend that DU weapons qualify as such. The effects of such weapons very much depend on the way in which, where, and when they are used and the resulting toxicological dose, route and magnitude of exposure, and the location of embedded fragments. As with other conventional weapons, DU-weaponry may be used in the most diverse manner. For example, it is conceivable that it may be used in a relatively isolated precision attack against a legitimate target on a battlefield far away from any potentially affected civilian population or civilian objects. In this case, the toxicological dose and other factors may not result in adverse health effects on the civilian population which reach the magnitude which is required in order for Article 51(4)(c) and the corresponding rule of customary international law to be violated. It may be different where DU weaponry is used on a larger scale in an urban environment or near comparably densely populated areas. Such uses are more likely to produce toxicological doses and magnitudes of exposure which have severe adverse health effects on the civilian population and individual civilians due to heavy metal poisoning, especially if the uranium is in soluble form.

In sum, whether and to what extent the use of DU ammunition violates the prohibition of indiscriminate attacks under Article 51(4)(c) and the corresponding rule of customary international humanitarian law will have to be determined on a case-by-case basis. While it does not do so under all circumstances, it may do so in some.

3.3.4 *Depleted uranium and the prohibition of disproportionate attacks*

As argued in the preceding section, a legal assessment of the use of DU weaponry in relation to the prohibition of indiscriminate attacks does not end if and when such use fails to produce effects which cannot be limited to military objectives and which expose civilians to significant dangers. Rather, in such a situation, an attack also needs to comply with the requirements of the principle of proportionality. In

---

[83] Cf., Customary Rule 71. For supporting practice, see Henckaerts and Doswald-Beck, *supra* n. 6, p. 248. For instance, the new military manual from New Zealand holds that 'Any weapon or ammunition which is of a nature to strike both military objectives and civilians or civilian objects without distinction is inherently indiscriminate.' *Supra* n. 25, § 7.2.8. Many scholars articulate the discrimination principle by distinguishing between those two approaches. For example, Schmitt holds that 'the principle of discrimination is bifurcated'. This author goes on to state: 'On the one hand, it limits the use of weapons that are by nature indiscriminate (…). By contrast, the second facet of the principle precludes indiscriminate use of weapons (…).' M. Schmitt, 'The principle of discrimination in 21st century warfare', 2 *Yale Human Rights & Development Law Journal* (1999) pp. 147-148.

[84] For contrary practice, see the statements of Canada, the Federal Republic of Germany, Italy and the United Kingdom made during the Diplomatic Conference on the Reaffirmation and Development of International Humanitarian Law Applicable in Armed Conflicts (Geneva, 1974-1977), reproduced in Henckaerts and Doswald-Beck, *supra* n. 6, Vol. 2, pp. 287-288. For a summary of this debate at the Diplomatic Conference, see F. Kalshoven, 'Arms, armaments and international law', 191 *RCADI* (1985-II) p. 236.

other words, such an attack would nevertheless be illegal if it entailed effects which may be expected to cause incidental loss of civilian life, injury to civilians, damage to civilian objects, or a combination thereof, which would be excessive in relation to the concrete and direct military advantage anticipated.[85]

The proportionality principle in international humanitarian law may be analysed from two different angles. Firstly, one can see it as a fundamental and leading principle, which materialised in different norms of the law applicable in armed conflicts.[86] Thus, the prohibition of indiscriminate attacks would be a facet of this principle. However, proportionality is often construed as merely expressing the specific rule banning attacks that have excessive effects in relation to the concrete and direct anticipated military advantage. Such limited interpretation stems from the status of Article 51(5)(b) AP I, which is viewed as the first clear codification of the proportionality rule.[87] Therefore, many scholars consider proportionality to amount to 'the issue of collateral damage to civilians resulting from attacks against military objectives'.[88] Although it might be described separately from the principle of distinction, the two are closely linked as part of the same legal regime protecting civilians and because they articulate one another. On the one hand, the proportionality test only applies when the principle of distinction has been respected and the attack is carried out against a legitimate target.[89] On the other hand, protection of civilians cannot be conceived in absolute terms without making war impossible. Therefore, if by aiming an attack at a military objective civilians are affected, it will not render such attack illegal *per se*.[90] Rather, international humanitarian law encapsulates the fact that collateral casualties are inherent in the nature of war-

---

[85] Cf., Art. 51(5)(b) AP I; Henckaerts and Doswald-Beck, *supra* n. 6, Rule 14, p. 46.

[86] For example, the Netherlands considers that the principle of proportionality occurs in IHL in two primary forms, one being the prohibition of disproportionate attacks, the other being the prohibition of superfluous injury and unnecessary suffering. See in the context of the working group on explosive remnants of war, 'Responses to document CCW/GGE/X/WG.1/WP.2, entitled IHL and ERW, dated 8 March 2005', *Response from The Netherlands*, *CCW/GGE/XII/XG.1/WP.4*, 7 November 2005, p. 3, § 1 (iii). For a similar view, J. Gardam, 'Necessity and proportionality in *jus as bellum* and *jus in bello*', in L. Boisson de Chazournes and Ph. Sands, eds., *International Law, the International Court of Justice and Nuclear Weapons* (Cambridge, Cambridge University Press 1999) p. 276.

[87] See United Kingdom, *The Manual of the Law of Armed Conflict*, *supra* n. 71 p. 25, para. 2.6.1. Another reference to proportionality in almost similar terms appears in Art. 57 of AP I concerning precautions in attack. The ICRC customary law study identifies the proportionality rule as a customary norm in the following terms: 'Launching an attack which may be expected to cause incidental loss of civilian life, injury to civilians, damage to civilian objects, or a combination thereof, which would be excessive in relation to the concrete and direct military advantage anticipated, is prohibited' (rule 14).

[88] Dinstein, *supra* n. 6, p. 59. See also H. Meyrowitz, 'The principle of superfluous injury or unnecessary suffering from the Declaration of St Petersburg of 1868 to Additional Protocol I of 1977', 299 *IRRC* (1994) pp. 109-110.

[89] See, for example, 'Responses to document CCW/GGE/X/WG.1/WP.2, entitled IHL and ERW, dated 8 March 2005', *Response from the Federal Republic of Germany*, *CCW/GGE/XI/WG.1/WP.9*, 29 July 2005, p. 4, para. 10 and 'Responses to document CCW/GGE/X/WG.1/WP.2, entitled IHL and ERW, dated 8 March 2005', *Response from Canada*, *CCW/GGE/XI/WG.1/WP.2*, 29 June 2005, p. 3, para. 12. See also Dinstein, *supra* n. 6, p. 119 and Schmitt, *supra* n. 83, p. 149.

[90] See Canada, Law of Armed Conflict at the Operational and tactical Levels, B-GJ-005-104/FP-021, Office of the Judge Advocate-General, 2004, p. 4-3, para. 413.

fare.[91] However, the immunity of civilians remains the rule and the doctrine of collateral damage must be read as a limited exception to this principle.[92] According to AP I, disproportionate attacks constitute an example of indiscriminate attacks 'among other' types of attacks. Attacks are regarded as indiscriminate and therefore prohibited only if they have excessive effects in relation to a certain military advantage. Although the test of proportionality to be applied refers to the term 'excessive' instead of 'disproportionate' to establish a violation of the prohibition, the question remains one of a balancing of two elements.[93] In contrast to the absolute prohibitions of paragraphs 4 and 5(a), the prohibition contained in subparagraph (b) is thus relative, involving a most difficult and ambivalent balancing act between human lives, damage to and destruction of civilian objects, on the one hand, and a military advantage, on the other hand. In its report to assess the NATO bombing campaign, the Committee appointed by the Office of the Prosecutor of the ICTY (hereafter, OTP) pointed at the difficulties raised by the application of this test:

> 'It is much easier to formulate the principle of proportionality in general terms than it is to apply it to a particular set of circumstances because the comparison is often between unlike quantities and values. One cannot easily assess the value of innocent human lives as opposed to capturing a particular military objective.'[94]

The difficult formulation of the proportionality rule and its inherent subjectivity derives from the compromise between the two fundamental humanitarian law principles of humanity and military necessity.[95] The imprecise wording and terminology further add to these difficulties and leave a number of questions unanswered.[96] The principle of proportionality requires that those who plan or decide upon attacks must take into account the effects of the attacks on the civilian population. They must determine whether those effects are excessive in relation to the concrete and direct military advantage anticipated and must thus balance the foreseeable extent

---

[91] Bothe, et al., *supra* n. 28, pp. 296-318, p. 300, at 2.3., fn. 4 and accompanying text. This is notwithstanding the prohibition of disproportionate collateral damage (see Art. 51(5)(b)).

[92] ICTY, *Kupreskić et al.*, Trial Ch. II, Case No IT-95-16, Judgment, 14 January 2000, para. 522.

[93] On the question of the choice of 'excessive' instead of 'disproportionate', see, J-F. Quéguiner, *Le principe de distinction dans le droit de la conduite des hostilités – Un principe traditionnel confronté à des défis actuels*, Ph.D. Thesis n°706, IUHEI, 2006, p. 367. See also W.J. Fenrick, 'The rule of proportionality and Protocol I in conventional warfare', 98 *Military Law Review* (1982) p. 106.

[94] Final Report to the Prosecutor by the Committee Established to Review the NATO Bombing Campaign Against the Federal Republic of Yugoslavia, 13 June 2000, 39(5) *ILM* (2000), para. 48.

[95] See on this question, E. Jaworski, '"Military necessity" and "civilian immunity": where is the balance?', 2 *Chinese YIL* (2003) p 186.

[96] Sandoz, et al., *supra* n. 6, p. 625, paras. 1977-1978. As noted by the authors of the OTP Report, one needs to answer difficult question for the principle of proportionality to be applied such as: 'a) What are the relative values to be assigned to the military advantage gained and the injury to non-combatants and or the damage to civilian objects? b) What do you include or exclude in totaling your sums? c) What is the standard of measurement in time or space? and d) To what extent is a military commander obligated to expose his own forces to danger in order to limit civilian casualties or damage to civilian objects?'. *Supra* n. 94, para. 49.

of incidental or collateral civilian casualties or damage and the relative importance of the military objective as a target. While in some situations there will be no room for doubt, there may be reason for hesitation in others, particularly due to the fact that the Protocol does not define what is 'excessive'.

The proportionality test must be made on a case-by-case basis because of the relative elements that must be balanced. Unsurprisingly, a legal evaluation may lead to different conclusions depending on the interpretation of each value at stake. It makes the issue of collateral damage one of the most controversial questions when assessing the legality of an attack.[97] In principle, this is no different when discussing the use of DU weaponry to target a legitimate military objective under the prohibition of disproportionate attacks. However, DU weapons trigger particular problems in as much as their secondary, long-term, effects on the civilian population raise some particular issues in relation to the temporal scope of the prohibition of disproportionate attacks. One may ask to what extent those effects, which are more distant from the initial use of DU weaponry, are taken into account in the proportionality test.

As the prohibition of disproportionate attacks implies a comparison between military advantage gained from an attack and the civilian casualties, defining the content and the boundaries of the former is of paramount importance. Measuring the excessive nature of collateral damage will mainly depend on the scope of the military advantage. Considering the military benefit from the destruction of a single target could be insufficient to meet the proportionality test whereas a broader definition of military advantage, which encompasses several targets, could make the losses acceptable.[98] State practice would seem to support the second, wider approach. Several states made declarations and statements in similar terms when signing or ratifying AP I. The Netherlands holds, for example: 'military advantage refers to the advantage anticipated from the attack considered as a whole and not only from isolated or particular parts of the attack.'[99] Such interpretation remains vague and difficult to apply as it seems to vary considerably according to the circumstances and to the context.[100] Despite this 'illusive nature'[101] of the proportionality rule, very high civilian losses and damages may not be justified even if the military advantage at stake is of great importance.[102]

---

[97] Whereas the Committee appointed by the Office of the Prosecutor of the ICTY concluded that the attacks conducted by NATO forces in Kosovo did not require an investigation, Amnesty international made a different analysis of the contested events. Cf., *NATO/Federal Republic of Yugoslavia: 'Collateral Damage' or Unlawful Killings?*, Amnesty International (2000).

[98] Jaworski, *supra* n. 95, p. 193.

[99] Declaration made at the time of the ratification, 26 June 1986, para. 5. See also Declarations and statements by Canada, United Kingdom and Italy. On that question see Bothe, et al., *supra* n. 28, pp. 309-311, at 2.6.2-2.6.3.

[100] Fenrick underlines that drawback by noting: 'Unfortunately, although these statements indicate the standard applies to military operations of a relatively broad scope, they do not indicate how the boundaries are to be set geographically or chronologically so that a determination of military advantage can be made.' *Supra* n. 93, p. 107.

[101] Jaworski, *supra* n. 95, p. 194.

[102] Sandoz, et al., *supra* n. 6, p. 626, para. 1980.

Under either of the two aforementioned approaches to the issue of what can count as 'military advantage' for the purposes of the proportionality test, the use of DU weapons does not raise specific issues. Being mainly armour piercing weapons, they provide a military benefit which can be assessed for a single target as well as for the whole attack carried out with DU weapons. It will then be necessary to resolve this question on a case-by-case basis like for any other weapon.

On the other hand, measuring the side of the equation opposite to the military advantage, i.e., the collateral damage resulting from the use of DU weaponry, proves to be more problematic. The key question in that regard is *which* consequences of an attack on the civilian population and objects are to be considered under the proportionality analysis. Article 51(5)(b) of AP I refers to the expression 'an attack *which may be expected to cause incidental* loss (...)', a wording similar to the ICRC study's customary rule on proportionality. This wording suggests that, while collateral damage is by definition incidental and secondary,[103] a sufficient degree of expected causation between the attack and its effects is required. However, an attack can result in a range of consequences whose link to the initial act and predictability may vary. Yet, what is to be considered 'sufficient', is not easily established. In state practice, expressions to describe this standard of expectation include 'foreseeable effects of an attack', 'foreseeable risk'[104] or 'likely to cause'.[105] However, they fail to provide conclusive answers to the question. As DU weapons impact on civilians include potential (long-term) effects, the question is to what extent they enter into the equation when assessing proportionality.

This issue arose with a similar sharpness in the context of the ongoing discussions on explosive remnants of war (ERW). Summarising the core aspect of the problem, Christopher Greenwood underlines that 'the question (...) is whether, when a weapon is properly targeted against a military objective, the risk that munitions will fail to explode when intended and thereby create a potential danger to civilians is sufficient to render indiscriminate or disproportionate an attack which would not otherwise be so regarded.'[106] It is therefore necessary to assess the scope of the criteria set up in the proportionality test to apply it to DU weapons.

The Commentary to AP I explains that, during the negotiations, some expressed a preference for the words 'which risks causing' rather than 'which may be expected to cause', but the former was rejected.[107] First of all, it is well-established in state practice that the determination of the notions of 'anticipated' for the military advantage and of 'expected' for the collateral damage needs to be done on the

---

[103] Cf., 'Collateral nature of effects', Joint Forces Command Glossary, available at <www.jfcom.mil/about/glossary.htm>.

[104] United Kingdom, *The Manual of the Law of Armed Conflict*, supra n. 71, para. 5.33.4.

[105] Germany, Military Manual, reproduced and commented in Dieter Fleck, ed., *The Handbook of Humanitarian Law of Armed Conflicts* (Oxford, Oxford University Press 2000) p. 177.

[106] See C. Greenwood, 'Legal Issues Regarding Explosive Remnants of War', Working Paper, *Group of Governmental Experts of the States Parties to the Convention on Prohibitions or Restrictions on the Use of Certain Conventional Weapons which may be deemed to be Excessively Injurious or to Have Indiscriminate Effects*, CCW/GGE/I/WP.10, 22 May 2002, p. 8, § 18.

[107] Sandoz, et al., *supra* n. 6, p. 684, para. 2209.

basis of the information reasonably available at the time of the attack.[108] This standard functions as a primary limitation on the relevant harmful effects to be considered. It is within this specific framework that one has to interpret the causation requirement deriving from the word 'expected'. As noticed by Christopher Greenwood, '[i]n its normal meaning, a consequence is said to be expected if it is thought more likely than not that that consequence will result.'[109] In that respect, two questions have to be distinguished. First, whether the consequences which enter the equation of proportionality extend to long-term effects. Second, whether long-term effects fall short of meeting the threshold of foreseeability inherent in the proportionality rule.

As to the first question whether and to what extent long-term effects following an attack must be taken into consideration while applying the proportionality principle, the answer must be sought in a balance between humanitarian considerations, which would favour a broad interpretation of effects on civilians, and a realistic application of the proportionality rule. The explosive remnants of war issue exemplifies the difficulties in striking that balance.

On one hand, some argue that only harmful effects on civilians from ERW in the immediate aftermath of an attack can be seen as 'expected' as required by the proportionality rule. Later, too many factors would intervene to make any assessment on predictability with available information at the time of the attack.[110] On the other hand, others consider it necessary to also assess long-term effects.[111] Having in mind, thanks to recent studies, the certainty of some weapons to turn into ERW, they reject the assertion that long-term effects could not be anticipated.[112] The latter approach then raises the question of how remote, and then less expected,

---

[108] See, for example, United Kingdom Declaration at the time of ratification, 28 January 1998. This State holds that 'Military commanders and others responsible for planning, deciding upon, or executing attacks necessarily have to reach decisions on the basis of their assessment of the information from all sources which is reasonably available to them at the relevant time.' See for similar practice, Henckaerts and Doswald-Beck, *supra* n. 6, p. 50.

[109] See C. Greenwood, *Observations on the Statement by Ambassador Tim Caughley of New Zealand to the Group of Governmental Experts, 20 June 2003*, Circulated by the UK Delegation to the Group of Governmental Experts to the CCW, 17-24 November 2003, pp. 2-3, § 5 (on file with the authors).

[110] See Greenwood, *supra* n. 106, pp. 9-10 and, from the same author, ibid., para. 6-7. It is worth noting that this expert dismisses long-term effects not as a matter of principle but because they cannot be, 'in most cases', expected. See also Gardam, *supra* n. 19, p. 119, asserting generally that 'State practice [...] appears to indicate that this type of damage [i.e. longer-term and more remote damage] is discounted' from the proportionality assessment.

[111] Tim McCormack, after having underlined the ongoing debate among states with regard to this issue, notes that 'the balancing test requires commanders and planners to take into account the expected damage to civilian property and the expected loss of civilian life, it should be both the short-term as well as the longer-term expectation that ought to be part of the equation'. T. McCormack, 'International Humanitarian Law Principles and Explosive Remnants of War', Working Group on Explosive Remnants of War, *CCW/GGE/XI/WG.1/WP.19*, 25 August 2005, p. 3, para. 9.

[112] Ibid., p. 4, para. 11, and International Committee of the Red Cross, 'Existing Principles and Rules of International Humanitarian Law applicable to Munitions that may become Explosive Remnants of War', *CCW/GGE/XI/WG.1/WP.7*, 28 July 2005, p. 4, para. 21.

the collateral damage has to be to fall outside the proportionality equation.[113] The challenging element would be to determine on a scale of effects which ones are relevant. The 'dud' rate of some weapons may be predictable, whereas the collateral damage to civilians may remain difficult to foresee, particularly when considering the time after the conflict. It is therefore possible to state that, although a margin of appreciation does exist and there is no clear-cut answer, some long-term effects are to be considered when it can be reasonably determined that there is a serious risk for civilians. Indeed, several states explicitly hold that ERW's long-term effects have to be taken into account under the proportionality test.[114] Specific situations raise special concerns and would prove to meet this threshold of predictability, such as use in populated areas[115] or against particular targets like electricity,[116] making long-term effects more expected. Without asking the commanders who are planning an attack to foresee the 'unknowable', the conditions set in the proportionality principle do not prevent planners from considering certain long-term effects, inasmuch as a broad interpretation of anticipated military advantage is favoured.[117] However, the determination of the excessive character of collateral

---

[113] C. Garraway, 'How Does Existing International Law Address the Issue of Explosive Remnants of War?', *CCW/GGE/XII/WG.1/WP.15*, 15 December 2005, p. 5.

[114] For example, Switzerland made the following statement: 'The military commander's proportionality assessment with regard to the choice and use of a particular means or method of warfare must also take into account the foreseeable incidental long-term effects of an attack such as the humanitarian costs caused by duds becoming ERW. Thus, ammunitions with high dud rates will influence the proportionality balance negatively and diminish the options of their use against legitimate military objectives.' See 'Responses to document CCW/GGE/X/WG.1/WP.2, entitled IHL and ERW, dated 8 March 2005', *Response from Switzerland, CCW/GGE/XI/WG.1/WP.13*, 3 August 2005, p. 3, para. 15. Austria interprets the words 'incidental' and 'expected' as covering also long-term effects of ERW 'Responses to document CCW/GGE/X/WG.1/WP.2, entitled IHL and ERW, dated 8 March 2005', *Response from Austria, CCW/GGE/XI/WG.1/WP.14*, 4 August 2005, p. 3, para. 9-10. See also Norway, which highlights the ongoing international debate on this issue, 'Responses to document CCW/GGE/X/WG.1/WP.2, entitled IHL and ERW, dated 8 March 2005', *Response from Norway, CCW/GGE/XI/WG.1/WP.5*, 29 July 2005, p. 5, para. 18-19. For a general assessment on the long-term effect question in the context of the ERW discussion, see T.L.H. McCormack, P.B. Mtharu, and S. Finnin, 'Report on States Parties' Responses to the Questionnaire: International Humanitarian Law and Explosive Remnants of War', Asia Pacific Centre for Military Law, March 2006, pp. 19-21.

[115] T.J. Herthel, 'On the chopping block: cluster munitions and the laws of war', 51 *Air Force Law Review* (2001) p. 268.

[116] J.W. Crawford, 'The law of noncombatant immunity and the targeting of national electrical power systems', 21 *Fletcher Forum of World Affairs* (1997) pp. 114-115, quoted by W.J. Fenrick, 'The law applicable to targeting and proportionality after Operation Allied Force: a view from the outside', 3 *YIHL* (2000) p. 69.

[117] Drawing a parallel between the two elements of the proportionality equation, the ICRC notes: 'If the concept of military advantage were to be enlarged, it seems only logical to also consider such 'knock-on effects', i.e., those effects not directly and immediately caused by the attack, but which are nevertheless the product thereof. In the ICRC's view, the same scale has to be applied with regard to both the military advantage and the corresponding civilian casualties. This means that the foreseeable military advantage of a particular military operation must be weighed against the foreseeable incidental civilian casualties or damage of such an operation, which include knock-on effects.' 'International humanitarian law and the challenges of contemporary armed conflicts', Report prepared by the International Committee of the Red Cross for the 28th International Conference of the Red Cross and Red Crescent, Geneva, September 2003, pp. 12-13.

damage will remain a question to be resolved on a case-by-case basis, as Oeter rightly notes that '[o]bjective standards for the appraisal of expected collateral damage and intended military advantage are virtually non-existent.'[118]

Those remarks shed some light on the DU weapons' effects issue under the prohibition of disproportionate attacks. As pointed out earlier, adverse consequences of DU on civilians result from the heavy-metal toxicity and radiation rather than blast and heat, which will primarily affect the legitimate target of the attack. Apparently, these effects should be analysed under the proportionality test. The US position mentioned above goes down this line although concluding that in most cases they would not be disproportionate.[119] However, by nature, such effects are time delayed and it may be challenged whether or not they fit within the definition of 'expected' collateral damage. The current uncertainty surrounding the danger of DU renders any proportionality assessment difficult.[120] More important is the link between the attack carried out with a DU weapon and the actual toxic effects on civilians. The contamination by DU dust as well as the level of exposure and the ensuing health risk depend largely on a number of variables and factors such as meteorological conditions, the type of the target or the nature of exposure. Assuming that DU poses a danger to civilians, the determination of the foreseeable risk stemming from such attack seems to be everything but easy. This is particularly true as, in general, such evaluation already appears difficult as 'the characteristics of the weapon in relation to the specific target and its surroundings' will intervene for any means of warfare.[121] In the case of DU's effects, many parameters beyond the control of the commander planning the attack come into play, blurring the causation link between the initial act and the potential harmful consequences. Moreover, at the time of the attack, it is doubtful that, on the basis of the information reasonably available to the planner, the collateral damage deriving from the use of DU weapons could be seriously predicted. However, one may suggest that in particular situations, like urban areas, the inherent toxicity of DU could give rise to serious concern and could have more certain adverse effects. In such cases, long-term effects of DU would have to be factored into the equation as meeting the 'expected' criteria of the proportionality rule. The foregoing nevertheless leads us to conclude that an application of the principle of proportionality to DU weaponry will be particularly difficult in relation to long-term effects.

---

[118] Oeter, *supra* n. 8, p. 179.

[119] See Department of the Air Force, HQUSAF (JA) Legal Memorandum concerning review by Harold R. Vague, *supra* n. 80.

[120] A. McDonald, 'The International Legality of Depleted Uranium Weapons', Background paper, *The international legal ramifications of the use of DU weapons*, Symposium on the Health Impact of Depleted Uranium Munitions, New York Academy of Medicine, 14 June 2003, available at <www.nuclearpolicy.org/files/nuclear/mcdonald_jun_14_03.pdf>, p. 14.

[121] F. Hampson, 'Means and methods of warfare', in P. Rowe, ed., *The Gulf War 1990-91 in International and English Law* (London, Sweet & Maxwell 1993) p. 89 at p. 92.

4.   DEPLETED URANIUM AND THE PROTECTION OF OBJECTS
      INDISPENSABLE TO THE SURVIVAL OF THE CIVILIAN POPULATION

The rule which prohibits attacking, destroying, removing or rendering useless objects indispensable to the survival of the civilian population[122] is seen as a corollary to the prohibition of starvation.[123] It goes without saying that any civilian object is already protected by the general rule discussed earlier. Nevertheless, far from being redundant, the special protection provided by this norm reflects the peculiar nature of the objects at stake and their importance for civilians. Within the context of DU, the protection of objects indispensable to the survival of the civilian population entails a number of issues. Firstly, it must be stressed that solely the prohibited action of rendering useless objects indispensable to the survival of the civilian population is relevant to DU's effects inasmuch as the other acts, such as attacking, may concern any type of weapon.[124] Secondly, the different forms of DU contamination appear to fall within the scope of this prohibition. The levels of contamination of food and drinking water may pose some serious threat to civilians' health 'where it is considered that there is a reasonable possibility of significant quantities of DU entering the ground water or food chain'.[125] The question is whether the secondary nature of DU toxic effects represents an obstacle in order to trigger the application of the rule prohibiting to render useless objects indispensable to the survival of the civilian population.

As mainly armour piercing projectiles, DU weapons' military utility stems from the heat, blast and penetrating capacity of heavy metal hitting the target. Even when used against non-armoured targets, such as soldiers, buildings, cars and trucks, the toxicity produced has no military function. In that respect, the earlier question resurfaces as to what extent the contamination of, for instance water, as an unintended consequence of DU use would constitute a violation of the rule on indispensable objects. Article 54 of AP I articulates this norm with the prohibition of starvation, describing the most usual ways in which this may be applied.[126] Therefore, the action of rendering useless seems to be behaviour carried out with a given purpose. Moreover, paragraph 2 refers to the clause 'for the specific purpose of denying them for their sustenance value to the civilian population'. The question of the intent or purpose has been discussed concerning the act of attack, several military manuals specifying that for the attack to be illegal, the intent has to be to

---

[122] Cf., Art. 54(2) to (5) AP I; Customary Rule 54.

[123] Cf., Art. 54 (1) AP I; Customary Rule 53.

[124] The Commentary of AP I explains that it 'should be noted that the verbs "attack", "destroy", "remove" and "render useless" are used in order to cover all possibilities, including pollution, by chemical or other agents, of water reservoirs, or destruction of crops by defoliants, and also because the verb "attack" refers, either in offence or defence, to acts of violence against the adversary, according to Article 49 § 1', Sandoz, et al., *supra* n. 6, p. 655, para. 2101.

[125] World Health Organization, Depleted uranium: Sources, Exposure and Health Effects (Geneva 2001) p. 147-148. See D. Fahey, Chapter 2, pp. 29 et seq.

[126] See Sandoz, et al., *supra* n. 6. p. 655, para. 2098.

prevent the civilian population from being supplied.[127] A mere objective observation of a particular result in terms of contamination is therefore insufficient and one must look for the additional requirement of the purpose of the act. It follows that adverse effects of DU could lead to a claim under this specific rule only once such a purpose can be established.

5.   DEPLETED URANIUM AND THE OBLIGATION TO TAKE PRECAUTIONARY
        MEASURES

The importance of precautionary measures in IHL is twofold. Firstly, they allow for full respect for other rules by imposing obligations of care before carrying out an action. Conceptually speaking, they act as a primary prerequisite in order to ensure respect of the norms regulating the conduct of hostilities and protecting civilians. There would be no relevance for the principles of distinction and proportionality if no precautions were taken to identify which objective is to be targeted.[128] Secondly, precautionary measures give rise to autonomous obligations in addition to the other rules of IHL. Therefore, one must analyse the compliance with norms requiring such measures in order to assess the legality of an attack.

Precautionary measures are codified in two provisions of AP I, one dealing with precautions in attacks,[129] the other addressing the reciprocal obligation concerning precautions against the effects of attacks.[130] The present contribution will focus on the former component as it is the most relevant with regard to DU weapons. Obligations to take precautionary measures in attacks are also part of customary international humanitarian law and belong to two main categories.[131] In addition to a general obligation of parties to a conflict to spare civilians and to take all feasible measures to avoid or minimise incidental harmful effects, several more specific obligations, aiming at ensuring constant care at each stage and concerning each aspect of an attack, rest upon those responsible for planning, deciding and

---

[127] The UK military manual specifies that 'the law is not violated if military operations are not intended to cause starvation but have that incidental effect'. See *The Manual of the Law of Armed Conflict, supra* n. 71, p. 74, para. 5.27.2. On the other hand, other manuals do not mention such a requirement. On this issue see Henckaerts and Doswald-Beck, *supra* n. 6, p. 190.

[128] Fenrick, *supra* n. 116, p. 57.

[129] Art. 57 encapsulates the second formulation of the principle of proportionality which is sometimes analysed in the context of the obligations regarding precautions in attack. See *The Manual of the Law of Armed Conflict, supra* n. 71, p. 86.

[130] Art. 58 holds:
'Precautions against the effects of attacks
The Parties to the conflict shall, to the maximum extent feasible:
(a) without prejudice to Article 49 of the Fourth Convention, endeavour to remove the civilian population, individual civilians and civilian objects under their control from the vicinity of military objectives;
(b) avoid locating military objectives within or near densely populated areas;
(c) take the other necessary precautions to protect the civilian population, individual civilians and civilian objects under their control against the dangers resulting from military operations.'

[131] See Henckaerts and Doswald-Beck, *supra* n. 6, pp. 51-67.

carrying out attacks.[132] This set of obligations is of tremendous importance for the legal analysis of DU weapons.[133] Regardless of whether these projectiles are indiscriminate or cause disproportionate collateral damage, commanders and states are legally bound to minimise adverse effects on civilians. This means that IHL may be violated in cases where precautionary measures are not taken, whatever the actual result of the attack.[134]

The general obligation to take precautionary measures to spare civilian persons and objects derives from a fundamental principle which inspires the whole body of IHL norms: the logic of the 'least harm' or 'the lesser of two evils'.[135] In that respect, precautions in attacks ensure that adverse effects that can be avoided will be. If not, there is also an obligation to minimise incidental harm. This duty of 'diligence' cannot be read in absolute terms and requires a standard of appreciation set in the formulas 'all feasible' and 'based on circumstances ruling at the time', which can be found in the relevant provisions.[136] This criterion is the same as the one relevant while applying the proportionality rule.[137] Assuming that the potential risk stemming from the use of DU weapons for civilian is recognised, long-term effects would have to be considered by planners and commanders. The recourse to DU weapons would require special care from those responsible. As highlighted by the new UK military manual, the general principle of precaution implies that 'the commander will have to bear in mind the effect on the civilian population of what

---

[132] On the difference between a general obligation and specific duties, see *The Manual of the Law of Armed Conflict, supra* n. 71, pp. 82-83, para. 5.32.1. See also Henckaerts and Doswald-Beck, *supra* n. 6, p. 51.

[133] It has been argued by some states that attacks using DU weapons comply with the rule of precaution as it concerns precaution in attacks rather than the specific means of making the attack. See McDonald, *supra* n. 120, p. 15. This is inconsistent with the fact that one of the specific obligations concerns the choice of means of warfare. Moreover, in the context of ERW discussions, several states recognised that the obligation to take precautions in the conduct of military operations should be applied to the use of munitions which could result in ERW. See McCormack, et al., *supra* n. 114, p. 21.

[134] In one of the arbitral decisions in the case between Eritrea and Ethiopia, although the Eritrea-Ethiopia Claims Commission (EECC) could not establish that Eritrea deliberately targeted a civilian neighbourhood, it held that this state failed to take all feasible precautions in its conduct of the air strike and therefore it found that 'Eritrea is liable for the deaths, wounds and physical damage to civilians and civilian objects caused in Mekele by the third and fourth sorties on June 5, 1998'. See EECC, Partial Award on Ethiopia's Central Front Claim 2, 28 April 2004, §§ 108-113.

[135] Sandoz, et al., *supra* n. 6, p. 705, para. 2226.

[136] See also Canada, Law of Armed Conflict at the Operational and Tactical Levels, *supra* n. 90, p. 4-4, para. 25-27. The manual states the following: 'Consideration must be paid to the honest judgement of responsible commanders, based on the information reasonably available to them at the relevant time, taking fully into account the urgent and difficult circumstances under which such judgements are usually made. (…) The test for determining whether the required standard of care has been met is an objective one: Did the commander, planner or staff officer do what a reasonable person would have done in the circumstances?'

[137] We refer here to the declarations and statements made by some states when signing or ratifying AP I discussed earlier. See Henckaerts and Doswald-Beck, *supra* n. 6, p. 52. The authors of the Study notes that 'the obligation to take all "feasible" precautions has been interpreted by many States as being limited to those precautions which are practicable or practically possible, taking into account all circumstances ruling at the time, including humanitarian and military considerations.'

he is planning to do and take steps to reduce that effect as much as possible.'[138] On the other hand, as stated above, one cannot ask a commander to predict all the knock-on effects of an attack, particularly when it concerns weapons whose toxicity is uncertain and depends on so many variables.

However, the general obligation of precaution would call at least for a minimum standard of care to take into account the potential risk for civilians arising from the DU weapons. Inasmuch as 'all feasible' precaution have to be taken, it can be asserted that parties are required to take into account information on expected subsequent effects.[139] Moreover, the specific duties of precaution entail an obligation with regard to the choice of means of warfare. Article 57(2)(a)(ii) AP I requires Parties to take all feasible precautions in the choice of means and methods to avoid or minimise harmful effects on civilians. This is a very important obligation concerning DU weapons. Concretely, it obliges consideration of what alternative methods and means of attack are available in order to use the one that causes least damage. Elements such as precision, accuracy and range of the weapons, as well as their properties in relation to the nature and the location of the target come into play.[140] However, for the choice to be possible it needs to be made between weapons providing a similar military advantage. Without being explicitly mentioned in the relevant paragraph, compared to Article 57(3), which deals with the choice between different military objectives, this seems to derive from the very nature of the choice and the way states interpret the rule. Indeed, the decision to use a particular weapon factors in several intertwined elements as military advantage, ranging from the actual destruction or neutralisation of the target to considerations about the security of forces.[141] Although it does not appear to be a formal condition for the specific obligation to be applied, in reality, if the weapon does not provide the same military gain, planners will use another means of warfare.[142]

One may draw some conclusions concerning DU weapons from the foregoing. DU weaponry being highly efficient in armour piercing, it is doubtful that other weapons would be available to fulfil the same military function with an identical efficiency. But it might be argued that tungsten munitions, although less efficient, could prevent extensive harmful effects on civilians from occurring, those effects outweighing the military utility at stake.

Finally, some brief remarks on the warning duty. If precautionary logic finds one application at all regarding DU weapons, it is certainly this duty. Whatever the degree of certainty about the health danger posed by DU, the existence of a risk should call for a warning in case civilians may be affected. Such a requirement lies at the very core of the rules under consideration. Whereas for other specific norms to be violated a certain threshold has to be met, this obligation of warning

---

[138] *The Manual of the Law of Armed Conflict, supra* n. 71, p. 82.

[139] See McCormack, et al., *supra* n. 114, p. 23.

[140] A.P.V. Rogers, 'Zero casualty warfare', 82 (837) *IRRC* (2000), pp. 165-181, p. 176.

[141] See, for example, *The Manual of the Law of Armed Conflict, supra* n. 71, p. 83, para. 5.32.5.

[142] Schmitt, *supra* n. 83, p. 152.

seems to only depend on the fact that civilians 'may be' affected.[143] Such warning must be given in advance to be useful. As noted by McDonald, the current practice is not to warn civilians before (or after) an attack using DU weapons.[144] One could argue that this lack of information does not arise as a matter of law if those weapons are said not to be dangerous. However, the instructions given by some states to their staff on the field in areas where DU weapons have been used change the legal analysis as they demonstrate the concern for a risk.[145] Therefore, precautionary measures constitute one of the most relevant IHL obligations with regard to DU weapons, given the legal obstacles to assessing the potential long-term effects for civilians.

6. CONCLUSION

Some cautious conclusions can be drawn from the foregoing. The use of DU is not in contravention of the prohibition to make civilians the object of attack as long as it is not directly aimed at hitting them or is used as a tactical objective. Neither has DU ammunition analogous effects to blind weapons, which cannot be directed against military objectives with a reasonable accuracy, because DU ammunition is characterised by a high degree of precision. It thus does not fall foul of the prohibition of attacks which employ a method or means of combat which cannot be directed at a specific military objective according to Article 51(4)(b) and customary international humanitarian law.

On the other hand, even though DU weapons can very well be directed at a specific military objective, it cannot be excluded that they take on a life of their own and consequentially affect civilians to a significant degree under certain circumstances. While this does not lead us to submit that these weapons are unlawful *per se,* they may be used in a way that the prohibition of indiscriminate attack is violated when their immediate or consequential effects reach a certain intensity or severity. Furthermore, attacks involving DU have to be assessed under the principle of proportionality, which extends to both immediate and long-term effects. However, in both cases of the prohibition of indiscriminate attacks and of disproportionate attacks as one subspecies, difficulties in establishing a causal connection between initial uses of DU and more remote consequences will complicate an assessment of these legal parameters. This makes firm compliance with the duty to take precautionary measures all the more important.

---

[143] *The Manual of the Law of Armed Conflict, supra* n. 71, p. 84, para. 5.32.8, and Canada, Law of Armed Conflict at the Operational and tactical Levels, *supra* n. 90, p. 4-4, para. 29.

[144] McDonald, *supra* n. 120, p. 15.

[145] Ibid., pp. 15-16.

Chapter 7
THE USE OF DEPLETED URANIUM AND THE PROHIBITION AGAINST THE USE OF POISONED WEAPONS

Jann K. Kleffner

1. INTRODUCTION

The prohibition of poison has a long tradition in the law of armed conflict and can be traced back many centuries.[1] The Lieber Code of 1863, which is generally recognised as the origin of what has come to be known as 'Hague Law' regulating the use of certain weapons and methods of warfare, included that prohibition,[2] as do subsequent treaties[3] and customary international humanitarian law.[4] The underlying rationale of the prohibition are that poison and poisoned weapons inflict superfluous injury and/or unnecessary suffering on combatants, and that their effects may be indiscriminate in violation of the principle of distinction between combatants and civilians.[5] Yet, as much as this generic prohibition of the use of poison is generally accepted, its precise contours are a matter of dispute. What exactly is poison and when does the use of substances which may have adverse effects on human health violate the prohibition of poison? This chapter considers these questions in relation to depleted uranium (DU) weapons.

---

[1] Such early sources include the *Ramayana* and the *Mahabharata*, two Sanskrit texts composed in the third century B.C. and between 200 B.C. and 200 A.D., respectively. They provide measures of humanitarian concern by postulating a series of principles regulating conduct in war, including the rule: 'When he fights his foes in battle, let him not strike with weapons concealed in wood, nor with such as barbed, *poisoned*, or the points of which are blazing with fire. Neither *poisoned* nor barbed weapons should be used. These are weapons for the wicked.' (emphasis added), cited in L. Green, 'What is – why is there – the law of war?' in M.N. Schmitt and L.C. Green, eds., *The Law of Armed Conflict: Into the Next Millennium* (Vol. 71) (Newport, Rhode Island, Naval War College International Law Studies 1998) pp. 141-183, p. 147, fn. 27-29 and accompanying text.

[2] Cf., e.g., Arts. 16 and 70 of the 1863 Instructions for the Government of Armies of the United States in the Field (Lieber Code).

[3] Cf., Art. 23(a) of the 1907 Hague Regulations; Protocol for the Prohibition of the Use of Asphyxiating, Poisonous or Other Gases, and of Bacteriological Methods of Warfare. Geneva, 17 June 1925; Art. 8 (2)(b)(xvii) ICC Statute.

[4] J.M. Henckaerts and L. Doswald-Beck, eds., *Customary International Humanitarian Law* (Cambridge, Cambridge University Press 2005) Vol. I: Rule 72, p. 251.

[5] E. David, *Principes de Droit des Conflits Armés*, 3rd edn. (Brussels, Bruylant 2002) p. 334.

McDonald / Kleffner / Toebes (eds.), Depleted Uranium Weapons and International Law
© *2008, T·M·C·ASSER PRESS, The Hague, The Netherlands and the Authors*

2.   DEFINING POISON AND POISONED WEAPONS

Neither the 1907 Hague Regulations and the customary rule regarding poison, which
both prohibit 'poison or poisoned weapons', nor the prohibition of 'the use in war
of asphyxiating, poisonous or other gases, and of all analogous liquids, materials or
devices' contained in the 1925 Geneva Gas Protocol, provide a definition of what
constitutes 'poison', 'poisoned weapons', 'poisonous gases' and 'analogous liquids, materials or devices'. Indeed, the ICRC Customary Law Study acknowledges
that '[m]ost States indicate that poison or poisoned weapons are prohibited without
further detail.'[6] In general parlance, 'poison' means 'a substance that when introduced into or absorbed by a living organism causes death or injury.'[7] Thus understood, DU weaponry could be considered poison or poisonous if and when it causes
death or injury. However, it has been suggested that the law of armed conflict requires more than this in order for the use of a substance to fall foul of the prohibition of poison.

3.   THE 'PREMEDITATION' APPROACH

When the International Court of Justice (ICJ) had to answer the question whether
the use of nuclear weapons is compatible with international law, it did so, *inter alia*,
in light of the prohibition of poison and poisoned weapons in the 1907 Hague Regulations and the prohibition under the 1925 Geneva Gas Protocol, holding that the
prohibited weapons 'have been understood, in the practice of States, in their ordinary sense as covering weapons whose *prime, or even exclusive, effect* is to poison
or asphyxiate'.[8] The Court thus went a long way in accepting the arguments advanced by the United Kingdom and the United States in the course of the ICJ proceedings. The United Kingdom had argued in its written statement that the prohibition
of poison and poisoned weapons was 'intended to apply to weapons whose primary
effect was poisonous and not to those where poison was a secondary or incidental
effect'.[9] In a similar vein, the United States had argued that the prohibition is applicable only to weapons 'that are designed to kill or injure by the inhalation or other
absorption into the body of poisonous gases or analogous substances.'[10] Identical
reasoning underlies the conclusion of the 1975 US review of DU weapons that such
weapons do not violate the prohibition of poison or poisoned weapons. The review

---

[6] Henckaerts and Doswald-Beck, *supra* n. 4, p. 253.

[7] *The Oxford Concise Dictionary*, 9th edn. (Oxford, Oxford University Press 2002) p. 1055.

[8] *Legality of the Threat or Use of Nuclear Weapons*, Advisory Opinion of 8 July 1996, ICJ General List No. 95, 35 *ILM* 809 and 1343 (1996), para. 55, emphasis added.

[9] UK, Written statement submitted to the ICJ, Nuclear Weapons Advisory Opinion, 16 June 1995,
paras. 3.59 and 3.60.

[10] US, Written statement submitted to the ICJ, Nuclear Weapons Advisory Opinion, 20 June 1995,
p. 24.

noted that 'DU's toxic radiological and chemical properties are an inherent characteristic of the substance and not a *designed, added in*, characteristic.'[11]

The pleadings of the United Kingdom and the United States reflect the view that a substance which has poisonous effects when released in the course of using a weapon does not violate the prohibition of poison if these effects were secondary, incidental and unintended and the weapon was not specifically developed with a view to having these effects. Drawing an analogy to different forms of *mens rea* in criminal law, the poisonous effects must be premeditated. Such an understanding of the prohibition of poison and poisoned weapons is reminiscent of the definition of 'chemical weapons' prohibited under the Chemical Weapons Convention. That definition requires that munitions and devices must be *specifically designed* to cause death or other harm through the toxic properties of those chemicals which would be released as a result of the employment of such munitions and devices.[12] The position of the United Kingdom and the United States may also be said to find further support in Article 3(a) of the 1993 Statute of the ICTY, which extends the jurisdiction of the Tribunal to violations of the laws and customs of war consisting of the 'employment of poisonous weapons or other weapons *calculated to* cause unnecessary suffering'[13] and in the identical war crime included in the 1996 Draft Code of Crimes against the Peace and Security of Mankind.[14] This wording, which was included into the ICTY Statute on the assumption that it reflected customary international humanitarian law,[15] and especially the words 'or *other*', may be read so as to suggest that poisonous weapons fall into the broader category of weapons 'calculated to cause unnecessary suffering'. 'Calculated to', in turn, implies an element of premeditation; a criterion that would not be satisfied if causing unnecessary suffering was merely an incidental, unintended side-effect of the use of a particular weapon.

According to the aforementioned 'premeditation-based' views, the use of DU weaponry would not violate the prohibition of poison or poisonous weapons. Its primary, intended effect is to pierce armour and it has been developed specifically for that purpose. It is not designed to cause death or injury when introduced into or absorbed by the human organism. Whatever the possible adverse health

---

[11] Department of the Air Force, HQUSAF (JA) Legal Memorandum concerning review by Harold R. Vague, Major General, USAF, The Judge Advocate-General United States Air Force, of the High Explosive Incendiary and Armour Piercing Incendiary Munitions, 14 March 1975 (emphasis added).

[12] Cf., Art. II (1)(b) of the Convention on the Prohibition of the Development, Production, Stockpiling and Use of Chemical Weapons and on their Destruction.

[13] Statute of the International Tribunal for the Prosecution of Persons Responsible for Serious Violations of International Humanitarian Law Committed in the Territory of the Former Yugoslavia since 1991 (adopted on 25 May 1993) UNSC Res. 827, emphasis added.

[14] Art. 20(e)(i) of the Draft Code of Crimes against the Peace and Security of Mankind as adopted by the International Law Commission at its 48th Session, Report of the International Law Commission on the work of its 48th session, 6 May-26 July 1996, GAOR, 51st Session, Supplement No. 10 (A/51/10) *Yearbook of the International Law Commission, 1996*, Vol. II(2).

[15] Cf., United Nations Secretary-General Report on the establishment of the International Criminal Tribunal for the former Yugoslavia (Report pursuant to paragraph 2 of Resolution 808 (1993) of the Security Council, 3 May 1993), S/25704, paras. 34, 41-42.

effects of DU dust when DU ammunition and armour is shot or breached, corroding or consumed by fire, or through inhalation of DU dust, ingestion, dermal absorption or absorption through wounds, these effects are secondary and incidental. DU weaponry is not specifically developed with a view to having these effects, which precludes it from falling foul of the prohibition of poison or poisoned weapons as understood in the aforementioned way.

4.   THE 'EFFECTS-BASED' APPROACH

The 'premeditation-based' approach to the prohibition of poison is, however, not beyond dispute. Although the ICJ asserted that the aforementioned practice 'is clear',[16] the Court appeared to over-emphasise the UK and US positions while neglecting to mention that other states have not limited their understanding of the prohibition of poison and poisoned weapons to only cover weapons which have been primarily (or even exclusively) designed and intended to have this effect.[17] Iraq, for instance, implied in 1991, during a debate in the UN Security Council concerning the aftermath of the First Gulf War, that the use of DU ammunition falls foul of the prohibition, while not contending that these weapons were designed especially with a view to cause poisoning.[18] State practice, in other words, is diverse and does not seem to establish agreement regarding the interpretation[19] of the prohibition of poison or poisoned weapons in the sense suggested by the ICJ.

Furthermore, the war crimes provisions in the ICTY Statute and the Draft Code on Offences against the Peace and Security of Mankind do not coincide with the criminalisation of employing poison or poisonous weapons under the Rome Statute of the International Criminal Court (ICC). While Article 8(2)(b)(xvii) provides that doing so amounts to a war crime in international armed conflict, nothing in the provision itself or the Elements of Crimes indicates that a weapon falls outside the prohibition for the reason alone that its poisonous properties are 'merely' incidental consequences of employing a weapon that is primarily designed for other purposes. Instead, the Elements of Crimes adopt an 'effects-based' approach. The first element refers to employing 'a substance or a weapon that releases a substance as a result of its employment', while the second element stipulates that '[t]he substance was such that it causes death or serious damage to health in the ordinary course of events, through its toxic properties.'[20] In a very similar way, the Elements

---

[16] *Legality of the Threat or Use of Nuclear Weapons*, *supra* n. 8, para. 55.

[17] Cf., on the *opinio juris* and practice of states, Henckaerts and Doswald-Beck, *supra* n. 4, Vol. II: Practice, Chapter 21, pp. 1590-1603. See also E. David, 'The Opinion of the International Court of Justice on the legality of the use of nuclear weapons', *IRRC* (1997) No. 316, pp. 21 at 27.

[18] Statement before the UN Security Council, UN Doc. S/PV.2981, 3 April 1991, pp. 29-30.

[19] Cf., Art. 31(3)(b) of the 1969 Vienna Convention on the Law of Treaties.

[20] Elements to Art. 8(2)(b)(xvii), Elements of Crimes, adopted on 9 September 2002, Doc. ICC-ASP/1/3 (part II-B). The third and fourth element of the war crime of employing poison or poisoned weapons are that the conduct took place in the context of and was associated with an international armed conflict, and that the perpetrator was aware of factual circumstances that established the existence of an armed conflict.

of the war crime of employing asphyxiating, poisonous or other gases, and all analogous liquids, materials or devices, under Article 8(2)(b)(xviii) of the ICC Statute, only require that '[t]he perpetrator employed a gas or other analogous substance or device' and that '[t]he gas, substance or device was such that it causes death or serious damage to health in the ordinary course of events, through its asphyxiating or toxic properties.'[21] The wording of the second Elements of Article 8(2)(b)(xvii) and (xviii) thus indicates that the weapon in question need not be employed with the premeditation to poison. Once the material elements of the prohibition are committed with intent and knowledge, a war crime according to Articles 8(2)(b)(xvii) or (xviii) is committed. In accordance with Article 30 of the Statute, in turn, it would suffice for establishing intent that a person is aware that the consequence of causing death or serious damage to health through the toxic properties of a given substance will occur in the ordinary course of events.[22] Consequently, weapons, which may not be *calculated* to have poisonous effects, but nevertheless release a substance which is said to cause death or serious damage to health in the ordinary course of events through its toxic properties, may fall under the prohibition of Article 8(2)(b)(xvii) and (xviii), while such conduct would not amount to a war crime under the 'premeditation-based' approach.[23]

It is only a short step from the 'effects-based' approach in the Rome Statute, which concerns itself with war crimes as secondary norms, to the same approach in relation to the underlying primary norm of the law of armed conflict, which prohibits poison or poisoned weapons. After all, a war crime provision may be narrower than the primary norm from which it derives, but the reverse does not hold true: primary norms cannot prohibit less than what entails consequences under the body of secondary norms governing the violation of that prohibition. In other words, one can derive from the war crimes provisions in Article 8(2)(b)(xvii) and (xviii) of the Rome Statute that the underlying primary norm is presumed to be not the 'premeditation-based' approach but the 'effects-based' approach. That primary norm thus imposes on states a prohibition to use weapons which release substances that cause death or serious damage to health in the ordinary course of events, through their toxic properties.

---

[21] Ibid. The second element is accompanied by a footnote, which clarifies that nothing in that second element 'shall be interpreted as limiting or prejudicing in any way existing or developing rules of international law with respect to development, production, stockpiling and use of chemical weapons'. The other two elements are identical to the third and fourth element of the war crime of employing poison or poisoned weapons under Art. 8(2)(b)(xvii). On the Elements of both war crimes, see C. Garraway, 'Article 8(2)(b)(xvii)', in R.S. Lee, ed., *The International Criminal Court – Elements of Crimes and Rules of Procedure and Evidence* (Ardsley, NY, Transnational Publishers 2001) pp. 178-180; K. Dörmann, *Elements of War Crimes under the Rome Statute of the International Criminal Court – Sources and Commentary* (Cambridge, Cambridge University Press 2003) pp. 281-291.

[22] Cf., Rome Statute, Art. 30(2)(b). See also Garraway, *supra* n. 21, p. 178: 'It was considered unnecessary to include any additional mental element, as no deviation was required from the *mens rea* requirement set forth in article 30.'

[23] In this vein, H. Fischer, 'The jurisdiction of the International Criminal Court for war crimes: Some observations concerning differences between the Statute of the Court and war crimes provisions in other treaties', in V. Epping, H. Fischer, W. Heintschel von Heinegg, eds., *Brücken bauen und begehen: Festschrift für Knut Ipsen zum 65 Geburtstag* (München, Beck 2000) pp. 77-101, 93.

5.   APPLYING THE 'EFFECTS-BASED' APPROACH TO DEPLETED URANIUM
        WEAPONS

Even if one were to adopt the 'effects-based' approach it is unlikely that the use of
DU weapons would violate the prohibition of poison and poisoned weapons. Skep-
tics would probably argue that the scientific uncertainty surrounding the health
effects of DU weapons makes it impossible to know whether these effects 'cause
death or serious damage to health in the ordinary course of events'. And yet, even if
one adopted a precautionary approach in light of the significant evidence which
suggest that DU residues have adverse health effects, a case for considering the use
of DU weapons as poison or poisoned weapons would be hard to make. The adverse effects are very much dependent on the specific circumstances of the use of
DU. As shown elsewhere in this book, the type of ammunition; the type of target
and soil; the oxygen content of surroundings of DU rounds or fragments; the intensity and duration of fires; the force of explosions; wind; time; and other factors,
including age, sex, diet, family history, health status, and lifestyle of those exposed,
affect the overall health consequences of exposure.[24] These context-specific aspects make an assessment whether DU residues cause death or serious damage to
health in the ordinary course of events, through its toxic properties, a highly complex undertaking. This is not to suggest that further research is unlikely to enhance
the quality of such assessments. Indeed, it is submitted that more research into the
effects of the use of DU is urgently called for and that the knowledge gained needs
to be translated into operational doctrines, which provide guidance to the military
on when, where and how to use DU weapons so as to avoid a potential violation of
the prohibition of poison and poisoned weapons under the given circumstances.
However, at this stage, it appears to the present author that one cannot conclude that
DU weapons violate that prohibition *per se*, or that the scientific knowledge about
the adverse health effects of the use of DU in a given situation provides sufficient
guidance to know whether the toxic properties of DU residues cause death or serious damage to health in the ordinary course of events.

---

[24] See on these factors, D. Fahey, Chapter 2 of this book.

Chapter 8
THE USE OF DEPLETED URANIUM AND THE DIRECT PROTECTION OF THE ENVIRONMENT UNDER *JUS IN BELLO*

Erik V. Koppe[1]

1. INTRODUCTION

In view of growing concern about the effects of the use of depleted uranium (DU) on man and his environment,[2] it is worth assessing whether its use is consistent with international rules that are intended to protect the environment during armed conflict. Such protection is embodied in three sets of rules: *jus in bello*, *jus pacis* and *jus ad bellum*, or the law of war or armed conflict, the law of peace, and the law with respect to the use of force. Because protection under *jus ad bellum* seems to be rather unique[3] and the protection under *jus pacis* depends on the applicability of peacetime international environmental law in times of armed conflict and appears to be subsidiary in character,[4] focus will be on the protection of the environment during armed conflict under *jus in bello*, which is primarily applicable during armed conflict. Furthermore, because *jus in bello* provides protection of the environment at various levels and space herein is limited, the frame of reference is further lim-

---

[1] The author would like to thank Dr. A.J.J. de Hoogh, Prof. Dr. H.H.G. Post, and Prof. Dr. W.D. Verwey for their valuable comments.

[2] See D. Fahey, Chapter 2 of this book.

[3] The Security Council reaffirmed in para. 16 of Resolution 687 of 3 April 1991 'that Iraq, without prejudice to the debts and obligations of Iraq arising prior to 2 August 1990, which will be addressed through the normal mechanisms, is liable under international law for any direct loss, damage – including environmental damage and the depletion of natural resources – or injury to foreign Governments, nationals and corporations as a result of its unlawful invasion and occupation of Kuwait.' UN Doc. S/Res/687 (1991), adopted on 3 April 1991, by 12 votes to 1, with 2 abstentions, on the situation between Iraq and Kuwait, para. 16. Resolution 687 seems to entail a form of strict liability for all environmental damage resulting from its invasion of Kuwait, including damage resulting from actions that are not illegal under the law of armed conflict and for damage resulting from actions by other belligerents and which therefore cannot be attributed to Iraq.

[4] On the effects of armed conflicts on treaties, see: A/CN.4/550, 'The effect of armed conflict on treaties: an examination of practice and doctrine'; Memorandum by the Secretariat, 1 February 2005; International Law Commission, 57th session, Geneva, 2 May-3 June 2005 and 4 July-5 August 2005; A/CN.4/552, First Report on the Effects of Armed Conflicts on Treaties, by Mr I. Brownlie, Special Rapporteur, of 21 April 2005; International Law Commission, 57th session, Geneva, 2 May-3 June and 4 July-5 August 2005, <http://www.un.org/law/ilc/sessions/57/57sess.htm>.

McDonald/Kleffner/Toebes (eds.), Depleted Uranium Weapons and International Law
© 2008, T·M·C·Asser press, *The Hague, The Netherlands and the Authors*

ited to those rules that are directly intended to protect the environment during international armed conflict, both under treaty (section 2) and under customary law (section 3).

2.   TREATY LAW

2.1   Introduction

Direct protection of the environment during armed conflict under treaty law is provided only by four relatively recent conventions. These are, in chronological order, the 1977 Convention on the Prohibition of Military or Any Other Hostile Use of Environmental Modification Techniques (ENMOD),[5] the 1977 Additional Protocol I to the 1949 Geneva Conventions (hereinafter, Additional Protocol I or AP I),[6] the 1981 Certain Conventional Weapons Convention, in particular its third Protocol on Incendiary Weapons (the Incendiary Weapons Protocol),[7] and the 1998 Statute of the International Criminal Court.[8] Since ENMOD, the Incendiary Weapons Protocol and the Statute of the International Criminal Court are not directly relevant to the topic under discussion, discussion herein will be limited to AP I.[9]

2.2   Additional Protocol I

Additional Protocol I was negotiated and adopted by a Diplomatic Conference in Geneva in 1977,[10] and consists of 102 articles. Of these, there are two that directly

---

[5] Convention on the Prohibition of Military or any Other Hostile Use of Environmental Modification Techniques, opened for signature on 18 May 1977, entered into force on 5 October 1978, 1108 *UNTS*, No. 17119.

[6] Protocol Additional to the Geneva Conventions of 12 August 1949, and Relating to the Protection of Victims of International Armed Conflicts, opened for signature on 12 December 1977, entered into force on 7 December 1978, 1125 *UNTS*, No. 17512.

[7] Protocol on Prohibitions or Restrictions on the Use of Incendiary Weapons (Protocol III) to the Convention on Prohibitions or Restrictions on the Use of Certain Conventional Weapons Which May be Deemed To Be Excessively Injurious or To Have Indiscriminate Effects, opened for signature on 10 April 1981, entered into force on 2 December 1983, 1342 *UNTS*, No. 22495.

[8] The Rome Statute opened for signature on 17 July 1998 and entered into force on 1 July 2002, 2187 *UNTS* No. 38544.

[9] Art. I ENMOD prohibits the use of environmental modification techniques as weapons of warfare, while Art. II defines environmental modification techniques as techniques 'for changing – through the deliberate manipulation of natural processes – the dynamics, composition or structure of the Earth, including its biota, lithosphere, hydrosphere and atmosphere, or of outerspace.' Art. 2(4) of the Incendiary Weapons Protocol prohibits making 'forests or other kinds of plant cover the object of attack by incendiary weapons except when such natural elements are used to cover, conceal or camouflage combatants or other military objectives, or are themselves military objectives.' And Art. 8(2)(b)(iv) of the Statute of the International Criminal Court holds individuals responsible when they are involved in 'launching an attack in the knowledge that such attack will cause incidental loss of life or injury to civilians or damage to civilian objects or widespread, long-term and severe damage to the natural environment which would be clearly excessive in relation to the concrete and direct overall military advantage anticipated.'

[10] As of October 2007, AP I had 167 States Parties and five signatory states, <http://www.icrc.org/> and <http://www.eda.admin.ch/>.

protect the environment during armed conflict: Articles 35 and 55. Article 35 is included in Section I, Part III of the Protocol and deals with 'Basic Rules' of Methods and Means of Warfare, while Article 55 deals with the 'Protection of the Environment' in the context of Chapter II ('Civilian Objects') of Section I ('General Protection Against the Effects of Hostilities') of Part IV dealing with the Civilian Population.

Both articles use similar terminology but are formulated differently. Article 35(3) states that

> '[i]t is prohibited to employ methods or means of warfare which are intended or may be expected, to cause widespread, long-term and severe damage to the natural environment.'

Article 55(1) provides that

> '[c]are shall be taken in warfare to protect the natural environment against widespread, long-term and severe damage. This protection includes a prohibition of the use of methods or means of warfare which are intended or may be expected to cause such damage to the natural environment and thereby to prejudice the health or survival of the population.'

Both provisions intend to protect the environment from the destructive effects of warfare but do so in different ways. Article 35(3) lays down a general prohibition against using certain means or methods of warfare that damage or could damage the environment, whereas Article 55(1) introduces a general duty of care in the first sentence[11] and then specifies this duty by an explicit prohibition in the second sentence. That prohibition repeats the general injunction of Article 35(3) but makes its application conditional on the effects on the civilian population. The choice of the word 'includes' in the second sentence implies that the prohibition of means and methods of warfare that prejudice the health or survival of the civilian population is illustrative and exemplary of the duty of care in Article 55(1), first sentence, and apparently also of the prohibition in Article 35(3). It seems, therefore, that activities that cause widespread, long-term and severe damage to the environment but that do not prejudice the health and survival of the civilian population are nevertheless prohibited both by Article 55(1), first sentence, and Article 35(3).[12]

Both Articles 35 and 55 prohibit the use of means and methods of warfare that are either intended or expected to damage the environment. This means that not

---

[11] Cf., K. Hulme, *War Torn Environment: Interpreting the Legal Threshold* (Leiden, Martinus Nijhoff Publishers 2004) pp. 80-81.

[12] See also Report to the Third Committee on the Work of the Working Group, Committee III, 3 April 1975 (CDDH/III/275), on Art. 48 *bis* [present Art. 55] and Art. 33 [present Art. 35], para. 3, of Protocol I, in H.S. Levie, *Protection of War Victims: Protocol 1 to the 1949 Geneva Conventions*, Vol. 2 (Dobbs Ferry, NY, Oceana Publications 1980) p. 270, and in H.S. Levie, *Protection of War Victims: Protocol 1 to the 1949 Geneva Conventions*; Vol. 3 (Dobbs Ferry, NY, Oceana Publications 1980) p. 270. Similarly, Y. Dinstein, *The Conduct of Hostilities under the Law of International Armed Conflict* (Cambridge, Cambridge University Press 2004) p. 182.

only deliberate or direct attacks on the environment are prohibited, but also attacks of which it is reasonably foreseeable that they will lead to collateral environmental damage.[13] This is irrespective of the weapons used and requires that those who deploy these means or methods of warfare must know beforehand to a certain extent that they will have detrimental effects on the environment. Conversely, if they do not reasonably know what shall be the consequences of the use of certain means and methods of warfare, and the damage is only accidental, no violation of either article can be established.[14]

Problematic in this context is that certain means and methods may cause harm that is not directly visible or demonstrable, and that it is still extremely difficult to analyse natural processes and to understand how certain activities will impact the environment in the long-term.[15] Therefore, although the Protocol does incorporate a precautionary element through the phrase 'or may be expected', the protection provided is less far-reaching than the precautionary approach sometimes taken by the international community of states in the framework of peacetime international environmental law,[16] such as in the case of the Montreal Protocol on Substances that Deplete the Ozone Layer.[17]

---

[13] M.N. Schmitt, 'Green war: An assessment of the environmental law of international armed conflict', 22 *Yale JIL* (1997) p. 72.

[14] M. Bothe, 'War and environment', in R. Bernhardt, ed., 4 *Encyclopaedia of Public International Law* (Elsevier, Amsterdam 2000) p. 1344; Dinstein, *supra* n. 12, p. 183. This interpretation is supported by declarations made by the United Kingdom and France upon ratification of AP I, respectively on 28 January 1998 and 11 April 2001. Both states said that the risk of environmental damage as a result of the use of means and methods of warfare must be assessed 'objectively on the basis of information available at the time', <http://www.icrc.org/ihl.nsf/>. See also J.-M. Henckaerts and L. Doswald-Beck, eds., *Customary International Humanitarian Law*, Vol. II: Practice, Part 1 (Cambridge, International Committee of the Red Cross/Cambridge University Press 2005) p. 877; UK Ministry of Defence, *The Manual of the Law of Armed Conflict* (Oxford, Oxford University Press 2004) p. 76 (hereinafter, UK Military Manual). France made the same declaration upon ratification with respect to Art. 8(2)(b)(iv) of the Statute of the International Criminal Court, which will be discussed further below, <http://untreaty.un.org/>.

[15] '[M]any interactive natural processes have not yet been (fully) understood, resulting in the fact that harmful effects which are not (yet) recognized or expected may occur now or in the future. Only quite recently, science has become able to demonstrate that even apparently restricted, relatively short-term and seemingly insignificant forms of environmental impact may subsequently turn out to have triggered serious or significant ecological disruption.' W.D. Verwey, 'Protection of the environment in times of armed conflict: In search of a new legal perspective', 8 *Leiden JIL* (1995) p. 12 (hereinafter, Verwey, Protection). Also W.D. Verwey, 'Observations on the legal protection of the environment in times of international armed conflict', 7 *Hague YIL* (1994) p. 37 (hereinafter, Verwey, Observations); W.D. Verwey, 'Comment: Protection of the environment in times of armed conflict – Do we need additional rules?' in R.J. Grunawalt, J.E. King and R.S. McClain, eds., *Protection of the Environment during Armed Conflict*, International Law Studies 1996, Vol. 69 (Newport, RI, Naval War College 1996) p. 561 (hereinafter, Verwey, Comment).

[16] A.P.V. Rogers, *Law on the Battlefield* (Manchester, Manchester University Press 2004) p. 169.

[17] Montreal Protocol on Substances that Deplete the Ozone Layer, opened for signature on 16 September 1987, entered into force on 1 January 1989, as amended in London (27-29 June 1990), Nairobi (19-21 June 1991) and Copenhagen (23-24 November 1992), Protocol to the Vienna Convention for the Protection of the Ozone Layer of 22 March 1985, 1522 *UNTS*, No. 26369.

Although definitions carry an inherent danger[18] and are not even always necessary,[19] it may still be useful to clarify the phrase 'natural environment' to a certain extent. Collins Dictionary describes 'the environment' in the ecological sense as 'the external surroundings in which a plant or animal lives, which tend to influence its development and behaviour'.[20] Collins Cobuild defines 'the environment' as 'the natural world or land, sea, air, plants, and animals'.[21] And according to the ICRC Commentary on the Additional Protocols

> '[t]he concept of the natural environment should be understood in the widest sense to cover the biological environment in which a population is living. It does not consist merely of the objects indispensable to survival (…) but also includes forests and other vegetation (…), as well as fauna, flora and other biological or climatic elements.'[22]

Although these descriptions seem rather obvious, it is not *prima facie* clear whether or not the phrase 'natural environment' in Articles 35(3) and 55 AP I includes the air and the marine environment. This would depend on the scope of AP I that appears from its language and from the intention of the drafters. The laws of war have traditionally distinguished between land, naval and aerial warfare[23] and it seems that as far as the carrying out of hostilities is concerned, AP I is primarily aimed at

---

[18] *Javolenus, omnis definitio iniure civili periculosa est, parum est enim ut nonsubverti possit*, 'every definition in the civil law is dangerous, for there is hardly one which cannot be undermined', Digest (of Justinian), 50.17.202, quoted from 'Final Report of the International Committee on the Formation of Customary (General) International Law; Principles Applicable to the Formation of General Customary International Law', in *The International Law Association, Report of the Sixty-Ninth Conference* (London 2000) p. 719, fn. 20.

[19] When Justice Stewart of the United States Supreme Court was confronted in 1964 with a case that involved the showing of an obscene motion picture and the freedom of speech in the Constitution's First Amendment and had to interpret the term pornography, he said: 'I shall not today attempt further to define the kinds of material I understand to be embraced within that shorthand description; and perhaps I could never succeed in intelligibly doing so. But I know it when I see it, and the motion picture involved in this case is not that.' Mr. Justice Stewart, concurring, in US Supreme Court, *Jacobellis v. Ohio*, 378 U.S. 184 (1964).

[20] *Collins Dictionary* (Glasgow, HarperCollins Publishers 1998) p. 517.

[21] *Collins Cobuild English Dictionary* (Glasgow, HarperCollins Publishers 1995) p. 555.

[22] Y. Sandoz, C. Swinaraski and B. Zimmerman, *Commentary on the Additional Protocols of 8 June 1977 to the Geneva Conventions of 12 August 1949* (Geneva, International Committee of the Red Cross 1987) p. 662. The Commentary is also available through <http://www.icrc.org/>.

[23] Each type of warfare had different purposes and required different techniques. Naval warfare was not intended to subjugate the enemy but rather to gain control over the oceans. I. Detter, *The Law of War* (Cambridge, Cambridge University Press 2000) p. 308; W. Heintschel von Heinegg, 'The law of armed conflict at sea', in D. Fleck, ed., *The Handbook of Humanitarian Law in Armed Conflicts* (Oxford, Oxford University Press 1995) p. 405. Greenwood observes that naval warfare also has larger scope for affecting the rights of neutrals. C. Greenwood, 'Historical development and legal basis', in Fleck, ibid., p. 11. See also Rauch on the specific and peculiar difficulties of the law of naval warfare: E. Rauch, *The Protocol Additional to the Geneva Conventions for the Protection of Victims of International Armed Conflicts and the United Nations Convention on the Law of the Sea: Repercussions on the Law of Naval Warfare; Report to the Committee for the Protection of Human Life in Armed Conflict of the International Society for Military Law and Law of War* (Berlin, Duncker & Humblot 1984) p. 59.

the affirmation of the rules with respect to warfare on land that were based on the 1899 and 1907 Hague Regulations on Land Warfare.[24]

There are strong indications, however, that both Article 55 and 35 also apply to naval and aerial warfare and consequently also protect the atmosphere and the marine environment. Firstly, Article 49(3) provides specifically that the rules of Part IV, Section I on the protection of the civilian population from the effects of hostilities, which includes Article 55, also apply to naval and aerial warfare to the extent that they affect civilians or civilian objects on land. This means, therefore, that within the framework of these kinds of naval and aerial operations, care must be taken 'to protect the natural environment against widespread, long-term and severe damage' which may also include the marine environment and the atmosphere. And, secondly, there is convincing evidence that Part III, Section I on means and methods of warfare, which includes Article 35(3), equally applies to land, naval and aerial warfare, despite the fact that most of its contents find their origin in the Hague Regulations on Land Warfare. Firstly, most provisions are generally considered to be rules of customary law applying to all forms of warfare; secondly, the title of Part II, Section I refers simply to 'warfare' in general; and thirdly, the ICRC's Draft Protocol which formed the basis of the negotiations at the Diplomatic Conference envisaged this part of the Protocol to apply to 'military operations as a whole carried out within the general framework of land, air or sea warfare'.[25] This view on the applicability of Part III, Section I, including Article 35(3), to naval and air warfare is generally supported in literature but usually without supporting evidence.[26]

Irrespective of the scope of application of either article, AP I does not just prohibit any damage to the environment. Both Articles 35(3) and 55 require that the damage to the natural environment is 'widespread, long-term, and severe', which resembles a similar threshold in Article I of the Environmental Modification Convention.[27] The meaning of these terms is left unde-

---

[24] Hague Convention (II) with respect to the Laws and Customs of War on Land, with annexed Regulations, signed on 29 July 1899, entered into force on 4 September 1900, supplemented by Hague Convention (IV) Respecting the Laws and Customs of War on Land, with annexed Regulations, signed on 18 October 1907, entered into force on 26 January 1910, in D. Schindler and J. Toman, eds., *The Laws of Armed Conflict: A Collection of Conventions, Resolutions and Other Documents* (Dordrecht, Martinus Nijhoff Publishers 1988).

[25] International Committee of the Red Cross, Draft Additional Protocols to the Geneva Conventions of 12 August 1949; Commentary, *supra* n. 22, p. 54.

[26] M. Bothe, 'Commentary – 1977 Geneva Protocol I Additional to the Geneva Conventions of 12 August 1949, and Relating to the Protection of Victims of International Armed Conflicts', in N. Ronzitti, ed., *The Law of Naval Warfare: A Collection of Agreements and Documents with Commentaries* (Dordrecht, Martinus Nijhoff Publishers 1988) pp. 761, 762; Dinstein, *supra* n. 12, p. 184; W. Heintschel von Heinegg, 'The law of armed conflict at sea', in Fleck, *supra* n. 23, p. 419; Rogers, *supra* n. 16, p. 168; Schmitt, *supra*, n. 13, p. 81; UK Military Manual, *supra* n. 14, p. 76. Only Simonds, Kiss and Shelton seem to exclude application of Art. 35(3) to naval and aerial warfare. A. Kiss and D. Shelton, *International Environmental Law* (Ardsley, NY, Transnational Publishers 2000) p. 562; S.N. Simonds, 'Conventional warfare and environmental protection: A proposal for international legal reform', 29 *Stanford JIL* (1992) p. 183.

[27] Art. I of this Convention prohibits the use of environmental modification techniques for hostile purposes if they cause 'widespread, long-lasting or severe effects'. ENMOD was negotiated in the

fined,[28] however, and must therefore be established in accordance with the general rules of treaty interpretation as reflected in Articles 31 and 32 of the Vienna Convention on the Law of Treaties.[29]

Because it is difficult to establish 'the ordinary meaning' of the terms of AP I, it is justified to refer to the preparatory works of the Protocol. The records of the Conference as collected by Howard Levie contain some references to the damage intended to be covered by the provisions and regarding the interpretation of the triple standard, both in individual statements from various delegations and in concluding reports. These range from claims that the threshold made sure that an individual tank commander who flattened a tree would be liable as a war criminal,[30] to a statement from the chairman of the Biotope Group, a specific working group on environmental protection during armed conflict, that short-term damage to the environment, such as artillery bombardment, was not intended to be covered by the environmental protection provisions of Protocol I. Disturbance had to be significant 'perhaps for ten years or more'.[31] The Report of the Second Session of Committee III, finally, contains the most extensive account on the interpretation of the damage threshold. According to this Report

---

same city and at the same time as AP I, so cross-reference was almost inevitable. Note that the damage threshold under ENMOD is alternative rather than the cumulative standard under AP ('widespread, long-lasting or severe' versus 'widespread, long-term, and severe').

[28] ENMOD, on the other hand, contains a number of understandings related to the Convention, which include specific definitions of the terms widespread, long-lasting, and severe. The Understandings were included in the report that was sent by the Conference of the Committee on Disarmament to the General Assembly. A/31/27, Report of the Conference of the Committee on Disarmament; Vol. I and II, United Nations, New York. United States Arms Control and Disarmament Agency, Arms Control and Disarmament Agreements; Texts and Histories of the Negotiations, p. 158, <http://disarmament2.un.org/TreatyStatus.nsf>.

[29] Vienna Convention on the Law of Treaties, signed on 23 May 1969, entered into force on 27 January 1980, 1155 *UNTS*, No. 18232. Note that the Vienna Convention is in principle not applicable to AP I because the Convention only entered into force on 27 January 1980 and AP I entered into force on 7 December 1978. However, the customary status of Arts. 31 and 32 has been confirmed by the International Court of Justice on various occasions, most recently in 2002 in the case between Indonesia and Malaysia on the sovereignty over Pulau Ligitan and Pulau Sipadan. *Case Concerning Sovereignty over Pulau Ligitan and Pulau Sipadan* (Indonesia/Malaysia), Merits, Judgment, 17 December 2002, *ICJ Rep.* 2002, p. 23, para. 37.

[30] Statement of the Representative of the United Kingdom in Committee III on 10 April 1975, Levie, Vol. 3, *supra* n. 12, p. 272. According to Kalshoven, '[i]t seems clear, therefore, that the man in the field will not easily come into conflict with this provision; rather it is addressed to higher levels of authority where the major decisions about the use of particular means and methods of warfare are taken.' F. Kalshoven, 'Reaffirmation and development of international humanitarian law applicable in armed conflicts: the Diplomatic Conference, Geneva, 1974-1977; Part II', 9 *NYIL* (1978) p. 130. Solf writes that both articles seem 'primarily directed to high level policy decisionmakers'. W.A. Solf, 'Article 55 – Protection of the natural environment', in M. Bothe, K.J. Partsch and W.A. Solf, *New Rules for Victims of Armed Conflicts; Commentary on the Two 1977 Protocols Additional to the Geneva Conventions of 1949* (The Hague, Martinus Nijhoff Publishers 1982) p. 348. Also R.G. Tarasofsky, 'Legal protection of the environment during international armed conflict', 24 *NYIL* (1993) p. 52.

[31] Report of the Chairman of the Group 'Biotope', Committee III, 11 March 1975 (CDDH/III/GT/35), in Levie, Vol. 2, *supra* n. 12, pp. 267-268, and Levie, Vol. 3, *supra* n. 12, p. 267.

'[t]he time or duration required (i.e., long-term) was considered by some to be measured in decades. References to twenty or thirty years were made by some representatives as being a minimum. Others referred to battlefield destruction in France in the First World War as being outside the scope of the prohibition. (...) However, it is impossible to say with certainty what period of time might be involved. It appeared to be a widely shared assumption [though] that battlefield damage incidental to conventional warfare would not normally be proscribed by this provision.'[32]

It seems, therefore, that the drafters intended to raise a significant threshold, and this interpretation of the damage threshold of Articles 35(3) and 55 by reference to the preparatory works has been well-established.[33]

  This does not mean that focus on the preparatory works for the interpretation of the threshold is without criticism. Verwey is skeptical because the terms have not been authoritatively defined and the preparatory works only clarify the word 'long-term',[34] while Bothe warns that too much emphasis on the preparatory works is dangerous because the drafters had only limited experience in the 1970s and only a few examples in mind.[35]

  In view of the almost inviolable standard under the preparatory works, it is certainly desirable that this interpretation changes over time. *Tempora mutanturnos etmutamurin illis.*[36] Our knowledge of the environment and of environmental problems has increased tremendously since the 1970s and as well as our concern for and appreciation of the environment. Peacetime international environmental law has

---

[32] Report of Committee III, Second Session (CDDH/215/Rev.1), in Levie, Vol. 2, *supra* n. 12, pp. 276-277.

[33] Neither the damage to the environment during the 1990-1991 Gulf War nor the damage resulting from the 1999 NATO bombing campaign were considered to fall under the damage threshold of AP I, even if the Protocol had been applicable. The Report of the US Department of Defense to Congress of 1992 refers to the conclusions of the Conference of Experts invited to Ottawa by the Canadian Ministry of Foreign Affairs from 9 to 12 July 1992. United States Department of Defense, *Conduct of the Persian Gulf War; Final Report to Congress; Pursuant to Title V of the Persian Gulf Conflict Supplemental Authorization and Personnel Benefits Act of 1991 (Public Law 102-25)* (Washington, D.C., US Government Printing Office 1992) pp. 624-625; United States Department of Defense, *Report to Congress on the Conduct of the Persian Gulf War—Appendix on the Role of the Law of War*, in 31 *ILM* (1992) pp. 636-637; International Criminal Tribunal for the Former Yugoslavia (ICTY), *Final Report to the Prosecutor by the Committee Established to Review the NATO Bombing Campaign against the Federal Republic of Yugoslavia*, 39 *ILM* (2000) p. 1262 (hereinafter, ICTY, Final Report).

[34] Verwey, Observations, *supra* n. 15, p. 36; Verwey, Protection, *supra* n. 15, p. 10; Verwey, Comment, *supra* n. 15, p. 560.

[35] M. Bothe, 'The protection of the environment in times of armed conflict; legal rules, uncertainty, deficiencies, and possible developments', 34 *GYIL* (1991) p. 56 and M. Bothe, 'Protection of the environment in times of armed conflict', in N. Al-Nauimi and R. Meese, eds., *International Legal Issues Arising Under the United Nations Decade of International Law; Proceedings of the Qatar International Law Conference' 94* (The Hague, Martinus Nijhoff 1995) p. 100.

[36] Times change and we change with them. According to the Van Dale Dictionary, the origin of the proverb is not clear, but some attribute it to Emperor Lotharius (795-855). G. Geerts and T. den Boon, et al., eds., *Van Dale; Groot Woordenboek der Nederlandse Taal* (Utrecht, Van Dale Lexicografie 1999), p. 4227. Lotharius was the eldest son of Louis the Pious, who was the son and heir to the throne of Charles the Great.

been one of the fastest growing fields in international law and plays an increasingly important role in international relations and international law. It is not unthinkable that these developments will lead to a reinterpretation of these terms and a lowering of standards.[37] After all, reference to the preparatory works is only a subsidiary method of treaty interpretation and is referred to in this context for want of anything better.

3. CUSTOMARY LAW

3.1 Introduction

In addition to the protection of the environment during armed conflict provided by the explicit references in AP I, the environment may also be protected by unwritten rules of customary international law, evidenced by a general practice[38] accepted as law. There are two possibilities. Firstly, it is possible that the explicit treaty provisions discussed above reflect customary international law, either because they were declaratory of a pre-existing norm of customary international law, or because they have developed into equivalent norms of customary international law (section 3.2). Secondly, it is possible that there are new customary rules of international law that directly protect the environment during armed conflict independently from these

---

[37] Bothe, *supra* n. 35, pp. 56-58; R. Desgagné, 'The prevention of environmental damage in time of armed conflict: Proportionality and precautionary measures, in 3 *YIHL* (2000) pp. 112-113; Hulme, *supra* n. 11, pp. 99-100; Simonds, *supra* n. 26, p. 174. Desgagné refers to a contribution of the ICRC to the 1992 Rio Conference on Environment and Development (UNCED) on environmental protection during armed conflict and evolving expectations: 'The question as to what constitutes (prohibited) "widespread, long-term and severe" damage and what is acceptable damage to the environment is open to interpretation. Such interpretation has to take the whole context into account, and will vary with changes in expectations with regard to the general need to protect the environment. Of course, the "travaux préparatoires" have also to be taken into consideration where relevant.' Desgagné, ibid., p. 112. According to the Australian representative Crawford in the Sixth Committee of the General Assembly in 1991, there had been agreement at the Conference of Experts on the Use of the Environment as a Tool of Conventional Warfare that was hosted by the Canadian Government in Ottawa in July 1991 that 'the application and development of the law of armed conflict must take account of the evolution of environmental concerns generally'. A/C.6/46/SR.20, Summary Record of the 20th meeting of the Sixth Committee of the General Assembly on 24 October 1991, p. 3, para. 8.

[38] State practice is constituted by both physical and verbal acts. The former includes actual behaviour by states; the latter includes references in military manuals, legislation, case law and statements within the framework of intergovernmental organisations. Relevant practice also includes statements from intergovernmental organisations. Judgments of international courts and tribunals, on the other hand, do not constitute state practice but may be used as evidence of the existence of a rule of customary law. J.-M. Henckaerts and L. Doswald-Beck, *Customary International Humanitarian Law*; Vol. I: Rules (Cambridge, Cambridge University Press 2005) pp. xxxii-xxxvi. See also on the importance of military manuals H.H.G. Post, 'Some curiosities in the sources of the law of armed conflict conceived in a general international legal perspective', 25 *NYIL* (1994) pp. 97-101; H.H.G. Post, 'The role of state practice in the formation of customary international humanitarian law', in I.F. Dekker and H.H.G. Post, eds., *On the Foundations and Sources of International Law* (The Hague, T.M.C. Asser Press 2003) pp. 142-145.

explicit treaty provisions and based on the fundamental principles of *jus in bello*, necessity and proportionality (section 3.3).

3.2   The customary status of Articles 35(3) and 55 Additional Protocol I

As far as the environmental protection provisions of AP I are concerned, there is general agreement that, unlike many other provisions of the Protocol, neither Article 35(3) nor Article 55 reflect a pre-existing rule of customary international law in the form in which they were adopted.[39] Apart from the fact that international concern for the environment in general only stems from the early 1970s,[40] and before 1977 conventions on the laws of war were silent on the environment, this appears from the Protocol's and the provisions' legislative history,[41] and is confirmed by the International Court of Justice (ICJ)[42] and by the International Committee of the Red Cross (ICRC).[43]

It is furthermore doubtful whether subsequent practice as evidence of a general *opinio juris* has led to the development of a customary rule of international law as reflected in both articles. This is only accepted when treaty provisions have a fundamentally norm creating character resulting in a general and widespread practice of non-party states and which is accompanied by a strong sense of *opinio juris*.[44] Although there are indications that both provisions are considered important

---

[39] Note that this does not preclude the existence of an earlier general principle of international law protecting the environment during armed conflict to a certain extent. Compare W.A. Solf, 'Protection of civilians against the effects of hostilities under customary international law and under Protocol I', 1 *American University Journal of International Law & Policy* (1986) p. 134.

[40] The first major international conference on the environment was the 1972 United Nations Conference on the Human Environment in Stockholm. A/CONF.48/14/Rev.1, Report of the United Nations Conference on the Human Environment, Stockholm, 5-16 June 1972, Declaration of the United Nations Conference on the Human Environment, United Nations, New York, 1973.

[41] The ICRC Draft Protocols did not contain provisions on environmental protection and never during the negotiations did any delegate refer to preexisting customary international law. On the contrary. When the Australian Delegate introduced an environmental protection paragraph, he stated that 'adoption of the article might well fill a gap in humanitarian law applicable in armed conflicts'. And upon the plenary adoption of Art. 35(3) by the Diplomatic Conference, the Federal Republic of Germany stated that 'paragraph 3 of this Article is an important new contribution to the protection of the natural environment in times of international armed conflict'. Levie, Vol. 3, *supra* n. 12, p. 263, Levie, Vol. 2, *supra* n. 12, p. 279.

[42] The ICJ stated in its Advisory Opinion on the request from the General Assembly that Arts. 35(3) and 55 AP were 'powerful constraints for all the States having subscribed to these provisions', which implies that neither provision had a preexisting customary equivalent. *Legality of the Threat or Use of Nuclear Weapons,* Advisory Opinion, 8 July 1996, *ICJ Rep.* 1996, pp. 226 at 242, para. 31.

[43] The ICRC stated in its Commentary on the APs that both provisions were new and repeated this in its 2005 study on Customary International Humanitarian Law. Commentary, *supra* n. 22, pp. 387, 662; Henckaerts and Doswald-Beck, *supra* n. 38, p. 152. See also Report of the Secretary-General to the General Assembly on the Protection of the Environment in the Environment in Times of Armed Conflict, UN Doc. A/48/269 of 29 July 1993, with Annexed ICRC Guidelines for Military Manuals and Instructions on the Protection of the Environment in Times of Armed Conflict, at p. 7, para. 38, by implication.

[44] *North Sea Continental Shelf,* Judgment, *ICJ Rep.* 1969, p. 3, paras. 72, 73 and 77, pp. 41-44. M. Shaw, *International Law,* 5th edn. (Cambridge, Cambridge University Press 2003) pp. 90-91.

and that they might be developing into rules of customary international law,[45] and although the ICRC concluded in its 2005 study on Customary International Humanitarian Law that since 1977 'significant practice [had] emerged to the effect that this prohibition has become customary',[46] this does not yet seem to be the case. There is substantial evidence to the contrary and, considering the fact that restrictions upon the independence of states cannot be presumed,[47] the existence of a customary rule should be proved beyond reasonable doubt.

There are only few occasions on which both provisions were invoked during actual armed conflict by non-state parties, but their number is too few to be regarded as significant evidence and it is not certain whether they were invoked because they were accepted as law.[48] And although the ICRC study refers to environmental protection provisions in the military manuals of 19 states,[49] using similar language as AP I, this fails to convince, since 15 states were already bound to observe both obligations when they included them in their manuals, because they

---

[45] Explicit or implicit references to Arts. 35(3) and 55 AP I can be found in preambular para. 4 of the Certain Conventional Weapons Convention, which seems to refer to both provisions; Section 6.3 of the Secretary-General's Bulletin of 6 August 1999 on the 'Observance by United Nations forces of international humanitarian law', ST/SGB/1999/13; Art. 20(g) of the Draft Code of Crimes against the Peace and Security of Mankind of 1996, in A/51/10, Report of the International Law Commission to the General Assembly on the work of its 48th session (6 May-26 July 1996), *Yearbook of the International Law Commission* (1996) Vol. II, Part Two (Geneva, United Nations 1998), <http://www.un.org/law/ilc/>; Art. 8(2)(b)(iv) of the ICC Statute; Principle 11 of the ICRC Guidelines for Military Manuals and Instructions on the Protection of the Environment in Times of Armed Conflict of 1993; and para. 702(g) of the ICRC's Model Military Manual; A/48/269, *supra* n. 43; A/49/323, Report of the Secretary-General on the United Nations Decade of International Law, of 19 August 1994, Principle 11; A. Roberts and R. Guelff, *Documents on the Laws of War* (Oxford, Oxford University Press 2000) pp. 609-614; A.P.V. Rogers and P. Malherbe, *Fight it Right*; *Model Manual on the Law of Armed Conflict for Armed Forces* (Geneva, ICRC 1999).

[46] Henckaerts and Doswald-Beck, *supra* n. 38, pp. 151-152, 154. Rule 45 of the customary international humanitarian law study on 'The Natural Environment' reads: 'The use of methods or means or warfare that are intended, or may be expected, to cause widespread, long-term and severe damage to the natural environment is prohibited. (...) '. Note that the study regards the United States as a persistent objector to this rule, and the United States, the United Kingdom and France as persistent objectors with respect to the use of nuclear weapons.

[47] According to the Permanent Court of International Justice, '[r]estrictions upon the independence of States cannot (...) be presumed'. The Case of the S.S. 'Lotus', 7 September 1927, Publications of the Permanent Court of International Justice, Collection of Judgments, Series A – No. 10 (Leyden, A.W. Sijthoff's Publishing Company 1927) p. 18.

[48] Both provisions were invoked by Iran in the war between Iraq and Iran in the 1980s, although neither state was or still is a party to Protocol I and despite the fact that both states attacked environmentally sensitive targets whenever they had the chance. C. Greenwood, 'Customary law status of the 1977 Geneva Protocols', in A.J.M. Delissen and G.J. Tanja, eds., *Humanitarian Law of Armed Conflict: Challenges Ahead; Essays in Honour of Frits Kalshoven* (Dordrecht, Martinus Nijhoff Publishers 1991) p. 101. Both provisions were also invoked after the 1990-1991 Gulf War although Iraq was not a party to AP I; by belligerents at the beginning of the war in the former Yugoslavia in 1991 and 1992 and after the NATO bombing campaign over Kosovo in 1999. United States Department of Defense, Report to Congress on the Conduct of the Persian Gulf War – Appendix on the Role of the Law of War, pp. 636-637; Henckaerts and Doswald-Beck, *supra* n. 14, p. 879; ICTY, Final Report, *supra* n. 33, p. 1262.

[49] Henckaerts and Doswald-Beck, *supra* n. 14, pp. 879-883.

had become party to the Protocol,[50] and two states were bound because they had signed the Protocol.[51] Of these 19 states, only Kenya and Serbia and Montenegro had included environmental protection provisions in their manuals before they acceded to the Protocol.[52] The absence of references in the military manuals of the United States, which chose not to start the procedure for ratification of AP I in the 1980s and which is nowadays probably the strongest and most significant military power in the world, on the other hand, may be more meaningful.[53]

Apart from the military manuals, there is not much evidence that can be derived from the various statements made before the ICJ within the framework of the Nuclear Weapons Opinions. Of the 43 states that provided either written or oral comments on both requests from the World Health Organization and the General Assembly, only three explicitly stated that Articles 35(3) and 55 had customary equivalents (Solomon Islands, Malaysia and Nauru);[54] one state believed that only Article 55 had developed into customary law (Qatar);[55] four states insinuated that both provisions had a general customary equivalent (India, Samoa, Zimbabwe and New Zealand);[56] and one state explicitly left it to the Court to decide whether both provisions had developed into customary law (New Zealand). On the other hand,

---

[50] Henckaerts and Doswald-Beck, *supra* n. 38, p. 152; Henckaerts and Doswald-Beck, *supra* n. 14, pp. 879-883. Ratifications can be viewed at <http://www.icrc.org/>.

[51] Both provisions were included in the 1983 Belgian Military Manual and the 1981 British Military Manual when both states were still only signatories. Henckaerts and Doswald-Beck, *supra* n. 38, p. 152. Both states signed the Protocol on 12 December 1977, <http://www.icrc.org/>.

[52] Henckaerts and Doswald-Beck, *supra* n. 38, p. 152; Henckaerts and Doswald-Beck, *supra* n. 14, pp. 879-883. Ratifications update is available at <http://www.icrc.org/>.

[53] The Army Field Manual of 1976, the Air Force Pamphlet of 1976 and the Air Force's 'Military Commander and the Law' are silent on environmental protection. While the Commander's Handbook on the Law of Naval Operations of 1995 does refer to environmental protection during armed conflict, it is embedded in an application of the principles of necessity and proportionality. Additionally, the US Operational Law Handbook of 2004 and the 'Law of War Handbook' state that Arts. 35(3) and 55 go beyond classic international humanitarian law and that the United States does not regard them as customary law. FM 27-10, *The Law of Land Warfare* (Washington, D.C., Department of the Army 1956), as changed in 1976, <http://www.afsc.army.mil/>; Air Force Pamphlet 110-31, *International Law – The Conduct of Armed Conflict and Air Operations* (Washington, D.C., Department of the Air Force 1976); T.L. Strand, M.W. Goldman, et al., eds., *The Military Commander and the Law* (Maxwell, AL, Air Force Judge Advocate-General School 2004), <http://milcom.jag.af.mil/>; NWP1-14M, *The Commander's Handbook on the Law of Naval Operations* (Norfolk, VA, Department of the Navy 1995), <http://www.nwc.navy.mil/>. See also A.R. Thomas and J.C. Duncan, eds., *Annotated Supplement to The Commander's Handbook on the Law of Naval Operations*, International Law Studies 1999, Vol. 73 (Newport, RI, Naval War College 1999); J.B. Berger III, D. Grimes, E.T. Jensen, et al., eds., *Operational Law Handbook* (Charlottesville, VA, International and Operational Law Department, The Judge Advocate-General's Legal Center and School 2004), <https://www.jagcnet.army.mil/>; *Law of War Handbook* (Charlottesville, VA, Center for Law and Military Operations (CLAMO) 2005), <https://www.jagcnet.army.mil/>.

[54] The Solomon Islands (WHO-WS, p. 62; WHO-WC, p. 67; GA-WS, p. 63); Malaysia (WHO-WS, p. 11); and Nauru (WHO-WS, Memorial III, p. 22), <http://www.icj-cij.org/>.

[55] Only Art. 55 had a customary equivalent according to Qatar. Art. 35(3) was only 'treaty-based'. (WHO/GA-OP, pp. 32 and 36), <http://www.icj-cij.org/>.

[56] India (WHO-WC, pp. 6, 12); Samoa (WHO-WS, p. 3); Zimbabwe (WHO/GA-OP, p. 27); and New Zealand (GA-WS, pp. 17-18, but more clearly in WHO/GA-OP, p. 27), <http://www.icj-cij.org/>.

three states adamantly denied the existence of customary counterparts of both provisions (the United Kingdom, the United States and France),[57] and the rest remained silent on the issue. This means that, apart from the differences between both camps, the limited number of supporters, the explicit opposition by some states and the silence of the vast majority do not justify the conclusion that both provisions have seen a widespread practice accepted as law.

Furthermore, both the ICJ and a Special Committee of the International Criminal Tribunal for the Former Yugoslavia (ICTY), as well as the large majority of authors, have either denied or expressed uncertainty on the customary status of both provisions. The ICJ stated in its 1996 Nuclear Weapons Opinion upon request of the General Assembly that Articles 35(3) and 55 provided additional protection for the environment and that they provided 'powerful constraints for all the States having subscribed to these provisions'.[58] The Special ICTY Committee[59] stated in 2000 that Article 55 'may (...) reflect current customary law'.[60] And authors such as Greenwood, Bothe, Spieker, Lijnzaad and Tanja, Verwey, Schmitt and Dinstein have all stated at some point in the 1990s or at the beginning of the 21st century that both provisions have not yet crystallised into customary law.[61]

3.3 Other customary rules directly protecting the environment during armed conflict

3.3.1 *Introduction*

In view of the lack of interest of the international community of states in the environment in the past, it is highly unlikely that there were rules of customary international law directly protecting the environment during armed conflict before the 1970s. There may have been rules before then that did protect the environment to a certain extent, but this was more by accident than by intention and would therefore have to count as indirect protection and is not under discussion here. After the Stockholm Conference of 1972,[62] however, and the entry into force of AP I and ENMOD, concern for the protection of the environment in general and for the pro-

---

[57] The United Kingdom (WHO-WS, p. 91; WHO-WC, p. 57; GA-WS, p. 57; WHO/GA-OP, pp. 36-37); the United States (WHO-WS, p. 30; WHO-WC, pp. 23-24; GA-WS, p. 25; WHO/GA-OP, p. 73); and France (GA-WS, p. 41), <http://www.icjcij.org/>.

[58] *Legality of the Threat or Use of Nuclear Weapons*, *supra* n. 42, p. 242, para. 31

[59] The Committee was established by the Prosecutor of the ICTY in 1999 to investigate NATO's bombing campaign over Kosovo.

[60] *Supra* n. 33, p. 1262.

[61] Greenwood, *supra* n. 48, p. 105, Bothe, *supra* n. 35, p. 56; H. Spieker, *Völkergewohnheitsrechtlicher Schutz der Natürlichen Umwelt im internationalen bewaffneten Konflikt; Waffenwirkung und Umwelt I* (Bochum, Universitätsverlag Brockmeyer 1992) pp. 458-459; Verwey, Observations, *supra* n. 15, p. 39; Verwey, Protection, *supra* n. 15, p. 15; Verwey, Comment, *supra* n. 15, p. 563; Schmitt, *supra* n. 13, p. 76; Dinstein, *supra* n. 12, pp. 185, 193.

[62] A/CONF.48/14/Rev.1, Report of the United Nations Conference on the Human Environment, Stockholm, 5-16 June 1972, Declaration of the United Nations Conference on the Human Environment, United Nations, New York, 1973.

tection of the environment during armed conflict in particular may have given rise to a sense of obligation and a customary protection of the environment during armed conflict, independently of the treaty provisions.

There are certainly indications that such rules have actually developed or have been in development since the late 1970s. In the first place, it is likely that there is a general customary duty of care for the environment during armed conflict. This appears from the development of international regulations protecting the environment in general and from repeated expressions of concern that will be discussed below, in particular. In the second place, it is not unlikely that, in view of the frequent references to the principles of necessity and proportionality in the context of environmental protection during armed conflict, both principles have found new manifestations in the form of two new rules of customary international law: the prohibition to cause wanton or wilful damage to the environment not justified by military necessity[63] and the prohibition to cause excessive collateral damage to the environment. The former is a reflection of the principle of necessity, in particular the principle of distinction or discrimination and is similar to the customary prohibition 'to destroy or seize the enemy's property, unless such destruction or seizure be imperatively demanded by the necessities of war' as laid down in Article 23(g) of the Hague Regulations of 1899 and 1907.[64] The latter is also a reflection of the principle of necessity, more particularly the principle of proportionality, and is related to the customary prohibition as laid down in Article 51(5)(b) AP I to launch 'an attack which may be expected to cause incidental loss of civilian life, injury to civilians, damage to civilian objects, or a combination thereof, which would be excessive in relation to the concrete and direct military advantage anticipated.'[65]

Rather than using the principle of necessity and its sub-principles of discrimination and proportionality as separate customary rules of international law, it is preferable to use them as framework concepts or umbrella norms from which specific rules of international law are derived, both conventional and customary,

---

[63] Note that the word 'wilful' and in particular the word 'wanton' already imply the absence of military necessity.

[64] The prohibition has also been recognised in Art. 53 of Geneva Convention (IV) relative to the Protection of Civilian Persons in Time of War of 1949 and in Art. 48 of AP I of 1977. Violation of both provisions partly entails individual criminal responsibility. Art. 147 of the Convention states that 'extensive destruction and appropriation of property, not justified by military necessity and carried out unlawfully and wantonly' is considered a grave breach and, in addition, Art. 85(3)(a) regards 'making the civilian population or individual civilians the object of attack' as a grave breach. More recently, individual criminal responsibility for extensive destruction of property not justified by military necessity was confirmed in Art. 8(2)(a)(iv) of the Statute of the International Criminal Court and in Arts. 20(a)(iv) and 20(e)(ii) of the 1996 Final Draft Code of Crimes against the Peace and Security of Mankind.

[65] According to Dinstein, in the past 'once an attack was directed at an indisputable military objective, any unavoidable injury or damage caused to civilians or civilian objects was accepted as 'collateral damage'. Dinstein, *supra* n. 12, p. 119. Failure to refrain from launching an indiscriminate attack which may be expected to cause excessive collateral damage is considered a grave breach of the Protocol according to Art. 85(3)(b) AP I and has been branded a war crime in the Statute of the International Criminal Court.

and to see them as the cornerstones and origin of most of the corpus of the laws of war. This is not only more proper in view of the differences in meaning between 'principles' and 'rules',[66] but it also provides clarity and keeps the customary rules of the law of armed conflict flexible. The beauty of principles is that they are dynamic and can adapt to new general perspectives, new general practices and new general concerns.

3.3.2 Duty of care

As far as the existence of a general and customary duty of care for the environment during armed conflict is concerned, there is probably sufficient evidence to support this conclusion, both in international and in national practice. It is arguable that a duty of care appears from the treaty provisions that have been concluded since 1977 directly protecting the environment during armed conflict, most explicitly from the first sentence of Article 55(1) of AP I, which states: 'Care shall be taken in warfare to protect the natural environment against widespread, long-term and severe damage.'

Apart from this, a customary duty of care for the environment during armed conflict is also reflected in a number of other, non-binding, international instruments. These include principle 26 of the 1972 Stockholm Declaration,[67] paragraph 5 of the 1982 World Charter for Nature,[68] principle 24 of the 1992 Rio Declaration,[69] paragraph 39.6 of Rio's Agenda 21,[70] and, indirectly, paragraph 14 of the

---

[66] Collins Dictionary defines 'principle', among other things, as 'a fundamental or general truth or law', 'the essence of something', 'a source or fundamental cause; origin', and as 'an underlying or guiding theory or belief'. Black's Law Dictionary defines 'principle' as follows: 'A fundamental truth or doctrine, as of law; a comprehensive rule or doctrine which furnishes a basis or origin for others; (…) that which constitutes the essence of a body or its constituent parts. (…).' A 'rule' is primarily defined by Black's Law Dictionary as '[a]n established standard, guide, or regulation' and by Collins Dictionary as 'an authoritative regulation or direction concerning method or procedure (…)'. *Collins Dictionary, supra* n. 20, pp. 1229, 1345; H.C. Black, et al., *Black's Law Dictionary* (St. Paul, MN, West Publishing 1991) pp. 828, 925.

[67] Report of the United Nations Conference on the Human Environment, A/CONF.48/14/Rev.1, Stockholm, 5-16 June 1972; Declaration of the United Nations Conference on the Human Environment, United Nations, New York, 1973; Principle 26 of the non-binding Stockholm Declaration that concluded the United Nations Conference on the Human Environment of Stockholm of 1972 stated: 'Man and his environment must be spared the effects of nuclear weapons and all other means of mass destruction. (…).'

[68] A/Res/37/7, adopted on 28 October 1982, by 111 to 1, with 18 abstentions; World Charter for Nature; Annex: World Charter for Nature. Para. 5 states that '[n]ature shall be secured against degradation caused by warfare or other hostile activities.'

[69] A/CONF.151/26/Rev.1 (Vol. I), Report of the United Nations Conference on Environment and Development, Rio de Janeiro, 3-14 June 1992, Vol. I, Resolutions Adopted by the Conference; Resolution 1, Adoption of Texts on Environment and Development; Annex I, Rio Declaration on Environment and Development, United Nations, New York, 1993. Principle 24 states: 'Warfare is inherently destructive of sustainable development. States shall therefore respect international law providing protection for the environment in times of armed conflict and cooperate in its further development, as necessary.'

[70] Ibid.; Annex II, Agenda 21, United Nations, New York, 1993. Para. 39.6 of Agenda 21 states: 'Measures in accordance with international law should be considered to address, in times of armed conflict, large-scale destruction of the environment that cannot be justified under international law.'

2002 Johannesburg Declaration.[71] Furthermore, the General Assembly declared 6 November of each year to be the 'International Day for Preventing the Exploitation of the Environment in War and Armed Conflict', which implies genuine concern and perhaps an international duty of care.[72]

Additional evidence can be found in the United States' Commander's Handbook on the Law of Naval Operations, the British Military Manual in the context of air operations and the non-binding San Remo Manual on International Law Applicable to Armed Conflicts at Sea, each of which states that operations should be carried out with due regard to the protection of the environment,[73] and by a number of verbal statements from government officials. Apart from the fact that a large number of states acknowledged before the ICJ that the use of nuclear weapons might have serious consequences for the environment, a number of states expressed specific concern for the environment during armed conflict. These were Sri Lanka, which referred to the protection of the environment as an established principle of international law,[74] and Iran, Sweden, and New Zealand, each of which explicitly expressed a concern for the environment during armed conflict.[75] Moreover, a number of states claimed the existence of a general customary principle of environmental security or safety, which combines elements of both international environmental law and the law of armed conflict.[76]

Finally, it seems that the existence of a duty of care for the environment during armed conflict has been confirmed by the ICJ in its 1996 Nuclear Weapons Advisory Opinion. The Court stated that environmental factors and considerations must play an important role in the implementation of the law of armed conflict,[77] which suggests the existence of a general duty of care for the environment. The ICRC stated in its Model Military Manual that '[i]n the conduct of military opera-

---

[71] A/CONF.199/20, Report of the World Summit on Sustainable Development, Johannesburg, South Africa, 26 August-4 September 2002, Resolution 1; Annex: Johannesburg Declaration on Sustainable Development, United Nations, New York, 2002. Para. 19 in the Declaration's section on commitments to sustainable development says: 'We reaffirm our pledge to place particular focus on, and give priority attention to, the fight against the worldwide conditions that pose severe threats to the sustainable development of our people, which include: (...); armed conflict; (...).'

[72] 6 November 2001 marked the 10th anniversary of the extinguishing of the last oil well fire. W.J. Hybl, Representative of the United States, in Press Release GA/9946 of 5 November 2001, <http://www.unis.unvienna.org/>.

[73] NWP1-14M, *supra* n. 53, p. 8-2; UK Military Manual, *supra* n. 14, p. 315, para. 12.24; San Remo Manual on International Law Applicable to Armed Conflicts at Sea, in *IRRC*, No. 309 (1995); and in Roberts and Guelff, *supra* n. 45, pp. 574-606, paras. 44, 11, 46(c) and 13(c). Note that although the San Remo Manual was drafted on a sub-state level, it is nevertheless authoritative, since it is believed to reflect 'the law which is currently applicable'.

[74] Sri Lanka (WHO-WS, p. 3), <http://www.icj-cij.org/>.

[75] Iran (GA-WS, p. 4); Sweden (WHO-WS, p. 5; GA-WS, p. 5); New Zealand (GA-WS, pp. 17-18), <http://www.icjcij.org/>.

[76] These were Nauru (WHO-WS, Memorial III, pp. 22-23; WHO-WC, Memorial II, pp. 27-28), Malaysia (WHO-WS, pp. 10-11; WHO-WC, pp. 27-29) and India (WHO-WC, pp. 12-13; GA-WS, p. 5), <http://www.icj-cij.org/>.

[77] *Legality of the Threat or Use of Nuclear Weapons*, *supra* n. 42, pp. 242-243, paras. 30, 32, 33.

tions, care must be taken to spare the environment',[78] and concluded in its 2005 study on Customary International Humanitarian Law that:

> 'Methods and means of warfare must be employed with due regard to the protection and preservation of the natural environment. In the conduct of military operations, all feasible precautions must be taken to avoid, and in any event to minimise, incidental damage to the environment. (...).'[79]

3.3.3 Prohibition of wanton destruction of and excessive collateral damage to the environment during armed conflict

3.3.3.1 Introduction

In addition to the existence of an international duty of care for the environment during armed conflict, it is not unlikely that from the principle of necessity and its sub-principles of discrimination and proportionality two new customary rules of international law have developed. The ICJ hinted at that possibility when it generally held that

> 'States must take environmental considerations into account when assessing what is necessary and proportionate in the pursuit of legitimate military objectives. Respect for the environment is one of the elements that go to assessing whether an action is in conformity with the principles of necessity and proportionality.'[80]

In the first place, this would be the prohibition of wanton or wilful destruction of the environment not justified by military necessity, i.e., the prohibition to attack the environment per se without military justification; in the second place, this would be the prohibition to cause disproportionate or excessive collateral damage to the environment. Although these customary rules are strongly related to the well-known customary prohibitions to wilfully and wantonly destroy property and to cause excessive collateral damage to civilian objects, both rules are new and must be distinguished in view of their focus on the environment.

    As was mentioned above, these more general and related customary rules of international law, that find their origin in the same principles, are generally recognised in case law and in literature and have been laid down in various conventions. Evidence of the existence of both new customary prohibitions that directly protect the environment is of relatively recent origin and can be found in international legal instruments, national legal instruments, and authoritative documents. Both rules are strongly related and are often mentioned together. Therefore, the evidence for their existence found in public statements and in literature will be

---

[78] Rogers and Malherbe, *supra* n. 45, p. 37, para. 702(c).

[79] Henckaerts and Doswald-Beck, *supra* n. 38, Rule 44, p. 147.

[80] *Legality of the Threat or Use of Nuclear Weapons*, *supra* n. 42, p. 242, para. 30.

discussed together (sections 3.3.3.4 and 3.3.3.5); the evidence found in international instruments and military manuals warrant separate discussion (sections 3.3.3.2 and 3.3.3.3).

3.3.3.2   The prohibition of wanton destruction of the environment

As far as the prohibition of wilful or wanton destruction of the environment is concerned, strong evidence is provided by General Assembly Resolution 47/37 of 25 November 1992,[81] the Assembly's first resolution that specifically dealt with the protection of the environment in times of armed conflict. Adopted without a vote, it stressed in preambular paragraph 5 'that destruction of the environment, not justified by military necessity and carried out wantonly, is clearly contrary to existing international law.' Although the resolution implied that this was already a generally accepted notion in international law, this statement found recognition and implicit confirmation by members of the International Law Commission[82] and by the ICJ in the 1996 Nuclear Weapons Advisory Opinion.[83]

Furthermore, the United States' Commander's Handbook on the Law of Naval Operations, the British Military Manual with respect to air warfare, the San Remo Manual on naval warfare, the ICRC Guidelines and the ICRC Model Manual all accept that damage to the natural environment that is not justified by military necessity and that is wanton is prohibited.[84]

Further evidence of this new customary prohibition can be found in the existence and object and purpose of Article 2(4) of the Incendiary Weapons Protocol, which prohibits the use of incendiary weapons without military justification against forests and plant cover, and in the discussions within the framework of the International Law Commission's drafting of the Draft Code of Crimes against the Peace and Security of Mankind.[85] Furthermore, the United States' Commander's

---

[81] A/Res/47/37, adopted without a vote on 25 November 1992, on the protection of the environment in times of armed conflict.

[82] Rosenstock at the 2448th meeting of the International Law Commission on 26 June 1996. Summary Records of the 2448th meeting, Wednesday, 26 June 1996; Draft Code of Crimes against the Peace and Security of Mankind, in A/CN.4/SER.A/1996, *Yearbook of the International Law Commission* (1996), Vol. I; Summary records of the meetings of the 48th session 6 May-26 July 1996, United Nations, Geneva, 1998, para. 26, pp. 110-111.

[83] In para. 32 the Court refers to Resolution 47/37, quotes preambular para. 5 and states that the resolution 'affirms the general view according to which environmental considerations constitute one of the elements to be taken into account in the implementation of the principles of the law applicable in armed conflict'. *Supra* n. 42, p. 242. Further evidence of this new customary prohibition can be found in the existence and object and purpose of Art. 2(4) of the Incendiary Weapons Protocol, which prohibits the use of incendiary weapons without military justification against forests and plant cover, and in the discussions within the framework of the drafting of the Draft Code of Crimes against the Peace and Security of Mankind by the International Law Commission.

[84] NWP1-14M, *supra* n. 53, p. 8-2; UK Military Manual, *supra* n. 14, p. 315, para. 12.24; San Remo Manual, *supra* n. 73, and in Roberts and Guelff, *supra* n. 45, pp. 574-606, para. 44; A/48/269, *supra* n. 43, principle 8; Rogers and Malherbe, *supra* n. 45, p. 37, paras. 702(b) and (d).

[85] Compare Art. 26 of the 1991 Draft Code which provides: 'An individual who wilfully causes or orders the causing of widespread, long-term and severe damage to the natural environment shall, on

Handbook on the Law of Naval Operations, the British Military Manual with respect to air warfare, the San Remo Manual on naval warfare, the ICRC Guidelines and the ICRC Model Manual all accept that damage to the natural environment that is not justified by military necessity and that is carried out wantonly is prohibited.

3.3.3.3 Prohibition of excessive collateral damage to the environment

As far as evidence for the existence of a customary prohibition of excessive collateral damage to the environment is concerned, strong evidence is provided by the formulation of Article 8(2)(b)(iv) of the 1998 Statute of the International Criminal Court. Although the addition of consideration of the environment to the proportionality obligation is new, and although there was no such thing as direct individual criminal responsibility for damage to the environment before 1998, the *chapeau* of Article 8(2)(b) seems to suggest that the provisions of Article 8(2)(b), including paragraph (iv), find their basis in pre-existing rules of international law. The *chapeau* of Article 8(2)(b) says:

> 'For the purpose of this statute, "war crimes" means: (...) Other serious violations of the laws and customs applicable in international armed conflict, within the established framework of international law.'

The 'established framework of international law' includes both conventional and customary international law and seems to exclude the possibility of progressive development. This is confirmed by the United Kingdom, which declared upon ratification of the Statute that it 'understands the term "the established framework of international law", used in Article 8(2)(b) and (e), to include customary international law as established by State practice and opinio iuris.'[86] None of the current States Parties has adopted a declaration or reservation questioning the legal validity of an environmental war crime clause in Article 8(2)(b)(iv) of the Statute, which you would expect in case of progressive development of international law.

---

conviction thereof, be sentenced [to ...]' and the discussions in the ILC on the proposals of the 1996 Working Group chaired by Tomuschat. The deliberations eventually led to the adoption of the current Art. 20(g) of the 1996 Draft Code, which dropped the reference to wilful destruction but maintained the reference to military necessity. The 1991 Draft Code: A/46/10, Report of the International Law Commission to the General Assembly on the work of its 43rd session, 29 April to 19 July 1991, *Yearbook of the International Law Commission*, Vol. II, Part Two (1994) (Geneva, United Nations 1994). The Tomuschat Working Group proposals: ILC (XLVIII)/DC/CRD.3. The text of this document was supposed to be reproduced in Vol. II, Part I of the *Yearbook of the International Law Commission*, 1996; unfortunately, Part I has been forthcoming since 1998 and may never be published despite having an ISBN and a UN Sales Number: ISBN: 92-1-133598-1; Sales No.: E.98.V.9. The discussions on the proposals from the Working Group: A/CN.4/SER.A/1996, *Yearbook of the International Law Commission*, Vol. I (1996); Summary records of the meetings of the 48th session 6 May-26 July 1996, United Nations, Geneva, 1998.

[86] Declaration upon ratification by the United Kingdom of Great Britain and Northern Ireland of the Statute of the International Criminal Court, available at <http://untreaty.un.org/>.

Apart from Article 8(2)(b)(iv) of the Statute, evidence and indications of state practice can be found in the United States' Commander's Handbook on the Law of Naval Operations,[87] in the 1994 San Remo Manual on International Law Applicable to Armed Conflicts at Sea,[88] the 1993 ICRC Guidelines,[89] and the 1999 ICRC Model Manual on the Law of Armed Conflict.[90] Furthermore, the prohibition of excessive collateral damage to the environment seems to be confirmed by the above-mentioned ICTY Committee. The Committee stated that

> '[e]ven when targeting admittedly legitimate military objectives, there is a need to avoid excessive long-term damage to the economic infrastructure and natural environment with a consequential adverse effect on the civilian population. Indeed, military objectives should not be targeted if the attack is likely to cause collateral environmental damage which would be excessive in relation to the direct military advantage which the attack is expected to produce.'[91]

3.3.3.4 Public statements

Evidence for both propositions can also be found in a number of public statements made by governmental officials and official representatives within the framework of the Sixth Committee of the General Assembly in 1991 and 1992[92] and before the ICJ within the framework of both requests for an Advisory Opinion on the legality of the use of nuclear weapons. In the Sixth Committee, a number of delegates discussed damage to the environment by reference to the principles of necessity and proportionality,[93] while Canada and Iran explicitly recognised the existence of a customary prohibition to cause unnecessary damage to the environment,[94] and Uru-

---

[87] NWP1-14M, *supra* n. 53, p. 8-2.

[88] San Remo Manual, *supra* n. 73, Roberts and Guelff, *supra* n. 45, pp. 574-606.

[89] A/48/269, *supra* n. 43.

[90] NWP1-14M, *supra* n. 53, p. 8-2; San Remo Manual, *supra* n. 73; Roberts and Guelff, *supra* n. 45, pp. 574-606, principles 13(c), 46(d), 51(d), 52(d) and 57(d); A/48/269, *supra* n. 43, principles 4, 8 and 9; Rogers and Malherbe, *supra* n. 45, paras. 702(e) and (f), p. 37.

[91] ICTY, Final Report, *supra* n. 33, pp. 1262-1263, paras. 15 and 18.

[92] Discussions were instigated by a proposal from Jordan to include an agenda item on the 'Exploitation of the Environment as a Weapon in Times of Armed Conflict and the taking of Practical Measures to Prevent Such Exploitation'. A/46/141, 8 July 1991, Request for the inclusion of an additional item in the provisional agenda of the 46th session; Exploitation of the Environment as a Weapon in Times of Armed Conflict and the Taking of Practical Measures to Prevent Such Exploitation; Annex: Explanatory Memorandum.

[93] In addition to Canada (A/C.6/46/SR.18, para. 12 and in particular para. 14 and A/C.6/47/SR.8, para. 20), the United States (A/C.6/46/SR.18, paras. 36-37 and A/C.6/47/SR.9, paras. 50-51); and the ICRC (Observer, A/C.6/46/SR.18, para. 49 and A/C.6/47/SR.8, para. 7) were most outspoken. See also Australia (A/C.6/46/SR.20, para. 7); Austria (A/C.6/46/SR.19, para. 5 and A/C.6/47/SR.8, para. 37); Iran (A/C.6/46/SR.18, paras. 27-28); Jordan (A/C.6/46/SR.18, para. 5; Nepal (A/C.6/46/SR.20, para. 28. Only reference to the violation of customary law by Iraq.); the Netherlands, on behalf of the European Communities, (A/C.6/46/SR.20, para. 1. Implicitly.); New Zealand (A/C.6/46/SR.20, para. 18); Sweden (A/C.6/46/SR.20, para. 22); and Russia (A/C.6/47/SR.9, para. 16).

[94] Canada (A/C.6/46/SR.18, para. 14 and A/C.6/47/SR.8, para. 20) and Iran (A/C.6/46/SR.18, para. 32).

guay, the United Arab Emirates and Brazil generally referred to customary law that protects the environment.[95] Japan, on the other hand, stated that 'legal rules aimed at protecting the environment had not yet been established under customary international law'.[96]

Before the Court, Egypt, India, Ireland, the Marshall Islands, Nauru, New Zealand, the Solomon Islands, the United States and, in particular, Iran referred to the protection of the environment in the context of the principles of necessity and proportionality.[97] Iran stated in its oral plea that

> '[a]s far as the law of armed conflict is concerned, both the customary rules and the provisions of treaty law prohibit belligerent parties, directly or indirectly, from inflicting unnecessary damage on the environment. Parties to the armed conflict are obliged, in accordance with well-established rules of customary law pertaining to armed conflict, to protect the environment in time of armed conflict. These rules include proportionality and the prohibition on military operations not directed against legitimate military targets, as well as the prohibition of destruction of enemy property not imperatively demanded by the necessities of war.'[98]

3.3.3.5 Experts and literature

The existence of two customary rules directly protecting the environment by prohibiting its willful and wanton destruction as well as excessive damage to it has been suggested by the ICRC in its 1993 Guidelines and 1999 Model Military Manual,[99] as mentioned above; by experts in the early 1990s;[100] and by experts in literature.[101] Furthermore, their existence is confirmed by the ICRC's 2005 study on Customary International Humanitarian Law. The ICRC concludes with respect to the protection of the natural environment in Rule 43:

---

[95] Uruguay and the UAE (A/C.6/46/SR.20, paras. 26 and 47) and Brazil (A/C.6/47/SR.9, para. 12).

[96] Japan (A/C.6/46/SR.20, para. 16). Note that Sweden stated that 'the existing rules of international law were not without shortcomings in that they did not make specific reference to environmental damage' which could refer to customary international law (A/C.6/46/SR.20, para. 22).

[97] Egypt (GA-WC, pp. 22-24, in particular paras. 53 and 55); India (WHO-WS, p. 2); Ireland (WHO-WS, p. 1, para. 2); the Marshall Islands (WHO/GA-OP, pp. 23-24); Nauru (WHO-WS, p. 36. According to Nauru, the environment enjoyed protection as a 'civilian object' by the customary rule of proportionality. Also WHO-WC, pp. 17-18); New Zealand (GA-WS, p. 17, n. 70); The Solomon Islands (WHO-WS, p. 52, para. 3.50; GA-WS, p. 55, para. 3.63); The United States (WHO/GA-OP, pp. 70-71), <http://www.icj-cij.org/>.

[98] Iran (WHO/GA-OP, p. 34), <http://www.icj-cij.org/>.

[99] A/48/269, *supra* n. 43, principles 4 and 8; Rogers and Malherbe, *supra* n. 45, p. 37, para. 702(b), (d), (e) and (f).

[100] Ibid., para. 74, p. 14.

[101] Compare for example Desgagné, *supra* n. 37, pp. 115-116; J.-M. Henckaerts, 'Towards better protection for the environment in armed conflict: recent developments in international humanitarian law', 9 *Review of European Community and International Environmental Law* (2000) p. 18; L. Lijnzaad and G.J. Tanja, 'Protection of the environment in times of armed conflict: the Iraq-Kuwait War', 40 *NILR* (1993) p. 184; Simonds, *supra* n. 26, pp. 169-170.

'The general principles on the conduct of hostilities apply to the natural environment: A. No part of the natural environment may be attacked, unless it is a military objective. B. Destruction of any part of the natural environment is prohibited, unless required by imperative military necessity. C. Launching an attack against a military objective which may be expected to cause incidental damage to the environment which would be excessive in relation to the concrete and direct military advantage anticipated is prohibited.'

The second part of Rule 45 provides: 'Destruction of the natural environment may not be used as a weapon.'[102]

3.3.3.6 Consequences

The likely existence of two additional customary rules that provide direct protection to the environment during armed conflict is important because it means not only more and stronger protection for the environment during armed conflict but also protection at two different levels: protection under treaty law applicable for States Parties and protection under customary law applicable for all states. Because both customary rules stem from a general international concern for the environment during armed conflict, on the one hand, and two general principles of the law of war that form the cornerstones of the *jus in bello*, on the other hand, and because there is no indication that they are more specific than the treaty norms, they do not supplant or override the treaty provisions. Instead, both customary norms are additional, supplementary or complementary to AP I.[103]

However, since both customary norms emerged after 1977 and fully matured in the 1990s, they need to be addressed, first, in accordance with the basic legal principle that new law prevails over old law, or *lex posterior derogat legi priori*. Then, if no violation can be established under customary law, treaty law may provide subsidiary protection *against* states that have become party to these conventions. This seems to be confirmed by the ICJ in the Nuclear Weapons Advisory Opinion. In paragraph 31, the Court notes that 'Articles 35, paragraph 3, and 55 of Additional Protocol I provide additional protection for the environment.'[104]

---

[102] Henckaerts and Doswald-Beck, *supra* n. 38, p. 143; Henckaerts and Doswald-Beck, *supra* n. 14, pp. 844-859.

[103] Although a number of authors as well as the ICRC (customary Rule 43) similarly refer to the additional and supplementary protection provided by pre-1977 customary law, they do not provide a satisfactory explanation for the possibility referred to by Lijnzaad and Tanja and Verwey that AP I might very well have lowered the level of protection of the environment because of the high damage threshold of Arts. 35(3) and 55 AP I and the fact that both provisions would prevail over pre-existing customary law because of their specific character and in accordance with the maxim *lex specialis derogat legi generali*. Lijnzaad and Tanja, *supra* n. 101, p. 182; Verwey, Observations, *supra* n. 15, pp. 36-37; Verwey, Protection, *supra* n. 15, p. 11; Verwey, Comment, *supra* n. 15, p. 560. Because it is likely that both customary norms only emerged after 1977 this problem does not arise here.

[104] *Legality of the Threat or Use of Nuclear Weapons*, *supra* n. 42, p. 226, at p. 242, para. 31.

4.   ASSESSMENT AND CONCLUSION

After discussing in detail the legal merits of the direct protection of the environment under *jus in bello*, it is possible to assess the legality of the use of DU under this set of rules. Assuming AP I is formally applicable, it prohibits the use of means and methods of warfare that are intended or expected to cause widespread, long-term and severe damage to the natural environment, in particular when this prejudices the health and survival of the population. Whereas it is unlikely that the use of DU is intended to cause damage to the environment, it is arguable that damage to the natural environment is actually foreseeable to a certain extent. Although DU is significantly less radioactive than natural uranium, it is still harmful to the natural environment, especially when it is used in concentrated quantities. Besides, uranium remains a toxic heavy metal. This was known when it was first used for military purposes and it is still known today.

However, it is unlikely that the damage done to the natural environment will pass the damage threshold of Articles 35(3) and 55. As has been mentioned above, the harmful effects of DU manifest themselves in flora and fauna primarily after inhalation, absorption or ingestion. Inhalation is likely shortly after impact when most particles are still airborne and is most damaging in the close vicinity of the impact when there is still a high concentration of uranium particles in the surrounding cloud.[105] Inhalation is also possible after the aerosols have settled on the ground or after resuspension by wind or human activity.[106]

Absorption and ingestion are only likely after a longer period of time when the uranium particles have landed on the ground by gravitation or via precipitation. The former may occur after particles have sunk into the groundwater; the latter may occur accidentally, by eating plants or vegetables that contain loose particles or after particles have entered the food chain.[107] Note, however, that 'bio-accumulation of uranium in plants and animals is not very high';[108] that uranium is therefore not effectively transported in the food chain;[109] and that '98% of uranium entering the body via ingestion is not absorbed, but is eliminated via the faeces'.[110] Only prolonged skin contact with complete DU shells could be damaging to man, but most intact shells are buried deep in the ground.

Therefore, the doses to which both flora and fauna will be exposed to will generally be small and will become even smaller with time. There may be some so-called 'hot spots' where the concentration of DU particles will be higher than nor-

---

[105] Studies show that most DU particles settle on the ground within minutes after impact and in close proximity to the target, although smaller particles may travel several hundred metres by wind. International Atomic Energy Agency, Features: Depleted Uranium, para. 14. Note that is small compared with radioactive fallout after a nuclear explosion that may travel over hundreds of kilometres and may even enter the stratosphere where it may stay for years.

[106] Ibid., para. 12.

[107] Ibid., paras. 12 and 13.

[108] Ibid., para. 14.

[109] Ibid., paras. 12 and 14.

[110] World Health Organization, Fact Sheet 257; Depleted Uranium. Also ibid., paras. 2 and 9.

mal, for example because of a quick rainshower after impact, but in general the doses will probably not be high enough to meet the 'widespread, long-term and severe' standard of AP I. Even if the interpretation given to the terms during the Diplomatic Conference is not accepted and the terms are interpreted more in conformity with present standards, it is not likely that the triple standard would be met. This seems to be confirmed by the World Health Organization (WHO)[111] and the findings of three independent field studies of the United Nations Environment Programme (UNEP) in Kosovo in 2001,[112] Serbia and Montenegro in 2002[113] and Bosnia-Herzegovina in 2003[114] and by one study of the International Atomic Energy Agency (IAEA) in Kuwait in 2003.[115] The WHO concluded from various studies on uranium miners that the effects of uranium on human beings are limited, and the UNEP and IAEA studies concluded that there is generally no significant or widespread DU contamination of the ground surface of the sites investigated,[116] which means that it cannot be detected or differentiated from natural uranium concentrations and the radiological and toxicological risks are insignificant. There are some traces of DU in the air and in some living organisms, such as lichen, but they are small and insignificant as well. However, because it is possible that a large number of penetrators are still buried deep in the ground, there is a risk of groundwater contamination in the long-term that might exceed WHO[117] and also environmental standards.

In view of the above, it is unlikely that the use of DU would violate the customary prohibition to cause wanton or excessive collateral damage to the environment. The use of DU does not lead to either wanton destruction of the environment or to excessive collateral damage in view of the studies carried out by the WHO, UNEP and the IAEA and its use has a definite military justification. Armour penetrating ammunition is very useful in neutralising enemy tanks, and although other materials, such as tungsten, can be used as an alternative,[118] the availability of

---

[111] World Health Organization, ibid.

[112] United Nations Environmental Programme, *Depleted Uranium in Kosovo; Post-Conflict Environmental Assessment* (Nairobi, UNEP 2001), <http://postconflict.unep.ch/>.

[113] United Nations Environmental Programme, *Depleted Uranium in Serbia and Montenegro; Post-Conflict Environmental Assessment in the Federal Republic of Yugoslavia* (Nairobi, UNEP 2002) <http://postconflict.unep.ch/>.

[114] United Nations Environmental Programme, *Depleted Uranium in Bosnia and Herzegovina; Post-Conflict Environmental Assessment* (Nairobi, UNEP 2003), <http://postconflict.unep.ch/>.

[115] International Atomic Energy Agency, *The Radiological Conditions in Areas of Kuwait with Residues of Depleted Uranium*, summary, <http://www.iaea.org/NewsCenter/Features/DU/>.

[116] In Bosnia and Herzegovina, the UNEP team found that three of the 14 sites visited were contaminated.

[117] UNEP, Kosovo, *supra* n. 112, p. 35; UNEP, Serbia and Montenegro, *supra* n. 113, p. 33; UNEP, Bosnia and Herzegovina, *supra* n. 114, p. 49.

[118] See the discussion on the military necessity of DU and the presence of alternatives by B. Carnahan, Chapter 4 of this book, and A. McDonald, Chapter 12 of this book. Similarly, A. McDonald, 'The International Legality of Depleted Uranium Weapons', section 2.1.3.2; Background paper for presentation on 'The International legal ramifications of the use of DU weapons', Symposium on The Health Impact of Depleted Uranium Munitions', held at the New York Academy of Medicine, 14 June 2003, <http://www.nuclearpolicy.org/DUSymposium.html>.

alternatives does not necessarily invalidate a military necessity argument. Besides, tungsten is presumed to be less effective and although it is less radioactive than DU, it is nevertheless toxic.[119]

The only argument that can be advanced in this context and that may have a chance of success seems to be the presumed customary obligation to observe a duty of care for the environment during armed conflict. Although the reports of UNEP, the IAEA and the WHO conclude that the use of DU does not cause wide-spread contamination of the environment and does not have a significant impact on the health of human beings, there are still sites that are contaminated and there are localised hot spots with above average levels of radiation and pollution. While these levels may not be considerable and perhaps not reason for major concern, they were still enough for UNEP to recommend a large number of measures, such as continued research and investigation, testing of groundwater, decontamination if necessary, the positioning of warning signs at sites that have been subject to attack with DU, the careful removal of penetrators and fragments found, informing the local population, and investigation of all health claims.[120]

Apparently, the use of DU is not completely harmless and its long-term effects on the environment and on individuals are still uncertain. It is only 16 years since its first use during armed conflict. Although a duty of care only involves an obligation to avoid harm that is reasonably foreseeable, there is reason for concern and therefore also reason for a careful and perhaps even precautionary approach towards the use of DU.

---

[119] At <http://www.lenntech.com/Periodic-chart-elements/W -en.htm>.

[120] UNEP, Kosovo, *supra* n. 112, pp. 37-38; UNEP, Serbia and Montenegro, *supra* n. 113, p. 33; UNEP, Bosnia and Herzegovina, *supra* n. 114, pp. 36-37; United Nations Environmental Program, Depleted Uranium in Bosnia and Herzegovina; Post-Conflict Environmental Assessment, pp. 59-61, <http://postconflict.unep.ch/>.

Chapter 9
THE USE OF DEPLETED URANIUM
AS A POTENTIAL VIOLATION OF HUMAN RIGHTS

Brigit Toebes[1]

1.   INTRODUCTION

It has been suggested that depleted uranium (DU) munitions may be chemically and radiologically hazardous and may have negative effects on the health of combatants and the civilian population surrounding an impacted area, and on the environment.[2] Assuming that the concerns arising out of the use of DU can be supported by scientific evidence, this chapter explores whether the use of DU weapons could violate human rights law and, if so, which rights might be implicated. Since the effects of DU on health and the environment are still unproven, the conclusions drawn in this report can only be tentative.

Before identifying the human rights potentially implicated by the use of DU and elaborating on the content of these rights (section 3), and discussing their potential applicability to the case of DU (section 4), the chapter will briefly address some preliminary issues: the territorial, personal and temporal applicability of human rights law (section 2). Since, thus far, user states appear to have used DU weapons only extraterritorially during international armed conflicts, an important preliminary question is whether states can violate human rights outside their territory and, if so, under what conditions. If the answer is in the affirmative, the question of the personal applicability of human rights law arises. Of particular relevance is the question whether there is a distinction between combatants and civilians for the purpose of the applicability of human rights law. In fact, four main categories of persons are relevant in relation to the use of DU: a state's own armed forces, a state's own civilians, and the combatants and civilians of another state where DU is used. Section 2 also examines in brief the applicability of human rights law during armed conflict and its relationship with the law of armed conflict.

First, a brief word regarding the distinction and relationship between the two broad categories of human rights, namely, civil and political and economic, social and cultural rights.

---

[1] The author wishes to thank Avril McDonald for her valuable substantive and editorial comments. She also wishes to thank Christiaan Roorda and Jann Kleffner for their comments and advice.

[2] See D. Fahey, Chapter 2 of this book.

McDonald / Kleffner / Toebes (eds.), Depleted Uranium Weapons and International Law
© *2008, T·M·C·ASSER PRESS, The Hague, The Netherlands and the Authors*

1.1   Civil, political and economic, social and cultural rights

The core notion underlying human rights law is human dignity. States, as the duty holders of human rights, are required to respect and to enhance the dignity of individuals. This includes respecting their lives, physical integrity and health.

Human rights law distinguishes between civil and political human rights (e.g., rights to life, physical integrity, family life, and freedom of speech), on the one hand, and economic, social and cultural rights (e.g., rights to health, food, housing, and education), on the other. The more modern conception of human rights considers this distinction to be only a formal one and assumes that all human rights are interdependent, interrelated and of equal importance.[3]

As far as the state obligations resulting from human rights are concerned, it is assumed that civil and political rights as well as economic, social and cultural rights can embrace positive as well as negative obligations on the part of states. More precisely, current human rights doctrine distinguishes between obligations to respect, to protect and to fulfil human rights.[4] Duties to respect are (negative) obligations to refrain from action, e.g., obligations not to arbitrarily kill someone or to torture someone. Duties to protect and fulfil, on the other hand, are (positive) obligations that require states to protect individuals against certain acts by third parties, or to provide or facilitate access to certain services. Obligations to protect may imply, for example, the duty to take legislative and other measures in order to protect individuals against the detrimental health effects of the chemical industry. Obligations to fulfil may, for example, imply the provision or the facilitation of legal assistance and of medical and educational services.

This distinction between obligations to respect, protect and fulfil is important for the case of DU, since it clarifies what specific duties states may have with respect to the various rights. With regard to the rights to life and health, examples of obligations to respect, protect and fulfil will be given below. The distinction will also be referred to when the applicability of human rights law in the case of DU is discussed.

It should be noted that, in principle, human rights are only binding for states to the extent that they have ratified or acceded to a treaty in which these rights are set forth. However, it has been argued that a number of rights, the so-called 'core rights' of the Universal Declaration on Human Rights, have obtained the status of customary international law.[5] These rights are, therefore, considered to be

---

[3] See United Nations World Conference on Human Rights, *Vienna Declaration and Programme of Action*, UN Doc. A/CONF.157/23, 12 July 1993, para. 5.

[4] Regarding this tripartite typology of obligations, see the report of the Special Rapporteur on the Right to Food, A. Eide, UN Doc. E/CN.4/Sub.2/1987/23. See also G.J.H. van Hoof, 'The legal nature of economic, social and cultural rights: a rebuttal of some traditional views', in P. Alston and K. Tomasevski, eds., *The Right to Food* (Utrecht, SIM 1984) pp. 97-111. For an elaboration of this typology in relation to the right to health see International Committee on Economic, Social and Cultural Rights (CESCR), General Comment 14, UN Doc. E/C.12/2000/4, <http://www.unhchr.ch>.

[5] See, *inter alia*, International Law Association, *Report on the Status of the Universal Declaration of Human Rights in National and International Law*, Report of the 66th Conference, 1984, pp. 525-563.

binding upon all states. Moreover, norms that are similar to international human rights are found in national constitutions.

2. TERRITORIAL, PERSONAL AND TEMPORAL APPLICABILITY OF HUMAN RIGHTS LAW

2.1 Territorial applicability of human rights

Human rights law primarily addresses the relation between states and the individuals living on their territories. A factor that complicates the applicability of human rights law in the present case is that, thus far, states appear to have only used DU weapons extraterritorially, in the context of international armed conflicts. As a result, the question arises whether human rights law covers the relation between the state and individuals (combatants and/or civilians) of opposing states.

Article 2(1) of the International Covenant on Civil and Political Rights (ICCPR) determines that every State Party is to guarantee the rights recognised therein 'to all individuals within its territory and subject to its jurisdiction'. From the 'Uruguayan cases' before the United Nations Human Rights Committee (HRC), the treaty-monitoring body of the ICCPR, it can be deduced that decisive is not the place where the violation occurs but rather the relationship between the individual and the duty holder.[6] In the case of *Lopez Burgos* v. *Uruguay*, the HRC held that a State Party 'can be held accountable for violations of rights under the Covenant which its agents commit upon the territory of another State, whether with the acquiescence of the Government of the State or opposing it.'[7]

This position was confirmed by the HRC in General Comment No. 31 on the 'Nature of the General Legal Obligations'. In paragraph 10 of the General Comment the HRC states that

> '(...) a State party must respect and ensure the right laid down in the Covenant to anyone within the power or effective control of that State party, even if not situated within the territory of the State party...This principle also applies to those within the power or effective control of the forces of a State party acting outside its territory, regardless of the circumstances in which such power or effective control was obtained, such as forces constituting a national contingent of a State party assigned to an international peacekeeing or peace-enforcement operation.'[8]

The International Court of Justice (ICJ) follows the HRC's approach in its Advisory Opinion in the 'Wall case'.[9]

---

[6] For a discussion of these cases see D. McGoldrick, *The Human Rights Committee* (Oxford, Clarendon Press 1994) pp. 177-178.

[7] *Lopez Burgos* v. *Uruguay*, UN Doc. A/36/40, p. 76; Mc Goldrick, *supra* n. 6, pp. 179-180.

[8] UN Doc. CCPR/C/21/Rev.1/Add.13, 26 May 2004.

[9] See *Legal Consequences of the Construction of Wall in the Occupied Palestinian Territory*, Advisory Opinion of 9 July 2004, paras. 109-111.

Unlike the ICCPR, the International Covenant on Economic, Social and Cultural Rights (ICESCR) does not allude to its territorial applicability. However, as explained by Sepúlveda, it can be deduced from the practice of the Committee on Economic, Social and Cultural Rights (CESCR), the treaty-monitoring body of the ICESCR, that the ICESCR applies to everyone within the jurisdiction of the State Party. The term 'within the jurisdiction' embraces all territories in which the state exercises even *de facto* control.[10] With regard to the responsibility of Israel for the realisation of economic and social rights in the Occupied Territories, for example, the CESCR held that the state's obligations under the ICESCR apply in 'all territories and populations under its *effective control*'.[11] Nevertheless, as Sepúlveda explains, under the ICESCR, states do not assume obligations to protect individuals against violations committed abroad by another state.[12]

Article 1 of the European Convention of Human Rights (ECHR) stipulates that States Parties to the ECHR shall secure the rights in the convention *vis-à-vis* persons 'subject to their jurisdiction'. In the case of *Cyprus* v. *Turkey*, the European Court of Human Rights (ECtHR) decided that the ECHR applies 'to all persons under their actual authority and responsibility, whether that authority is exercised within their own territory or abroad.'[13] In the case of *Loizidou* v. *Turkey*, the Court held that the concept of jurisdiction is not restricted to the national territory of the state. The responsibility of the state

> 'could also arise when as a consequence of military action – whether lawful or unlawful – it exercises effective control of an area outside its national territory. The obligation to secure, in such an area, the rights and freedoms set out in the Convention, derives from the fact of such control whether it be exercised directly, through its armed forces, or through a subordinate local administration.'[14]

However, in the *Banković* case, the ECtHR stressed that 'its recognition of the exercise of extra-territorial jurisdiction by a Contracting State is exceptional'. It held that the Court recognises extraterritorial jurisdiction

> 'when the respondent State, through the effective control of the relevant territory and its inhabitants abroad as a consequence of military occupation or through consent, invitation or acquiescence of the Government of that territory, exercises

---

[10] M. Sepúlveda, *The Nature of the Obligations Under the International Covenant on Economic, Social and Cultural Rights* (Antwerp/Oxford/New York, Intersentia 2003) pp. 272-277.

[11] Concluding Observations with respect to Israel, UN Doc. E/1992/22, para. 234. See Sepúlveda, *supra* n. 10, p. 274. See also the International Court of Justice (ICJ), *Legal Consequences of the Construction of Wall in the Occupied Palestinian Territory*, *supra* n. 9, para. 112.

[12] Sepúlveda, *supra* n. 10, p. 275.

[13] ECtHR, 2 Rep., Judg. & Dec., pp. 125 at 136 (1975).

[14] *Loizidou* v. *Turkey*, judgment of 18 December 1996, (Appl. No. 15318/89), para. 52. See also ECtHR, *Drozd and Janousek* v. *France and Spain*, Judgment of 27 May 1992, ECHR Reports, Series A, No. 240, para. 91 ('the responsibility of States parties can be involved because of acts of their authorities producing effects outside their own territory'.).

all or some of the public powers normally to be exercised by that Government'.[15]

Furthermore, the Court held that Member States can only exert extraterritorial jurisdiction over the territory of other States Parties to the ECHR.[16]

Altogether, the case law of the various treaty monitoring bodies makes it clear that the human rights instruments concerned can apply extraterritorially in some circumstances,[17] although the ECtHR interprets its instrument more restrictively than the other bodies. Generally speaking, the ECtHR requires that the state concerned exercises effective control over the foreign territory. Effective control is without doubt exercised if a state has an occupying force controlling the foreign area. The situation will be different if a state 'only' bombs the alien territory. According to Kamminga, it is unlikely that under such circumstances the state exercises sufficient control to have positive (human rights) obligations in that area. Nevertheless, such a state is clearly required to comply with the negative obligations resulting from the right to life.[18]

2.2 Personal applicability of human rights

As we are dealing with the effects of DU on combatants as well as on civilians, the question arises whether combatants and civilians enjoy the same human rights.

In principle, human rights law, which is primarily designed to apply under peacetime conditions (see below section 2.3), applies equally to all individuals within the state's territory or subject to its jurisdiction.[19] Unlike international humanitarian law (IHL), human rights law does not treat combatants as a special group, nor does it guarantee them a different level of protection than it does to other individuals.

The applicability of human rights law to soldiers was addressed in a case before the ECtHR concerning the deprivation of liberty of soldiers. In *Engel and others* v. *the Netherlands*, the Court held that, in principle, the convention applies

---

[15] ECtHR, *Banković* v. *Belgium et al.*, Appl. No. 52207/99, decision as to the admissibility, 12 December 2001, para. 71. See also 'De extraterritoriale reikwijdte van het EVRM' [the extraterritorial scope of the ECHR], 27 *NJCM-Bull.* (2002) pp. 631-641 (note: M.T. Kamminga). For a more recent case see ECtHR, *ISSA and Others* v. *Turkey*, Appl. No. 31821/96, 16 November 2004, paras. 67-82.

[16] *Banković* case, *supra* n. 15, para. 80.

[17] Art. 1 American Convention on Human Rights speaks of the 'persons subject to their jurisdiction'. For an interpretation of the extraterritorial jurisdiction of the American Declaration, see the *Coard* case (Report No. 109/99, Case No. 10.951, 29 September 1999, para. 37 – 'In principle, the inquiry turns not on the presumed victim's nationality or presence within a particular geographic area, but on whether, under the specific circumstances, the State observed the rights of a person subject to its authority and control').

[18] See Kamminga, *supra* n. 15.

[19] An exception to this rule can be found in Art. 8(2) ICESCR, which allows for lawful restrictions on the exercise of the right to form trade unions by members of the armed forces, the police and the administration of the state. The existence of this restriction may underline the assumption that such restrictions are not allowed for the other rights in the ICESCR.

to members of the armed forces and not only to civilians. The Court nevertheless wanted to bear in mind 'the particular characteristics of military life and its effects on the situation of individual members of the armed forces'.[20] It should be noted, however, that this case concerned a peacetime situation. During an armed conflict, combatants have a different level of human rights protection than civilians, to the extent that they may lawfully be killed. This matter will be dealt with more extensively below and in section 4.1.

As to whether enemy armed forces can rely upon human rights, the answer, as for enemy civilians, will depend upon whether the state accused of breaching human rights extraterritorially exercises 'effective control' over the area concerned.

2.3   Temporal applicability of human rights

During peacetime, the applicable human rights will apply in their entirety, but during states of emergency, including armed conflict, states may suspend (derogate from) all but a hard core of (civil and political) human rights.[21] Once an armed conflict – whether international or non-international in character – comes into being, another body of law, IHL, kicks in and becomes the dominant law. In case of a conflict between the applicable human rights law (the *lex generalis*) and IHL (the *lex specialis*), IHL will trump.

To some extent, IHL can be seen as a species of human rights law,[22] which protects the rights of particular categories of human beings (principally, civilians taking no part in hostilities and fighters (combatants and civilians taking a direct part in hostilities) *hors de combat* through injury, shipwreck or detention) during periods of armed conflict.[23] This assumption is supported by international jurisprudence. In its Advisory Opinion on the *Legality of the Threat or Use of Nuclear Weapons*, the ICJ determined that

> 'In principle, the right not arbitrarily to be deprived of one's life applies also in hostilities. The test of what is an arbitrary deprivation of life, however, then falls to be determined by the applicable *lex specialis*, namely, the law applicable in armed conflict which is designed to regulate the conduct of hostilities. Thus whether a particular loss of life, through the use of a certain weapon in warfare, is to be considered an arbitrary deprivation of life contrary to Article 6 of the

---

[20] ECtHR, *Engel and others* v. *the Netherlands*, 8 June 1976, published in A22 (see para. 54). The Court decided that Mr Engel's provisional arrest violated Art. 5(1) ECHR.

[21] See, *inter alia*, T. Meron, 'Notes and comments on the inadequate reach of humanitarian law and human rights law and the need for a new instrument', 77 *AJIL* (1983) pp. 589 at 603.

[22] A.H. Robertson and J.G. Merrils, *Human Rights in the World*, 3rd edn. (New York, St. Martin's Press 1992) p. 277.

[23] What these people have in common is that they do not or no longer take part in hostilities. Of course, to some extent IHL also safeguards the interests of combatants by, for example, interdicting the use against them of weapons causing superfluous injury or necessary suffering.

> Covenant, can only be decided by reference to the law applicable in armed conflict and not deduced from the terms of the Covenant itself.'[24]

Similar conclusions were drawn by the Inter-American Commission on Human Rights (IACHR):

> '(...) Human rights instruments contain norms of a higher level (Article 27 of the American Convention), that, under circumstances such as those we are dealing with of armed conflict, need to be supplemented with interpretation in the light of humanitarian law.'[25]

And

> 'There is an integral linkage between the law of human rights and humanitarian law because they share a "common nucleus of non-derogable rights and a common purpose of protecting human life and dignity" and there may be a substantial overlap in the application of these bodies of law. Certain core guarantees apply in all circumstances, including situations of conflict (...).' [26]

In conclusion, the question whether the applicable human rights have been breached during an armed conflict can only be determined by IHL as the *lex specialis*. However, the relationship between human rights and IHL may not be exactly the same for economic, social and cultural rights as it is for civil and political rights, as discussed below in section 4, where the temporal applicability of human rights is analysed in more depth. With these points in mind, we will now turn to a discussion of the relevant human rights norms.

3.   THE RELEVANT HUMAN RIGHTS AND THE DUTIES THEY IMPOSE ON
     STATES

Given the potentially negative impact on people's lives and health of the use of DU weapons, the human rights that are likely to be of primary importance with regard to the use of DU weapons are the rights to life and health. Other human rights that may potentially be violated include the rights to food, a clean environment, family life, information, and the right to a remedy. In addition, the use of DU weapons could raise questions regarding the relation between combatants of user states and their own states in terms of international labour standards on safe and healthy working conditions.

---

[24] ICJ, *Legality of the Threat or Use of Nuclear Weapons, Advisory Opinion,* General List No. 95, 8 July 1996, p. 8, para. 25. See also T. Meron, 'The humanization of humanitarian law', 94 *AJIL* (2000) pp. 239 at 266.

[25] IACHR, Case 10.480, El Salvador, Report 1/99, 27 January [*Parada Cea et al.*], para 65. See also, *inter alia*, Case 11.142, Colombia Report No. 26/97, 30 September 1997 [*Ribon Avillan* Case], para. 132.

[26] IACHR, Case 10951, United States, Report 109/99, 29 September 1999 [*Coard and al.* – Invasion of Grenada].

3.1   Civil and political rights

3.1.1   *The right to life*

The right to life is recognised in every broad-based human rights treaty, including in Article 6 of the ICCPR and Article 2 of the ECHR.[27] This human right, which is considered to be the 'supreme human right',[28] primarily offers protection against arbitrary killing. Its scope has, however, been gradually expanded so as to include a broader range of guarantees, including protection from malnutrition, epidemics, and nuclear weapons.[29]

The meaning of Article 6 ICCPR has been elaborated in two so-called 'General Comments', non-binding pronouncements that seek to explain a certain aspect of the treaty concerned in order to assist states to fulfil their treaty obligations. General Comment 6 of the HRC explains the meaning of Article 6. Referring to armed conflicts, it declares that 'States have the supreme duty to prevent wars, acts of genocide and other acts of mass violence causing arbitrary loss of life'.[30] It also gives a broad interpretation of Article 6, by claiming that it 'cannot properly be interpreted in a restrictive manner, and the protection of this right requires that States adopt positive measures'.[31] As examples of such positive measures, the Committee refers to measures to reduce infant mortality and to increase life expectancy, especially in adopting measures to eliminate malnutrition and epidemics.[32]

General Comment 14 also deals with the dangers of nuclear weapons. It states that the 'designing, testing, manufacturing, possession and deployment of nuclear weapons is among the greatest threats to the right to life which confront mankind today' and that therefore states should 'take urgent steps (...) to rid the world of this menace'.[33] The HRC thus recognises that protection against nuclear weapons is part of the scope of the right to life.

Several complaints on this issue were submitted to the HRC within the context of its individual complaints procedure. For example, in *Vaihere Bordes* v. *France* the plaintiffs claimed that France's intention to conduct a series of underground nuclear tests in the South Pacific threatened their right to life and their right

---

[27] See also, e.g., Art. 5 of the Universal Declaration of Human Rights (UDHR) and Art. 4 of the American Convention on Human Rights (ACHR, Organization of American States – OAS).

[28] *Inter alia,* Art. 6(1) ICCPR. The term 'supreme human right' can be found in Human Rights Committee, *General Comment 6*, UN Doc. A/37/40, para. 1.

[29] See B.C.A. Toebes, *The Right to Health as a Human Right in International Law* (Antwerp/ Oxford/New York, Intersentia 1999) p. 261. See also Y. Dinstein, 'The right to life, physical integrity, and liberty', in L. Henkin, ed., *The International Bill of Rights: The Covenant on Civil and Political Rights* (New York, Colombia University Press 1981) p. 114 and B. Gammie, 'Human rights implications of the export of banned pesticides', 25 *Seton Hall Law Review* (1994) p. 585.

[30] HRC, *General Comment 6, supra* n. 28, para. 2.

[31] Ibid., para. 5.

[32] Idem.

[33] HRC, *General Comment 14*, adopted on 2 November 1984, published in UN Doc. A/40/ 40, HRC Annual Report 1985, Annex VI, pp. 162-163 (second general comment on Art. 6), paras. 4 and 7.

not to be subjected to arbitrary interference with their privacy and their family life.[34] In the case of *Port Hope* v. *Canada* the plaintiff alleged that the storage of radioactive waste in Port Hope threatened the life of present and future generations.[35] While both complaints were declared inadmissible on formal grounds,[36] one could argue that the right to life may require states to take positive measures to deal with the side effects of nuclear testing and nuclear waste.

As regards the ECtHR, noteworthy is the case of *L.C.B.* v. *the United Kingdom*. The applicant, who was diagnosed with leukaemia, alleged that the United Kingdom had failed to monitor the extent of her father's exposure to radiation on Christmas Island and to warn and advise her parents or monitor her health prior to her diagnosis. The Court held, *inter alia*, that since the radiation did not reach dangerous levels in the area concerned, there was no violation of the right to life in Article 2 ECHR.[37]

As noted in section 1.1, human rights may embrace state obligations to respect, as well as to protect and to fulfil. Sepúlveda has made an overview of state obligations resulting from the right to life. It is presented as a 'continuum of duties', starting with the negative end of the spectrum ('negative' obligations to respect) and ending at the positive end ('positive' obligations to fulfil). It reads as follows:

A. The duty to refrain from intentional and unlawful taking of life;
B. The duty to take positive measures to protect individuals under the jurisdiction of the state from being killed:
   1. To protect individuals from being killed by state agents;
   2. To protect individuals from being killed by third parties;
C. The duty to carry out an effective official investigation when individuals have been killed or have disappeared;
D. The duty to provide an effective remedy (duty to take measures as a consequence of a previous failure to protect the right to life);

---

[34] *Mrs Vaihere Bordes and Mr John Temeharo* v. *France,* Communication No. 645/1995, inadmissibility decision of 22 July 1996 (UN Doc. A/51/40, Vol. II, Annex IX 7. See also 18 *HRLJ* (29 August 1997) Nos. 1-4, pp. 36-39. See Toebes, *supra* n. 29, p. 264.

[35] *E.H.P. (name deleted) on her own behalf and, as chairperson of the Port Hope Environmental Group, on behalf of the present and future generations of Port Hope, Ontario, Canada, including 129 Port Hope residents who have specifically authorized the author to act on her behalf* v. *Canada,* Communication No. 67/1980, inadmissibility decision of 27 October 1982 (UN Doc. CCPR/C/OP/2), Selected Decisions of the Human Rights Committee under the Optional Protocol, Vol. 2, pp. 20-22. It concerned a communication submitted by a Canadian environmental group. The communication was declared inadmissible because of exhaustion of domestic remedies. See Toebes, *supra* n. 29, p. 263. See also the case of *E.W. et al.* v. *the Netherlands*, Communication No. 429/1990, inadmissibility decision of 8 April 1993, where a group of Dutch citizens complained that their government's agreement to allow a deployment of cruise missiles fitted with nuclear warheads was a violation of their right to life.

[36] As will be discussed below in section 3.1.2, the European Court of Human Rights is rather inclined to base similar deliberations on the right to family life in Art. 8 ECHR.

[37] ECHR, *L.C.B.* v. *the United Kingdom*, 9 June 1998, published in Reports 1998-III.

E. The duty to satisfy the basic needs necessary for sustaining the lives of those persons who cannot do so themselves:
    1. For individuals held in state custody;
    2. For deprived individuals under the jurisdiction of the state.[38]

Sepúlveda's analysis demonstrates that the obligations resulting from the right to life vary from the 'negative' obligation to refrain from arbitrary killing to the positive obligation to ensure the survival of individuals.

The right to life under human rights law continues to apply during armed conflict (it cannot be derogated from), although it is not absolute (combatants can lawfully be killed as can civilians who take a direct part in hostilities). Although during an armed conflict, IHL is the *lex specialis*, like human rights law, it recognises the right to protection from arbitrary killing (meaning that the core content of both bodies of law overlaps as far as the right to life is concerned). The most relevant IHL provisions in this regard are those designed to ensure protection from attack and humane treatment for persons protected by the four Geneva Conventions of 1949 and their two Additional Protocols (AP I and II) of 1977.[39] Regarding the right to life, the following provisions are of particular relevance. As far as the protection of civilians is concerned, Article 51 AP I provides, *inter alia,* that the civilian population shall not be the object of attack and that indiscriminate attacks are prohibited.[40] A comparable provision is Article 13 AP II, which applies during non-international armed conflicts. Similarly, the prohibition of unnecessary suffering and superfluous injury applies to combatants during situations of international armed conflict (Art. 35(2) AP I) and during non-international armed conflict under customary international law,[41] and any deprivation of life must not conflict with this principle. In addition, rules concerning the protection of the environment apply (Art. 35(3) AP I) during international armed conflict.[42] Other relevant provisions include Article 54 AP I (the protection of objects indispensable to the survival of the civilian population) and Articles 57 and 58 AP I (on precautionary measures). For the implications of these norms in relation to the use of DU weapons, readers are referred to Chapters 4 through 8 of this book.[43]

3.1.2   *The rights to family life and information*

Another potentially relevant provision in human rights law is the right to family life in Article 8 ECHR. In several decisions of the ECtHR, the right to family life has

---

[38] Sepúlveda, *supra* n. 10, pp. 140-155.

[39] Toebes, *supra*, n. 29, p. 78. See also D. Plattner, 'International humanitarian law and inalienable or non-derogable human rights', in D. Prémont, et al., eds., *Non-Derogable Rights and States of Emergency* (Brussels, Bruylant 1996) pp. 349 at 363.

[40] See J. Kleffner and T. Boutruche, Chapter 6.

[41] See Art. 35(2) AP I. See M. Zwanenburg, Chapter 5.

[42] Arts. 35(3) and 55 AP I. It should be noted, however, that it would be very difficult to link a specific right to life with an attack against the natural environment.

[43] See in particular M. Zwanenburg, Chapter 5, J. Kleffner and T. Boutruche, Chapter 6, and E. Koppe, Chapter 8.

been employed to offer a certain protection against environmental health threats. The case of *López Ostra* v. *Spain* concerned the nuisance caused by a waste treatment plant and its effects on the applicant's daughter's health in the town of Lorca in Spain. The Court opined that 'severe environmental pollution may affect individuals' well-being and prevent them from enjoying their homes in such a way as to affect their private and family life adversely'. It concluded that the municipality of Lorca had failed to take steps to protect the applicant's right to respect for her home and for her private life, and that Article 8 had accordingly been violated.[44] In fact, the municipalities had a so-called 'obligation to protect' citizens against the detrimental effects of the polluting factory (see section 1.1).

Furthermore, Article 8 is considered to include a right to be informed about such environmental health threats, i.e., a right to information.[45] In the *Anna Guerra* case the Court held that:

> 'In the instant case the applicants waited, right up until the production of fertilisers ceased in 1994, for essential information that would have enabled them to assess the risks they and their families might run if they continued to live at Manfredonia, a town particularly exposed to danger in the event of an accident at the factory.
> The Court holds, therefore, that the respondent State did not fulfil its obligation to secure the applicants' right to respect for their private and family life, in breach of Article 8 of the Convention.'[46]

Similarly, in another case, the applicants complained that they had been denied access to the records compiled in relation to the radiation levels and the medical treatment they had received following experimental nuclear explosions at Christmas Island carried out by the British Government in the 1950s. The Court held that:

> 'Where a Government engages in hazardous activities, such as those in issue in the present case, which might have hidden adverse consequences on the health of those involved in such activities, respect for private and family life under Article 8 requires that an effective and accessible procedure be established which enables such persons to seek all relevant and appropriate information.'[47]

---

[44] ECHR, *López Ostra* v. *Spain* (Judgment of 9 December 1994, A.303C (1995)).

[45] On the right to be informed about environmental threats, see L. Boisson de Chazournes, 'Nonderogable rights and the environment', in Prémont, et al., *supra* n. 39; A.C. Kiss and D. Shelton, *International Environmental Law* (New York, Transnational Publishers 2000) pp. 160-162.

[46] ECHR, *Guerra and others* v. *Italy*, Judgment of 19 February 1998, Rep., Judg. & Dec. (1998-1), para. 60. With regard to the freedom to receive information under Art. 10 ECHR, the Court held that this right 'cannot be construed as imposing on a State, in circumstances such as these of the present case, positive obligations to collect and disseminate information of its own motion' (para. 53). For a more recent decision see ECtHR, *Fadeyeva* v. *Russia*, Judgment 9 June 2005, (Appl. No. 55723/00), *inter alia*, paras. 88-93.

[47] ECHR, *McGinley and Egan* v. *United Kingdom*, Judgment of 9 June 1998, Rep., Judg. & Dec. (1998-III), para. 101. It should be noted that no violation was found since the applicants did not avail themselves of an existing procedure to request from the competent authorities the production of the documents in question.

It emerges that in the above-mentioned cases the government failed to comply with its 'obligation to fulfil', i.e., its obligation to establish a procedure in order to enable persons to seek information.

The case law of the ECtHR on Article 8 therefore provides some openings towards dealing with environmental health issues, including a right to be informed about such potential dangers.

3.1.3   *The right to a remedy*

Many human rights instruments guarantee a right to a remedy if the rights and freedoms set forth by those instruments are violated.[48] The Universal Declaration of Human Rights stipulates that 'everyone has the right to an effective remedy by the competent national tribunals for acts violating the fundamental rights granted him by the constitution or laws'.[49] A right to a remedy embraces a wide range of obligations, including to financially compensate the victim; to investigate the facts; to bring those found responsible for the violations to justice; to provide fair compensation for damages; to offer legal protection on privacy, family, or home; and to offer access to a tribunal for the determination of rights and duties.[50] The right to a remedy is also recognised in the ICCPR (Art. 2(3)) and the ECHR (Art. 6).

For our purposes, it is worth noting that the right to a remedy has been used to offer protection in cases where people's environmental health has been endangered. For example, in the case of *Zander* v. *Sweden*, the ECtHR held that Article 6 ECHR was violated since the applicants had been denied a remedy for their potential exposure to contaminated water.[51]

3.2   **Economic, social and cultural rights**

3.2.1   *The right to health*

The right to health is part of the category of economic, social and cultural rights. Several UN human rights conventions recognise the right to health, the most important of which are the ICESCR (Art. 12), the Convention on the Elimination of All Forms of Discrimination Against Women (CEDAW, Art. 12) and the Convention on the Rights of the Child (CRC, Art. 24). Similarly, several regional human rights conventions recognise the right to health.[52] The scope of the right has been

---

[48] E.g., Arts. 2(3) ICCPR, 6 ECHR, 8 ACHR, 7 and 26 African Charter. See also CESCR, General Comment No. 3, *The Nature of States Parties Obligation*, UN Doc. E/1991/23 (5th session, 1990), para. 5, <www.unhchr.ch>.

[49] Art. 8 UDHR.

[50] *Inter alia*, Arts. 2(3) ICCPR, 6 ECHR and XVII of the American Declaration.

[51] ECtHR, *Zander* v. *Sweden*, 279B Eur. Cf., H.R. (ser. A) (1993). See also ECtHR, *Zimmerman and Steiner* v. *Switzerland*, 66. Eur. Ct. H.R. (Ser. A) (1983). See Kiss and Shelton, *supra* n. 45, p. 165.

[52] Art. 11 European Social Charter (ESC, Council of Europe), Art. 16 African Charter on Human and Peoples' Rights (Organisation of African Unity), Art. XI American Declaration on the Rights and Duties of Man and Art. 10 Protocol of San Salvador (Organization of American States – OAS).

clarified recently by the adoption of a General Comment on the right to health in Article 12 ICESCR, which has been drawn up by the CESCR.[53] This chapter will take Article 12 ICESCR and General Comment 14 as a point of departure.

The right to health is not a right to be healthy.[54] Rather, it embraces a number of health services and freedoms for which the state can be held responsible. Roughly speaking, the right to health has two dimensions: it embraces a right to appropriate health care services as well as a right to a number of underlying determinants of health.

Regarding the right to health care facilities, Article 12(2)(d) recognises the obligation of states to 'create conditions which would assure to all medical service and medical attention in the event of sickness'. According to General Comment 14, this embraces states' duties to ensure the provision of 'timely access to basic curative, rehabilitative health services, and health education', and 'regular screening programmes (...)'.

Regarding the underlying determinants of health, Article 12(2)(b) refers to the obligation of states to 'improve all aspects of environmental and industrial hygiene'. General Comment 14 explains that this obligation includes, *inter alia*, 'preventive measures in respect of occupational accidents and diseases', and (...) 'the prevention and reduction of the population's exposure to harmful substances such as radiation and harmful chemicals'.[55]

General Comment 14 defines the so-called core content of the right to health, i.e., a minimum that is inherent in the right to health and which should apply under all circumstances.[56] A connection is made, *inter alia*, with the Primary Health Care Strategy of the World Health Organization (WHO), which enumerates a number of core health services that states should furnish under all circumstances.[57] States should provide access to the following specific services:

> 'health facilities, goods and services on a non-discriminatory basis, especially for vulnerable or marginalized groups; minimum essential food which is nutritionally adequate and safe, to ensure freedom from hunger to everyone; basic shelter, housing and sanitation, and an adequate supply of safe and potable water; essential drugs, as from time to time defined under the WHO Action Programme on Essential Drugs; and ensure equitable distribution of all health facilities, goods and services; adopt and implement a national public health strategy and plan of action (...),

and

> 'to ensure reproductive, maternal (pre-natal as well as post-natal) and child health care; to provide immunization against the major infectious diseases oc-

---

[53] CESCR, General Comment 14, *supra* n. 4.

[54] Ibid., para. 8.

[55] Ibid., para 15.

[56] Ibid., paras. 43-44. For a further explanation of the concept of core content, see section 4.2.2.

[57] WHO, *Primary Health Care, Report of the International Conference on Primary Health Care*, Alma-Ata, USSR, 6-12 September 1978, '*Health For All' Series No. 1* (Geneva/New York, WHO 1978), Chapter 3, para. 50.

curring in the community; to take measures to prevent, treat and control epidemic and endemic diseases; to provide education and access to information concerning the main health problems in the community, including methods of preventing and controlling them; to provide appropriate training for health personnel, including education on health and human rights.'[58]

For the right to health, the following state obligations to respect, protect and fulfil can thus be discerned:

Obligations to respect:

- The obligation to respect equal access to available health services and not to impede individuals or groups from their access to the available services;
- The obligation to refrain from acts which encroach upon people's health, such as activities causing environmental pollution.

Obligations to protect:

- The obligation to take legislative and other measures to assure that people have (equal) access to health services if provided by third parties;
- The obligation to take legislative and other measures to protect people from health infringements by third parties.

Obligations to fulfil:

- The obligation to adopt a national health policy and to devote a sufficient percentage of the available budget to health;
- The obligation to provide the necessary health services or create conditions under which individuals have adequate and sufficient access to health services, including, in particular, health care services as well as clean drinking water and adequate sanitation.[59]

This overview demonstrates that inherent in the right to health are not only positive obligations to provide health services and to protect people's health but also negative obligations to refrain from acts that encroach upon people's health.

In addition, reference can be made to a number of provisions in IHL which contain protection that is related to a right to health during armed conflicts. In fact, Geneva law contains numerous references to the above-mentioned dimensions of the right to health, in particular to the right to health care services. Regarding the right to health care facilities, Geneva law deals with the inviolability of medical services in situations of conflict. It purports to ensure the undisturbed and safe delivery of medical care during armed conflict and it prescribes that individuals who do not participate directly in hostilities and who are *hors de combat* be cared

---

[58] *Supra* n. 4, paras. 43-44.
[59] See B.C.A. Toebes, 'The right to health', in A. Eide, et al., eds., *Economic, Social and Cultural Rights,* 2nd edn. (The Hague, Kluwer Law International 2001) pp. 169-190.

for without discrimination.[60] For example, Geneva Convention IV provides that hospitals and safety zones shall be established so as to protect wounded, sick, elderly persons, children, and mothers.[61] Additional Protocol II requires that protection and care is given to the wounded, sick and shipwrecked.[62] For example, Article 7 AP II requires that all the wounded, sick and shipwrecked, whether or not they have taken part in the armed conflict, shall be respected and protected. Article 7(2) requires that they shall receive the medical care and attention required by their condition.

Regarding the second dimension of the right to health, the underlying preconditions for health, IHL contains a reference to environmental health. Article 35(3) AP I on the protection of victims of international armed conflicts prohibits the employment of methods of warfare which cause widespread damage to the environment.[63]

3.2.2   The right to food

Another economic and social right that is of importance in the context of DU is the right to food. It is recognised as part of the right to an adequate standard of living in Article 11 of the ICESCR. It can also be found in Article 24 of the CRC and, at the regional level, in Article 12(1) of the Protocol of San Salvador (Organization of American States, OAS).

In its General Comment No. 12 (1999), the CESCR has given its interpretation of the right to adequate food. It states: 'The right to adequate food is realized when every man, woman and child, alone or in community with others, has physical and economic access at all times to adequate food or means for its procurement.'[64] This General Comment also formulates a so-called 'core content' of the right to food, which is, as mentioned above, the minimum that is inherent in the right to food and which should be guaranteed under all circumstances. It embraces:

(a) 'the availability of food in a quantity and quality sufficient to satisfy the dietary needs of individuals, free from adverse substances, and acceptable within a given culture;

---

[60] See, *inter alia*, Arts. 12-15 and 18-37 Geneva Convention I (on the wounded and sick in international armed conflicts); Arts. 7 and 12-40 Geneva Convention II (wounded, sick and shipwrecked of the armed forces in international conflicts); Arts. 13, 15, 17, 20 and 29-33 Geneva Convention III (prisoners of war in international armed conflicts); Arts. 14, 76 and 91-92 Geneva Convention IV; Arts. 12 and 21-30 AP I; and Arts. 5, 7, 9 and 11 AP II.

[61] Art. 14 Geneva Convention IV.

[62] See Part III Protocol II.

[63] For an interpretation of this provision see E. Koppe, Chapter 8.

[64] CESCR, General Comment No. 12 on the right to adequate food, Report of the Committee on Economic, Social and Cultural Rights, UN Doc. E/2000/22, pp. 102-110, <www.unhchr.ch>. See also A. Eide, 'The right to an adequate standard of living including the right to food', in A. Eide, et al., eds., *Economic, Social and Cultural Rights,* 2nd edn. (The Hague, Kluwer Law International 2001) pp. 133-148, and in Annex 2.

(b) the accessibility of such food in ways that are sustainable and that do not inter-
     fere with the enjoyment of other human rights.'[65]

For our purposes, an important aspect of this core content is that food be 'free from
adverse substances'. According to General Comment 12, it:

> 'sets requirements for food safety and for a range of protective measures by both
> public and private means to prevent contamination of foodstuffs through adul-
> teration and/or through bad environmental hygiene or inappropriate handling at
> different stages throughout the food chain; care must be taken to identify and
> avoid or destroy naturally occurring toxins'.[66]

In addition, it can be observed that during armed conflicts IHL may offer some
protection of people's nutritional needs. Although more limited than with respect to
the right to health, Geneva law contains several references to the right to food. For
example, Geneva Convention III provides that prisoners of war must have access to
basic daily food rations and that their habitual diet must be taken into account.
Furthermore, canteens have to be installed in their camps.[67] Geneva Convention IV
provides that an occupying power is required not to hinder preferential measures
regarding food and other services adopted prior to an occupation in favour of children and mothers.[68] Furthermore, AP II requires that persons whose liberty has
been restricted have access to food and drinking water.[69] Most important for our
purposes, Article 54 of AP I prohibits starvation as a method of warfare. It also
prohibits parties to armed conflicts from attacking and destroying objects indispensable for the survival of individuals, including foodstuffs, agricultural areas,
crops, and drinking water installations.[70] The latter may also cover pollution of
water reservoirs, or destruction of crops by defoliants.[71]

3.2.3 *The right to a clean environment*

Relevant for the present case is also the right to a clean environment.[72] As it is not
specifically recognised in any UN human rights instrument, it is not firmly established under international law. A complicating factor is that it is not certain whether
the right to a clean environment is an individual or a collective right. The interests
protected by the right to environmental protection cannot easily be individualised,

---

[65] General Comment No. 12, ibid., para. 8.

[66] Ibid., para. 10.

[67] Arts. 26 and 27 Geneva Convention III.

[68] Art. 50 Geneva Convention IV.

[69] Art. 5(1)(b) AP II.

[70] Art. 54 AP I.

[71] ICRC, *Commentary on the Additional Protocols of 8 June 1977 to the Geneva Conventions of
12 August 1949* (Geneva, Martinus Nijhoff Publishers 1987) p. 655.

[72] Other terms that may be applied are the rights to a healthy environment, to environmental protection and simply the right to an environment. On this topic see Kiss and Shelton, *supra* n. 45, pp. 141-
187.

since the right aims at the protection of a public interest, i.e., a clean environment.[73] A number of openings towards dealing with the right to a clean environment can nevertheless be mentioned.

First, several UN bodies have proclaimed the right to a clean environment in non-binding instruments. For example, the Human Rights Sub-Commission dealt with the issue during its 1989 session,[74] when it appointed a Special Rapporteur to study the relation between the environment and human rights. The Rapporteur's final report contains a set of Draft Principles on Human Rights and the Environment, which recognise a right to a 'secure, healthy and ecologically-sound environment'.[75] In addition, the General Assembly recognised in a 1990 resolution that 'all individuals are entitled to live in an environment adequate for their health and well-being'.[76]

Furthermore, the right to a clean environment is recognised as a separate human right in two regional human rights instruments. The African Charter on Human and Peoples Rights and the Protocol on Economic, Social and Cultural Rights to the American Convention on Human Rights ('Protocol of San Salvador') both recognise the right to an environment.[77] In addition, the right to a clean environment is recognised in the constitutions of more than 60 countries, according to Kiss and Shelton.[78]

A national case in which a constitutional right to an environment was considered justiciable is the 1993 Philippine case of *Minors Oposa* v. *Secretary of the Department of Environmental and Natural Resources*.[79] The case involved an effort to have logging licenses revoked because of deforestation resulting from extensive logging. The plaintiffs, a group of 43 children who were represented by their

---

[73] The solution chosen by Galenkamp is to maintain that the rights remain vested in the individual, but to accept them as collective rights procedurally. According to Galenkamp, these rights 'are in fact rights that are materially vested in individuals, but at the same time aimed at the protection of some public interests so that collectivities should be able to act in law to protect them'. M. Galenkamp, 'Collective Rights', in *SIM Special No. 16* (Utrecht, Netherlands Institute of Human Rights 1995) pp. 53 at 91-92.

[74] It did so in response to two interventions by the Sierra Club Legal Defense Fund concerning aerial fumigation programmes carried out by Guatemala and the United States and the construction of a road in a national park in Ecuador, respectively. See also Toebes, *supra* n. 29, p. 172.

[75] Human Rights and the Environment, Final report prepared by Fatma Ksentini, Special rapporteur, UN Doc. E/CN.4/Sub.2/1994/9, 6 July 1994, Principle 2. See also Toebes, *supra* n. 29, at pp. 173-174 and Kiss and Shelton, *supra* n. 45, p. 177. See also a number of resolutions of the Human Rights Commission, *inter alia*, UN Docs. E/CN.4/RES/1996/13, more generally see the United Nations Website, <http://www.unhchr.ch>.

[76] Need to Ensure a Healthy Environment for the Well-Being of Individuals, G.A. Res. 45/94, UN GAOR, 45th Sess., UN Doc. A/RES/45/94 (1990).

[77] See Art. 24 of the African Charter on Human and Peoples' Rights and Art. 11 Protocol of San Salvador.

[78] They explain that such instruments '(...) either proclaim the principle that an environment of a specified quality constitutes a human right or impose duties upon the state'. Kiss and Shelton, *supra* n. 45, p. 175.

[79] Philippines Supreme Court Decision in *Minors Oposa* v. *Secretary of the Department of Environment and Natural Resources (DENR)*, 30 July 1993, 33 *ILM* (1994) p. 173.

parents, invoked parts 15 and 16 of the Philippine Constitution, recognising the rights to health and to a healthy environment.[80] The Supreme Court recognised that the 'right to ecology' was violated and concluded that all logging licenses were to be revoked or rescinded by executive action.[81]

International environmental law does not explicitly recognise the right to an environment. However, the Rio Declaration states that human beings are 'entitled to a healthy and productive life in harmony with nature'.[82] It also promotes a number of other rights, including the rights to information, participation and a remedy.[83]

In conclusion, since the right to an environment is not firmly embedded in the human rights system, it is doubtful whether it can currently be described as *lex lata*.[84] Given its weak status under international law, the approach that is usually taken is to seek protection of the environment through other existing human rights, including the rights to life, health, privacy and information.[85]

3.2.4   *The right to safe and healthy working conditions*

During and after an armed conflict, combatants and others may suffer harm from handling weapons or weapons platforms. Since this contact takes place within the framework of their work, it is relevant to identify international human rights standards on occupational health and safety.

A general provision can be found in Article 7 ICESCR, which recognises the right to the enjoyment of just and favourable conditions of work.[86] According to paragraph (b) of this provision, this includes safe and healthy working conditions. Since this provision has not been elaborated in a General Comment, it is difficult to pinpoint what exactly it requires.[87]

---

[80] Parts 15 and 16 of Art. II of the Declaration of Principles and State Policies of the 1987 Constitution. Part 16 reads as follows: 'The state shall protect and advance the right of the people to a balanced and healthful ecology in accord with the rhythm and harmony of nature.'

[81] *Minors Oposa* decision, *supra* n. 79, pp. 191 and 193.

[82] Rio Declaration and Programme of Action, adopted at Stockholm on 16 June 1972, Principle 1, <www.unep.org>.

[83] Ibid. See also Principle 1 of the 1972 Stockholm Declaration.

[84] Ibid., at p. 63. See also Kiss and Shelton, *supra* n. 45, pp. 145-146.

[85] See Kiss and Shelton, ibid. See also R. Picolotti and J.D. Taillant, *Linking Human Rights and the Environment* (Arizona, Arizona University Press 2003). In Chapter 1 of this study ('The Environmental Jurisprudence of International Human Rights Tribunals' at pp. 1-30) Shelton distinguishes between procedural rights (information, participation, and a remedy) and substantive rights (*inter alia*, life, private and family life) that can offer protection in relation to environmental harm.

[86] See also Art. 3 European Social Charter.

[87] The Committee's reporting Guidelines require that states indicate what legal, administrative or other provisions exist that prescribe minimum conditions of occupational health and safety. States are to provide statistical and other information on the number, nature and frequency of occupational accidents and diseases (Art. 7, question 3). See Revised Guidelines Regarding the Form and Contents of Reports to be Submitted by States Parties under Arts. 16 and 17 ICESCR, UN Doc. E/C.12/1991/1, 12 June 1991, <www.unhchr.ch>.

In addition, the International Labour Organization (ILO) has established a vast number of international standards on occupational health matters. General guiding policies are provided by the Occupational Safety and Health Convention (1981, No. 155), the Occupational Health Services Convention (1985, No. 161), and the Working Environment Convention (1977, No. 148). Protection against specific risks are, *inter alia*, provided by the Chemicals Convention (1990, No. 170), the Radiation Protection Convention (1960, No. 115), the Occupational Cancer Convention (1974, No. 139) and accompanying recommendations. Also of importance are the Labour Inspection Convention (1974, No. 81), the Workman's Compensation (Occupational Diseases) Convention (1934, No. 42) and the Workman's Compensation Accidents Convention (1925, No. 17).[88]

States Parties to these conventions are required to report regularly on their application, and the extent of compliance is subject to examination and public comment by the ILO machinery. Complaints about alleged non-compliance may be made by the governments of other ratifying states or by employers' or workers' organisations.[89]

Under IHL, labour conditions (for example, of POWs and other detainees) are to a certain extent protected but it is questionable whether a right to healthy labour conditions is included.[90]

A particular question is whether combatants are covered by the ILO Conventions. Combatancy has not yet been addressed as a special branch of economic activity under the ILO reporting procedures. Nevertheless, since combatants are not explicitly excluded by these conventions, in principle these standards should also apply to them.[91]

4.   THE USE OF DEPLETED URANIUM AS A POTENTIAL VIOLATION OF
      HUMAN RIGHTS

The human rights that were identified above as potentially relevant shall now be applied to the case of DU. A distinction is drawn between two phases: during the armed conflict, when both combatants and civilians may come into contact with DU, and after the armed conflict, when the long-term negative health effects of DU may emerge. Although DU weapons may be used during an armed conflict, the question arises at what point any potential violation of human rights occurs: during or after the armed conflict? For the sake of clarity, a distinction is drawn between obligations to respect human rights (i.e., not to cause harm by using DU) and obligations to fulfil human rights (i.e., to offer services that are needed to deal with the negative effects of DU).

---

[88] <www.ilo.org>. See also the accompanying recommendations.
[89] Ibid.
[90] E.g., GC III (Part III – Labour of Prisoners of War); Art. 39 GC IV.
[91] Art. 2 of the Occupational Safety and Health Convention (1981, 155), provides that States Parties may exclude from its application, in part or in whole, 'particular branches of economic activity'.

4.1   **During an armed conflict**

The first stage that needs to be examined is the period during which the exposure to DU takes place, i.e., the phase of the armed conflict. Persons potentially exposed during this phase include the user state's own soldiers who may suffer exposure through friendly fire or handling of DU-contaminated military hardware, and secondly, the enemy combatants and civilians who may be exposed to DU.

4.1.1   *Civil and political rights*

Regarding civil and political human rights, the non-derogability rule applies, which implies that a special set of core human rights cannot be derogated from during an armed conflict.[92] The right to life is one of the rights that can under no circumstances be derogated from.[93] The other civil and political rights can be limited in time of war or other public emergencies which 'threaten the life of the nation'.[94] For the case of DU, this implies that of the above-mentioned relevant civil and political rights, only the right to life can offer protection during armed conflicts in cases where a state has made a valid derogation from the relevant human rights treaty.

As noted above, individuals have a right not to be arbitrarily deprived of their right to life even during armed conflicts. However, during an armed conflict, the situation for combatants differs from that for civilians. Article 15(2) ECHR determines that the right to life in Article 2 ECHR allows for 'deaths resulting from lawful acts of war', meaning that the ECHR allows the killing of combatants. The ICCPR does not contain such a provision, but the use of the term 'arbitrarily' in Article 4 implies that there may be lawful violations of the right to life that are not arbitrary. In this case, the lawfulness or otherwise of the deprivation must be addressed by reference to the *lex specialis*, i.e., IHL.

As it is lawful to kill combatants under IHL, the main issue is whether the right to life of protected persons, namely, the civilians of the enemy state and persons *hors de combat*, has been lawfully deprived under the applicable law, which is both IHL and human rights. As discussed in Chapter 2, the local population is unlikely to experience short-term exposures to DU as high as soldiers on the battlefield.[95] Nevertheless, if an armed conflict turns out to be long lasting, the negative effects of DU could emerge in the local population during the armed conflict.

As discussed above, the rules of IHL enable a determination to be made whether there has been a violation of human rights. In the *Tablada* and *Las Palmeras*

---

[92] Art. 4(2) ICCPR: no derogation from Arts. 6 (life), 7, 8(1) and (2), 11, 15, 16 and 18; Art. 16(2) ECHR: no derogation from Arts. 2 (life – except in respect of deaths resulting from lawful acts of war), 3, 4(1) and 7. See, for example, Plattner, *supra* n. 39.

[93] Arts. 4 ICCPR and 15 ECHR. The other civil and political rights can be limited in times of war or other public emergencies which 'threaten the life of the nation'.

[94] Arts. 4(1) ICCPR and 16(1) ECHR. In addition, some of the provisions of the ICCPR and ECHR contain specific limitation clauses.

[95] See D. Fahey, Chapter 2 (Conclusions).

decisions, the Inter-American Commission and Court of Human Rights paved the way for interpreting the right to life in the light of IHL.[96] In the *Tablada* decision the Commission argued that its ability to resolve claimed violations of the right to life arising out of an armed conflict 'may not be possible in many cases by reference to [the right to life in] Article 4 of the American Convention alone'. The reason is, according to the Commission, that 'the American Convention contains no rules that either define or distinguish civilians from combatants and other military targets, much less specify when a civilian can be lawfully attacked or when civilian casualties are a lawful consequence of military operations'.[97]

As discussed above, a number of provisions of IHL offer protection that is comparable to a human right to life. Of particular importance are the prohibition of indiscriminate attacks and the protection of objects indispensable to the survival of the civilian population, as set forth in the APs. Also of relevance is the protection of objects indispensable to the survival of the civilian population.[98] These provisions clearly indicate the existence of a human right to life of civilians during armed conflict, albeit one that is not unlimited (for example, civilian deaths are permitted if incidental collateral damage during lawful attacks on military objectives) and that must – just as for combatants – be calibrated by reference to IHL as the *lex specialis*.

As far as a state's obligations in relation to the use of DU during an armed conflict are concerned, it is useful to distinguish between obligations to respect and obligations to fulfil human rights. Obligations to respect in this case would be the negative obligations of states to respect life, i.e., to avoid attacks on protected persons and to avoid or minimise the potentially negative effects of DU use. It can be argued that if civilians suffer harm from exposure to DU and find themselves in a life-threatening situation, they can rely on the right to life to claim that the opposing state has violated the obligation to refrain from life-threatening acts. As noted above, such negative obligations may also exist if the opposing state does not, strictly speaking, exercise total 'effective control' over the area concerned.

Regarding the obligation to fulfil, the question arises whether the opposing state has a positive obligation under human rights law to take measures to ensure the survival of the civilians. It is not certain whether this broader scope also applies during armed conflicts. As stated by Kamminga, whether a state can have such positive obligations depends on the extent to which it exercises control over the area.[99] If it has installed an occupying force in the area exercising 'effective control', it can be said to have such obligations, including the obligation to protect the lives of individuals from the negative effects of DU.

---

[96] IACHR Report No. 55/97, Case No. 11.137, Argentina, OEA/Ser/L/V/II.97, Doc. 38, 30 October 1997. See also L. Zegveld, 'The Inter-American Commission on Human Rights and international humanitarian law: A comment on the Tablada case', *International Review of the Red Cross* No. 324, (1998) pp. 505-511. See also the *Las Palmeras* case, Inter-American Court of Human Rights, Judgment of 4 February 2000.

[97] IACHR report, p. 44, para. 161.

[98] Arts. 51 AP I, 13 AP II, 54 AP I. See J. Kleffner and T. Boutruche, Chapter 6.

[99] Kamminga in *NJCM-Bull., supra* n. 15.

In Chapter 2, Fahey noted that incomplete and false information was provided about the health of veterans participating in the US Department of Defense's
DU Program by Pentagon officials seeking to downplay public concerns about DU.
Although not of direct relevance to US soldiers, it was noted above that Article 8 of
the ECHR requires that an effective and accessible procedure be established which
enables those exposed to hazardous activities to seek all relevant and appropriate
information, as confirmed by case law of the ECtHR.

Regarding the right to a remedy, this broad human right may in various
ways offer protection in the case of DU, e.g., the right to access to a court to seek
compensation for damages caused by the possible detrimental effects of DU.[100]

4.1.2   *Economic, social and cultural rights*

The ICESCR, unlike conventions recognising civil and political rights, does not
single out a set of non-derogable human rights.[101] It only contains limitation clauses
for the substantive rights in the Covenant. In the absence of a derogation clause, the
question arises whether and to what extent economic, social and cultural rights
apply during armed conflicts. Answering these questions enables us to assess to
what extent individuals have a right to protection of their health and food status
during armed conflicts. In other words, if negative effects of DU emerge during the
armed conflict, does this constitute a violation of individuals' right to protection of
their health and food status? Another question is whether soldiers are protected by
the right to healthy labour conditions.

According to Sepúlveda, the lack of a derogation clause in the ICESCR is
an appropriate omission. According to the author, 'the very nature of economic,
social and cultural rights does not allow for the kind of derogation contemplated in
civil and political rights instruments'. The author holds that 'it seems difficult to
imagine a circumstance in which a derogation from the rights contained in the
ICESCR would be necessary to maintain peace and order from the perspective of
human rights law'. Derogations from the rights to food or health could not 'assist in
resolving a conflict situation rather than worsening it'.[102]

Whether or not the absence of a derogation clause is an appropriate omission, it can be assumed that in the absence of such a clause the rights of the Covenant apply during armed conflict. The question is rather, in the present author's
view, whether the rights of the Covenant apply fully or to a limited extent.

In the ICESCR, two sets of limitations can be found, neither of which
specifically mentions war or any other type of emergency. Article 4 sets out a gen-

---

[100] For more on the right of individuals suffering ill effects as a possible result of exposure to DU
to a remedy under human rights law, see A. McDonald, Chapter 11.

[101] Contrary to the ICESCR, the ESC contains a general derogation clause which does not single
out specific rights as non-derogable but which applies to the entire Charter. In order not to complicate
the discussion, the focus in this part will be on the implications of the ICESCR.

[102] Sepúlveda, *supra* n. 10, p. 295. For a more sceptical view see M. Craven, *The International
Covenant on Economic, Social and Cultural Rights* (Oxford, Clarendon Press 1995) pp. 26-27.

eral limitations clause regarding the Covenant, and Article 2(1) sets out a clause based on limitations of available resources.[103]

Under Article 4 limitations can only be imposed if 'determined by law, compatible with the nature of rights, and solely for the purpose of protecting the general welfare in a democratic society'. Article 4 allows for limitations necessary for the purpose of 'protecting the general welfare', but it is not clear how the term 'general welfare' is to be interpreted; it is confusing and perhaps dangerous. It seems to leave some room for states to argue that certain cutbacks, due to economic recession, may be needed for the 'general welfare' in order to avoid excessive deficits in the state budget. Some authors have argued that Article 4 does not permit states to reduce the level of enjoyment of a right using the argument of resource scarcity.[104] According to Alston and Quinn, limitations for reasons of 'economic development' cannot be justified by reference to Article 4.[105] Rather, such reasons would have to be based on Article 2(1), which allows for resource-based restraints on the exercise of rights.

In situations of armed conflict, there seems to be some room for states to argue on the basis of Article 4 that they have to infringe upon people's housing, food or health facilities, not because of resource scarcity but in order to be able to pursue their military activities, which supposedly are necessary for the 'general welfare', including the restoration of peace. In fact, states could argue here that some measures are militarily necessary.[106] For example, a state could argue that it has to occupy a settlement in order to pursue its conduct of the war, or that it has to employ or test certain weapons which may be harmful to people's health.

According to the Limburg Principles, however, Article 4 was not meant 'to introduce limitations on rights affecting the subsistence or survival of the individual or integrity of the person'.[107] The terms 'survival of the individual' and 'integrity of the person' indicate that such infringements may not be of such gravity that they infringe upon people's subsistence, survival or integrity. It seems that infringements

---

[103] In addition, the ICESCR contains one rights-specific limitation clause in Art. 8 (right to form trade unions). In addition to the limitations mentioned in Art. 4 ICESCR, Art. 31(1) ESC allows for limitations for the 'protection of national security and public health or morals'). It should be noted that the revised European Social Charter (open for ratification) maintains the general derogation clause in Art. F of Part V. Furthermore, contrary to the American Convention, which includes a derogation clause in Art. 27, the AP to the American Convention on Human Rights in the Area of Economic, Social and Cultural Rights ('Protocol of San Salvador') does not contemplate the possibility of derogating from the rights set forth therein.

[104] P. Alston and G. Quinn, 'The nature and scope of States Parties' obligations under the International Covenant on Economic, Social and Cultural Rights', 9 *Human Rights Quarterly* (1987) pp. 156 at 205.

[105] Alston and Quinn, *supra* n. 104, p. 202, referring to the *travaux préparatoires*: remark made by the Chilean representative, UN Doc. E/CN.4/SR.235, p. 13.

[106] See B. Carnahan, Chapter 4 of this book.

[107] *The Limburg Principles on the Implementation of the International Covenant on Economic, Social and Cultural Rights,* UN Doc. E/CN.4/1987/17, principle 47, see also 9 *Human Rights Quarterly* (1987) pp. 122-135. The Limburg Principles are a set of non-binding principles which provide an explanation of the meaning and implications of the ICESCR.

on people's rights to health, food and housing are allowed to the extent that they leave their subsistence, survival and integrity intact. There appears to be a certain minimum level of protection inherent in economic, social and cultural rights that should remain intact under all circumstances. This would imply that the use of DU should not affect people's possibility of subsistence or survival.

This assumption is in line with the approach adopted with respect to Article 2(1), which requires states to realise the rights in the Covenant 'to the maximum of their available resources'. This leaves States Parties some room to argue that they do not have sufficient resources to realise the rights in the ICESCR.[108]

Attempts have been made to set a limit to the concept of progressive realisation by states. In its General Comment 3, the CESCR has argued that 'a minimum core obligation to ensure the satisfaction of, at the very least, minimum essential levels of each of the rights, is incumbent upon every state party'.[109] Accordingly, in spite of States Parties' resource scarcity, the Committee requires them to guarantee a certain minimum level of protection under all circumstances. This approach is usually referred to as the 'core content approach'.[110]

There is a discernible development towards a definition of inalienable and fairly concrete minimum standards in economic, social and cultural rights. In the view of the present author, there are strong reasons to assume that these minimum standards apply during armed conflicts. Underlining this assumption is the fact that IHL contains numerous references to economic, social and cultural rights.[111] The fact that IHL contains minimum standards that are comparable to the core contents of economic, social and cultural rights suggests that people have a right to minimum protection of their socio-economic position during armed conflicts.

Reference can also be made to the Declaration on Minimum Humanitarian Standards (Turko/Åbo, 1990),[112] which formulates a number of minimum stan-

---

[108] Similar escape hatches are provided by the terms 'to take steps...by all appropriate means' and 'progressive realization' in this provision. See CESCR, *General Comment No. 3, The Nature of States Parties Obligations*, UN Doc. E/1991/23 (5th session, 1990), para. 10, and the Limburg Principles, 1987, *supra* n. 107, paras. 16-25.

[109] *General Comment 3, supra* n. 108, para. 10; and *Limburg Principles, supra* n. 107, para. 25. See also CESCR, General Comment 8 on the relationship between economic sanctions and respect for economic, social and cultural rights, UN Doc. E/C.12/1997/8, para. 7 ('the state [targeted by the sanctions] and the international community should do everything possible to protect at least the core content of the economic, social and cultural rights of the affected peoples of that State'.) <www.unhchr.ch>.

[110] See, *inter alia*, A.P.M. Coomans, 'Clarifying the core elements of the right to education', in A.P.M. Coomans and G.J.H. van Hoof, eds., *The Right to Complain about Economic, Social and Cultural Rights*, SIM Special No. 18 (Utrecht, SIM 1995) pp. 11-27. For an overview of the discussion, see A.P.M. Coomans, 'Economic, social and cultural rights', SIM Special No. 16 (Utrecht, SIM 1995) pp. 17-19. For the rights to food and health, see sections 3.2.1 and 3.2.2 of this chapter.

[111] See also sections 3.2.1 and 3.2.2 *supra* on the rights to health and food.

[112] The text of the Declaration can be found in the *Report of the Sub-Commission on Prevention of Discrimination and Protection of Minorities* (report of the Secretary-General prepared pursuant to Commission Resolution 1995/29, UN Doc. E/CN.4/1996/80). This document was considered by the Sub-Commission on Prevention of Discrimination and Protection of Minorities in 1991 who transmitted it to the Commission on Human Rights in 1994. The Commission on Human Rights in turn requested the Secretary-General to prepare 'an analytical report on the issue of fundamental standards of

dards that apply in any type of situation, including one that cannot be characterised as an international or non-international conflict. The concern prompting this initiative was that 'it is often situations of internal violence that pose the greatest threat to human dignity and freedom'.[113] The declaration makes a number of implicit references to economic, social and cultural rights. Whereas Article 10 refers to the economic and social rights of children, Articles 12 and 13 refer to the right to health, by offering a right to medical services and general protection to the wounded and sick. Articles 14 and 15 implicitly refer to economic, social and cultural rights by stipulating that medical and humanitarian personnel should be in a position to carry out their work in an undisturbed manner.[114]

Soldiers and locals exposed to or suffering from the effects of DU during an armed conflict could rely on the minimum protection offered by the aforementioned relevant economic and social rights guaranteed in the ICESCR. They could rely on the right to health in Article 12, read with the general limitation clause in Article 4, to argue that the right to health can be limited only to the extent that DU should not affect their 'survival'. In addition, they could rely on Article 12, read with Article 2(1), to maintain that they have a right to the minimum health services necessary to deal with the effects of DU. This is to some extent underlined by IHL as the *lex specialis*, which guarantees the undisturbed delivery of medical services.[115]

It was noted above that the right to health has two dimensions: a right to health care facilities as well as a right to underlying conditions for health, including rights to safe drinking water and a healthy environment. Both dimensions are relevant for the present case. A right to health care facilities implies a right to adequate screening before and proper treatment after any negative health effects of the use of DU in weapons emerge. Accordingly, if the use of DU appears to be harmful, the right to health may provide a basis to argue that states have the obligation to prevent and treat such negative effects. A right to a healthy environment, as one of the determinants of health, implies a right to be free from exposure to any negative effects of DU in weapons.[116]

---

humanity', taking into consideration the issues raised in the report of the International Workshop on Minimum Humanitarian Standards held in Cape Town, South Africa in September 1996. Resolution 1997/21 entitled 'Minimum humanitarian standards', see UN Doc. E/CN.4/1998/87, 5 January 1998, UN Doc. E/CN.4/1998/87/Add.1, 12 January 1998, CHR Res. 1998/29, ESCOR Supp. (No. 3) at p. 113. See also the elaboration by a group of experts of the Declaration on Minimum Humanitarian Standards in Turku/Åbo, Finland in 1990 (see UN Doc. E/CN.4/1996/80. The concept of such a Declaration was first discussed by Meron, *supra* n. 21, and A. Eide, T. Meron and A. Rosas, 'Combating lawlessness in gray zone conflicts through minimum humanitarian standards', 89 *AJIL* (1995) p. 215.

[113] UN Doc. E/CN.4/1998/87, para. 8, *supra* n. 112.

[114] UN Doc. E/CN.4/1996/80 (Annex), *supra* n. 112.

[115] As mentioned in section 2.3, the mentioned provisions purport to ensure the undisturbed and safe delivery of medical care during armed conflict and prescribe that individuals not participating directly in hostilities and those who are wounded, sick or placed *hors de combat* be cared for without discrimination. See, *inter alia*, Arts. 12-15, 18-37 GC I; Arts. 7 and 12-40 GC II; Arts. 13, 15, 17, 20, 29-33 GC III; Arts. 14, 76, 91-92 GC IV; Arts. 12, 21-30 AP I; Arts. 5, 7, 9 and 11 AP II.

[116] The right to health provision that most explicitly refers to environmental protection is Art. 24 of the Convention on the Rights of the Child (CRC).

Regarding the right to safe and healthy working conditions for a state's own combatants during the armed conflict, one could argue that they have a right to a minimum standard of working conditions, and a right to minimum health services as part of these healthy labour conditions. The difficulty, of course, is that the profession of arms, by definition, is one that is inherently unsafe and frequently unhealthy.

If consumption of contaminated food takes place during armed conflict or leads to detrimental health effects during this phase, combatants and civilians could rely on the minimum protection offered by the right to food during armed conflict, which includes the obligation to ensure that food is free from adverse substances. As regards the right to food under IHL, it was noted that it does not embrace an obligation to ensure that food that is free from adverse substances, something that would be necessary for the case of DU.[117]

4.2   After the armed conflict: dealing with the long-term health effects

As any adverse health effects associated with exposure to DU will most often emerge after the armed conflict, the question arises whether the limitations and derogations that applied during the armed conflict remain intact. The rights implicated may have been derogated from for the duration of the conflict, when the contact with DU occurred, yet the effects of the exposure may appear at a time when the right is again fully recognised.

A two-tiered distinction is made below: first, it is argued that the minimum human rights obligations that apply during the armed conflict continue to apply after the armed conflict, and suggested that there is a 'continuing violation' of these minimum standards (section 4.2.1). When we refer to these so-called 'continuing minimum violations', we are, in fact, primarily referring to the quest for compensation for damage caused by DU during the armed conflict. Secondly, in addition to these continuing violations, after the armed conflict the full right to health and to other services necessary to address the negative effects caused by DU will be restored (section 4.2.2). In other words, on the basis of the aforementioned human rights, DU user states are under a positive 'obligation to fulfil' access to a remedy, to information, to health services, and other services needed to address the negative effects which may be caused by DU. It should be noted that this two-tiered distinction is not always clear-cut and that some confusion may arise from the question whether there exists a 'minimum violation' or a (violation of a) positive obligation to provide access to certain services.

4.2.1   Continuing violation of minimum standards that applied during the armed conflict

As far as negative health impacts sustained during an armed conflict are concerned, soldiers who have been exposed to DU as a result of the actions or omissions of

---

[117] Arts. 26 and 27 of GC III. For example, Geneva Convention III provides that prisoners of war in international armed conflicts must have access to basic daily food rations and that their habitual diet must be taken into account. Furthermore, canteens have to be installed in their camps.

their own DU-using state cannot seek compensation for damages on the basis of the right to life after the armed conflict. Nor can they rely on the other civil and political rights, for the same reasons. However, a state's own combatants might be able to rely on the fact that they are offered minimum protection by economic and social rights. During armed conflict, individuals, including friendly combatants, have a right to minimum protection of their health. The soldiers could argue that, since the use of DU may affect their chances of survival, their own DU-using state has a responsibility to compensate for the negative health impacts of DU caused during the armed conflict. As noted, however, their case may be (perhaps fatally) undermined by the fact that the profession of arms is inherently unsafe and often unhealthy. Indeed, it could be argued that when an individual joins the military he or she accepts limitations on his or her right to life and health, *inter alia*.

Where the DU-using state is an enemy state, the question needs to be tackled whether this state exercised 'effective control' during the armed conflict.

Secondly, the inhabitants of an area that has been exposed to DU and who are dealing with the long-term negative (health) effects of this exposure have a (continuing) right to compensation for damage on the basis of the minimum protection offered by the right to life, as applied during the armed conflict. Furthermore, like the soldiers, they can rely on the minimum protection offered by the rights to health and food in order to seek compensation for damage to their health and their environment. The inhabitants can address the state where the impact takes place as well as the state that used DU. Concerning the latter, again the key question is whether this state exercised 'effective control' during the armed conflict.

4.2.2 *Full right to health and other services after the armed conflict*

After the armed conflict, the soldiers of a DU-user state who are exposed and who suffer negative health effects that may be associated with DU have a full right to information about their exposure, a right to a remedy to address any damage, and a right to health services, including medical treatment and screening for the treatment of any (possible) negative health effects. Enemy combatants will need to establish that the user state exercised 'effective control' over the area where DU was used.

Regarding the civilian inhabitants of an impacted area, it can be argued that after the armed conflict the state where the impact takes place as well as the state that used DU bear responsibility for any negative effects caused by DU to civilians living in the area. Relevant human rights in this respect could be the rights to life, family life, information, a remedy and health.

On the basis of the right to information, the using state has the responsibility to indicate where it has dropped DU weapons, and to warn of the possible dangers to civilians.

Concerning the right to health, combatants as well as civilians have a right to appropriate screening for the possible negative effects of DU and a right to appropriate medical treatment once any (potential) negative health effects emerge. The above-mentioned General Comment on the Right to Health in Article 12 ICESCR mentions the following requirements for the adequate provisions of health services: they should be available, accessible, acceptable, and of good quality.[118] The accessibility of health care services requires equal accessibility (non-discrimination), physical accessibility and economic accessibility.[119] If DU has been deployed in far-off regions, the principle of physical accessibility requires that the needed health services are also available there. Furthermore, the principle of economic accessibility or affordability requires that if the health services are not affordable for those in need of them, financial arrangements should be made to enable them to pay for them.

Research has suggested that internal exposure to DU might potentially result in reproductive and developmental effects. Animal studies indicate that maternal exposure to uranium compounds has resulted in foetal effects, including reduced body weight and length.[120] If it can be proven that DU has such effects, Article 12 of CEDAW, guaranteeing the reproductive health of women, could become relevant. Children born with birth defects linked to DU could potentially invoke Article 24 of the Convention on the Rights of the Child (CRC), as could children who have been directly exposed, for example, by inadvertently playing with DU remnants. The obligation of states to take measures to provide clean drinking water and to take into consideration the risks of environmental pollution is explicitly recognised in Article 24(2)(c) CRC.

---

[118] General Comment 14, *supra* n. 4, para. 12.

[119] Ibid.

[120] See D. Fahey in Chapter 2 of this book.

Part Three
RESPONSIBILITY AND REMEDIES CONCERNING
THE USE OF DEPLETED URANIUM WEAPONS

Chapter 10
STATE RESPONSIBILITY FOR THE USE OF
DEPLETED URANIUM WEAPONS IN ARMED CONFLICT

Tobias Gries and Manfred Mohr

1.   INTRODUCTION

The previous chapters analysed the legality of the use of depleted uranium (DU) weapons. Once there is a breach of international law through a state's use of DU weapons, the injured state and its citizens may raise the issue of the responsibility of the injuring state.

The principle of state responsibility has been on the agenda of the International Law Commission (ILC) for over 46 years.[1] The commission finalised its work in 2001 by adopting the 'Draft Articles on Responsibility of States for Internationally Wrongful Acts'[2] (hereinafter, ILC Draft Articles). These articles reflect to a large extent recognised rules of customary international law. They will therefore serve as a basis for the present investigation of state responsibility with respect to the use of DU weapons. In addition, the authors will refer to the practice of states and the case law of international and national judicial bodies to determine the recognised rules of state responsibility.

The issue of international responsibility and its consequences arises in inter-state and state-individual relations. However, the principle of state responsibility applies only between states.[3] Any claim of an individual against a state other than the state of his or her nationality is not an issue of state responsibility as under-

---

[1] Following the request of the UN General Assembly, GA Resolution 799 (VIII) of 7 December 1953, the ILC decided at its 7th session in 1955 to investigate the issue of state responsibility.

[2] UN International Law Commission, *Report on the Work of its Fifty-third Session (23 April – 1 June and 2 July – 10 August 2001)*, General Assembly, Official Records, 55th Session, Supplement No. 10 (A/56/10), <http://www.un.org/law/ilc/reports/2001/2001report.htm>, pp. 29-365 (hereinafter, ILC Report 2001).

[3] M. Sassòli, 'State responsibility for violations of international humanitarian law', 84 *IRRC* (2002) pp. 401-434, 402; see also Art. 33 ILC Draft Articles stating: '1. The obligations of the responsible State set out in this Part may be owed to another State, to several States, or to the international community as a whole, depending in particular on the character and content of the international obligation and on the circumstances of the breach. 2. This Part is without prejudice to any right, arising from the international responsibility of a State, which may accrue directly to any person or entity other than a State.'

McDonald/Kleffner/Toebes (eds.), Depleted Uranium Weapons and International Law
© *2008, T·M·C·ASSER PRESS, The Hague, The Netherlands and the Authors*

stood by public international law. The availability of remedies for individuals with regard to DU is discussed elsewhere in this book.[4]

State responsibility is furthermore independent of any responsibility or liability established under the national law of the injuring or injured state.[5] The success of claims of individuals for compensation brought under national law against their own or foreign states based on their status as a victim of a violation of international humanitarian law is not contingent on a finding of state responsibility. Nevertheless, the establishment of state responsibility and with it the proven facts may support an individual claim under national law, if the international obligation which has been violated by the respective state corresponds to the claim of the individual under national law.

State responsibility is also to be distinguished from the criminal responsibility of individuals under international law. As a matter of course, there are cases where the act of an individual (committed on behalf of a state) gives rise to both state responsibility and individual criminal responsibility. In these cases, the act committed by the individual is attributable as an act of state (state responsibility) and constitutes at the same time a crime under international law.[6] However, state responsibility does not presuppose individual criminal responsibility.[7]

State responsibility establishes a legal relationship between the injuring and the injured state. This (secondary) legal relationship exists beside the primary obligations deriving from the international law recognised by the respective states.[8] The objective of the law on state responsibility is to react to a breach of a primary obligation. It aims towards the restoration of the lawful situation, including the removal of the consequences of the unlawful conduct as well as the prevention of a repetition of the violation. The injuring state continues to be bound by the primary

---

[4] See A. McDonald, Chapter 11 of this book.

[5] See for the parallel claims in the *Vavarin* case, German Supreme Court (Bundesgerichtshof), III ZR 190/05, judgment of 2 November 2006, in which the claimant refers to international as well as national claims.

[6] International crimes are listed in Art. 50 Geneva Convention for the Amelioration of the Condition of the Wounded and Sick in Armed Forces in the Field of 12 August 1949 (hereinafter, Geneva Convention I); Art. 51 Geneva Convention for the Amelioration of the Condition of the Wounded, Sick and Shipwrecked Members of Armed Forces at Sea of 12 August 1949 (hereinafter, Geneva Convention II); Art. 130 Geneva Convention relative to the Treatment of Prisoners of War of 12 August 1949 (hereinafter Geneva Convention III); Art. 147 Geneva Convention relative to the Protection of Civilian Persons in Time of War of 12 August 1949 (hereinafter, Geneva Convention IV); Art. 85 Protocol Additional to the Geneva Conventions of 12 August 1949, and Relating to the Protection of Victims of International Armed Conflicts of 8 June 1977 (hereinafter, Protocol I), all available at <http://www.icrc.org>; Art. 5(1)(c) Rome Statute of the International Criminal Court, 17 July 1998, UN Doc A/CONF.183/9 (hereinafter, ICC Statute); H.-H. Jescheck, 'War crimes', in R. Bernhardt, ed., 4 *Encyclopedia of Public International Law* (hereinafter, *EPIL*), (Amsterdam, et al., North Holland Publishing Company 2000) pp. 1349-1354.

[7] According to Art. 25(4) ICC Statute, criminal responsibility of an individual does not exclude the responsibility of the state to which the act of the individual may be attributable.

[8] R. Ago, 'Third report on state responsibility', II *Yb ILC* (Part One) (1971), Doc. A/CN.4/246 and Add.1-3, pp. 199, 206.

obligation.[9] The establishment of state responsibility does not extinguish the state's obligation to fulfil its primary obligation and with it the duty to end any wrong.[10]

1.1   State responsibility is not equivalent to strict liability

State responsibility, as a rule, is not strict liability. Strict liability establishes an obligation to pay compensation without fault or even without breaching an international obligation.[11] This liability has been recognised only for certain activities on the basis of international treaties and for ultra-hazardous activities causing transboundary harm on the basis of customary international law.[12] The ILC Draft Articles on Hazardous Activities[13] are – contrary to their original conception – irrelevant for the present investigation. The whole set of articles concentrates on prevention but not on liability or responsibility, although insufficient prevention might establish the responsibility of a state.[14]

Apart from the Draft Articles on Hazardous Activities, the general principles of law applicable to hazardous activities have never been applied to warlike activities. Even the ICJ in its Advisory Opinion on the *Legality of the Threat or Use of Nuclear Weapons* found that the obligation to prevent transboundary harm caused by the use of weapons in times of armed conflict is not 'intended to deprive a State of the exercise of its right of self-defense under international law because of its obligations to protect the environment'.[15] Consequently, any responsibility may only derive from a violation of rules of humanitarian law protecting the natural environment.[16] But humanitarian law does not provide for any strict liability, established independently from a violation of that body of law.

1.2   International humanitarian law as the *lex specialis*

The legality of military activities of states during armed conflicts is determined to a large extent by the rules of humanitarian law. However, the beginning of an (in-

---

[9] Art. 29 ILC Draft Articles states: 'The legal consequences of an internationally wrongful act under this Part do not affect the continued duty of the responsible State to perform the obligation breached.'

[10] Art. 29 ILC Draft Articles; K. Ipsen, 'Grundzüge der völkerrechtlichen verantwortlichkeit', in K. Ipsen, ed., *Völkerrecht*, 5th edn. (Munich, C.H. Beck 2004) p. 620.

[11] M. Bedjaoui, 'Responsibility of states: Fault and strict liability', *EPIL, supra* n. 6, pp. 212-216, 212.

[12] *Legality of the Threat or Use of Nuclear Weapons*, ICJ Advisory Opinion of 8 July 1996, *ICJ Rep.* (1996) pp. 226-593, 242, para. 29.

[13] Text of the Draft Articles on Prevention of Transboundary Harm from Hazardous Activities of 3 August 2001, *ILC Report 2001, supra* n. 2, pp. 370-377 (hereinafter, ILC Draft Articles on Hazardous Activities).

[14] General Comment, ILC Draft Articles on Hazardous Activities, *supra* n. 13, p. 377.

[15] *Supra* n. 12, p. 242, para. 30.

[16] For these rules see E. Koppe, Chapter 8 of this book. Guidelines for Military Manuals and Instructions on the Protection of the Environment in Times of Armed Conflict, 36 *IRRC* (1996) pp. 230-237.

ternational) armed conflict and with it the activation of humanitarian law does not automatically suspend or substitute rules of international law applicable in times of peace, including the general rules of international law.[17] The principle of state responsibility is recognised as such a general rule of public international law.[18] It applies therefore to conduct in or in connection with an armed conflict. Nevertheless, the general principle may be specified by the rules agreed upon by the involved states,[19] including those deriving from humanitarian law, in accordance with the maxim *lex specialis derogat lex generalis*.

In the field of international humanitarian law, two, almost identical, norms provide such more specific rules. As contained in Article 3 of the Fourth Hague Convention of 1907[20] and Article 91 of AP I,[21] they provide that parties to an armed conflict are 1) responsible for all acts committed by persons forming part of their armed forces, and 2) liable to pay compensation for violations of international humanitarian law.[22] In contrast to these two instruments, the four Geneva Conventions do not contain specific rules of state responsibility. Nevertheless, Articles 51 Geneva Convention I, 52 Geneva Convention II, 131 Geneva Convention III and 148 Geneva Convention IV recognise the obligation to pay compensation for severe breaches of humanitarian law. This *lex specialis* notwithstanding, the general rules on state responsibility are not necessarily excluded. In the words of the ILC:

> 'For the *lex specialis* principle to apply it is not enough that the same subject matter is dealt with by two provisions; there must be some actual inconsistency between them, or else a discernible intention that one provision is to exclude the other.'[23]

In the light of this argument of the ILC, Sassòli has raised the question[24] whether humanitarian law establishes a self-contained system as defined by the ICJ,[25]

---

[17] *ILC Report 2001, supra* n. 2, p. 62: 'On the other hand the present articles are concerned with the whole field of State responsibility. Thus they are not limited to breaches of obligations of a bilateral character, e.g. under a bilateral treaty with another State. They apply to the whole field of the international obligations of States, whether the obligation is owed to one or several States, to an individual or group, or to the international community as a whole.'

[18] Ago named the principle of state responsibility as 'one of the principles most deeply rooted in the doctrine of international law and most strongly upheld by State practice and judicial decisions'. *Supra* n. 8, p. 205.

[19] *ILC Report 2001, supra* n. 2, p. 62.

[20] Art. 3 Hague IV provides: 'A belligerent party which violates the provisions of the said Regulations shall, if the case demands, be liable to pay compensation. It shall be responsible for all acts committed by persons forming part of its armed forces.'

[21] Art. 91 AP I provides: 'A Party to the conflict which violates the provisions of the Conventions or of this Protocol shall, if the case demands, be liable to pay compensation. It shall be responsible for all acts committed by persons forming part of its armed forces.'

[22] J. de Preux, 'Article 91 – Responsibility', in C. Pilloud, J. de Preux, Y. Sandoz and B. Zimmermann, eds., *Commentary on the Additional Protocols of 8 June 1977 to the Geneva Conventions of 12 August 1949* (Geneva, ICRC 1987) (hereinafter, *ICRC Commentary Protocol I*) p. 1053, para. 3645.

[23] Commentary Art. 55 ILC Draft Articles, *ILC Report 2001, supra* n. 2, p. 358.

[24] Sassòli, *supra* n. 3, p. 404.

[25] Case concerning United States Diplomatic and Consular Staff in Tehran (*United States of America v. Iran*), ICJ Judgment, *ICJ Rep.* (1980) pp. 3-65, 40, para. 86.

referred to by the ILC as a system *lex specialis* to the Draft Articles.[26] After analysing particular elements of state responsibility, Sassòli concludes that there is no exclusion of the general rules of state responsibility by humanitarian law.[27] Humanitarian law is not an autonomous system uninfluenced by the general rules of state responsibility. The relevant rules of international humanitarian law are therefore *lex specialis* to the general principles and rules of state responsibility. Consequently, the general rules of state responsibility continue to apply as long as and to the extent[28] that there is no particular rule of humanitarian law substituting the general rules.[29] However, as far as state responsibility for the use of DU is concerned, other rules of a *lex specialis* nature may flow from environmental law and special international treaties regulating the status of forces in peacetime in foreign states.[30]

2. THE ELEMENTS OF STATE RESPONSIBILITY

It is generally recognised under customary international law that '[e]very internationally wrongful act of a State entails the international responsibility of that State'.[31] The same elements apply to rules of state responsibility established by Article 3 Hague Convention IV and Article 91 AP I. According to the general definition, state responsibility consists of three constitutive elements: (1) the act – action or omis-

---

[26] *ILC Report 2001, supra* n. 2, p. 358 and *Yb ILC* (1975), Vol. II, p. 69 with regard to the question of accountability.

[27] Sassòli, *supra* n. 3, p. 433.

[28] See Art. 55 ILC Draft Articles: '... where and to the extent...'; Commentary Art. 55 ILC Draft Articles, *supra* n. 2, p. 358: 'For the *lex specialis* principle to apply it is not enough that the same subject matter is dealt with by two provisions; there must be some actual inconsistency between them, or else a discernible intention that one provision is to exclude the other.'

[29] *ICRC Commentary Protocol I, supra* n. 22, p. 1053, paras. 3645-3646; W. Heintschel von Heinegg, 'Entschädigung für verletzungen des humanitären völkerrechts', in W. Heintschel von Heinegg, S. Kadelbach, B. Heß, M. Hilf, W. Benedek and W.-H. Roth, eds., *Entschädigung nach bewaffneten Konflikten. Die Konstitutionalisierung, der Welthandelsordnung, Berichte der Deutschen Gesellschaft für Völkerrecht*, Band 40 (Heidelberg, , C.F. Müller 2003) pp. 1-61, 35; F. Kalshoven, 'State responsibility for warlike acts of the armed forces', 40 *ICLQ* (1991) pp. 827-858, 838. Steinkamm characterises the rules of attribution under Arts. 91 and 3 as nothing more than the reflection of the general rule of state responsibility applicable in times of war and in times of peace. See A.A. Steinkamm, 'War damages', *EPIL, supra* n. 6, pp. 1354-1360, 1354.

[30] NATO Member States have adopted Status of Force Agreements (SOFA, Agreement between the Parties to the North Atlantic Treaty regarding the Status of their Forces (SOFA), 19 June 1951, 199 *UNTS* (1954) p. 67), regulating the privileges and liabilities of the members of the forces stationed abroad on the territory of another state. The NATO-SOFA, for instance, contains in Art. VIII(a) comprehensive exclusion of claims for damages arising out of any act conducted under the scope of the Alliance. Therefore, any claim for compensation resulting from a damage caused by the employment of DU weapons by forces of a NATO Member State during military exercises on the territory of another NATO Member State would be excluded by Art. VIII(a) NATO-SOFA.

[31] Art. 1 ILC Draft Articles. The Permanent Court of International Justice already stated in the *Chorzow Factory* case (Indemnity) (Merits) (PCIJ Ser. A., No. 17 (1928), p. 29) that '[it] is a principle of international law, and even a general conception of law, that any breach of an engagement involves an obligation to make reparations.'

sion – committed, (2) which is attributable to a state under international law, and (3) constitutes a breach of an international obligation of that state.[32]

It is widely accepted today that an injury or damage – apart from the violation of a right of the injured state as such – is not necessary to establish state responsibility.[33] The notion of 'injured state' in Article 42 ILC Draft Articles, defining the state entitled to invoke the responsibility of another state, refers to the state whose rights have been violated, without demanding any further injury or damage.[34] It should be clarified at this point that the exclusion of the criteria of 'injury' applies only to the establishment of state responsibility. The determination of the kind and amount of compensation as a legal consequence of that responsibility, on the other hand, depends on the injury.[35]

2.1   The act committed

The determination of state responsibility depends on a particular and clearly identified conduct attributable to a state. Depending on the nature of the conduct and the circumstances of the case, the relevant rules of international law determining the attribution and lawfulness of the conduct may differ. It is therefore not sufficient to claim a violation of international law because of the use of DU weapons in a general or abstract sense. The responsibility of states whose armed forces use DU weapons depends on their specific international obligations and the particular use of the weapons (in specific circumstances), at least as long as there is no general prohibition of the use of DU weapons.[36]

There are four different kinds of conduct involving the employment of DU weapons which may lead to the responsibility of a state if the particular conduct falls under a prohibition established by humanitarian law: (1) the direct use of DU in combat, (2) the assistance by allied forces of states using DU weapons, (3) omitting to conduct sufficient investigation before the employment of certain DU weapons as well as the omission of sufficient precaution before and during the deployment of DU weapons, and finally (4) omitting to clean-up known contaminated areas.

---

[32] See Art. 2 ILC Draft Articles, *supra* n. 2.

[33] Case concerning the Difference between New Zealand and France concerning the Interpretation or Application of Two Agreements, concluded on 9 July 1986 between the Two States and which relate to the Problems Arising from the Rainbow Warrior Affair, 30 April 1990, XX *RIAA* (1994) pp. 217, 267, paras. 106-110; Commentary Art. 2 ILC Draft Articles, *supra* n. 2, p. 73: 'It is sometimes said that international responsibility is not engaged by conduct of a State in disregard of its obligations unless some further element exists, in particular, "damage" to another State. But whether such elements are required depends on the content of the primary obligation, and there is no general rule in this respect.'

[34] Commentary Art. 42 ILC Draft Articles, *ILC Report 2001*, *supra* n. 2, p. 297.

[35] See below for the question of legal consequences arising out of responsibility, section 3 of this chapter.

[36] See G. den Dekker, Chapter 3, and M. Zwanenburg, Chapter 5 of this book.

2.1.1 Direct application in combat

The use of DU in combat includes its application as ammunition, as well as armour improving the defence capability of military objects.[37] The issue of responsibility may affect all states that contribute to the use of DU. This includes the contributing country of the person literally pushing the trigger of the weapon firing DU ammunition but also those countries contributing the military personnel ordering or controlling the combat mission. As it will be shown later, and from the perspective of the general rules of state responsibility, the decisive issue is that of attribution of the acts performed. Responsibility for the direct use of DU depends on a state's involvement in the chain of command, uninterrupted by autonomous operative decisions. We will discuss this in section 2.2 below.

The same is true regarding the use of DU as armour. Since armour is designed to protect against lethal impact by munitions, it is clear that those who reinforce military objects, such as tanks, using such armour employ DU as a means of warfare. There are no reasons to treat such usage of DU any differently from its use in munitions. In both cases, the release of contaminated dust is caused by the clash of the DU-modified armour or munitions with other objects. Consequently, those states employing DU armour cannot justify their action by blaming the states that have used conventional munitions to strike it for any release of DU-contaminated dust.

2.1.2 Assistance in combat

Besides the possibility of the direct responsibility of a state using DU weapons in combat for any resulting ill effects, the question arises whether allied forces which assist in combat wherein DU weapons are deployed are responsible for their contribution. This is an aspect of the larger question of the responsibility of coalition partners for the actions of other members of a military coalition. One can think of, for example, those allied forces that provided the necessary support for the military operations executed by US aircraft and US and UK tanks against Iraq in the second Gulf War ('Operation Iraqi Freedom') in 2003[38] or against the Federal Republic of Yugoslavia in 1999.[39] Their supporting acts included target identification, flight control, logistics supply in the air as well as on the ground, direct assistance during combat by their elimination of secondary targets, and assistance with defence against potential enemy attacks. The question arises whether such assistance runs contrary to the international obligations of these states and whether these states are interna-

---

[37] See D. Fahey, Chapter 1.

[38] Ibid., section 4.4.

[39] 'Allied pilots flew 37,465 sorties, of which over 14,006 were strike missions. By comparison with previous campaigns, support sorties outnumbered strike sorties. This campaign, facing unpredictable reactions from Yugoslav defences, required protective combat air patrols in multiple locations, on some days up to seven, around the area.' W.K. Clark, 'When force is necessary: NATO's military response to the Kosovo crisis', in 47 *NATO Review*, Web edition, pp. 14-18, <http://www.nato.int/docu/review/1999/9902-03.htm>.

tionally responsible for their contribution.[40] A further question is whether all members of a coalition are (equally) responsible for the actions of each other member, even where they have not participated in or supported such actions. The answers to these questions, discussed below in sections 2.2.2 and 2.2.3, depend on whether the conduct of an organ of one state or an international organisation can be attributed to another state or any individual member of the organisation.

2.1.3   Omission of sufficient investigation or precaution

Certain omissions might also give rise to state responsibility.[41] In principle, there is no difference between the legal responsibility for an act or an omission.[42] In particular cases it might be difficult to draw a clear dividing line between the two. But not every omission is equal to an act. In the *Diplomatic and Consular Staff* case, the International Court of Justice (ICJ) found Iran responsible for its inaction because it failed to take appropriate steps, in circumstances where such steps were evidently called for.[43] In the *Corfu Channel* case, the ICJ held that a sufficient basis for Albania's responsibility was the fact that it knew, or must have known, of the presence of the mines in its territorial waters and did not warn third states of their presence.[44] In the context of armed conflict, a state is responsible for the acts of the soldiers in the field as well as for the actions or inactions of commanders directing and controlling (or failing to direct and control) these soldiers.

International law also imposes on states a direct obligation to act in certain cases. One such rule is Article 36 Additional Protocol I (AP I). As discussed in Chapters 4 and 12 of this book, it requires states to investigate the lawfulness of (new)[45] weapons, means or methods of warfare. Consequently, not only the development, acquisition or use of these weapons but also the omission of sufficient investigation into the effects arising from their use has the potential to entail state responsibility.

A further example is Article 57 AP I, which clearly sets out the need to execute an attack with the necessary precaution, preventing the civilian population from any direct or indirect effects. The obligation to take precautions applies before and during the attack. It begins before the commencement of the attack, with the need for comprehensive investigation of the targets, their legitimacy as objects of

---

[40] The relevant criteria are defined by Art. 16 ILC Draft Articles, *supra* n. 2 and discussed in detail under section 2.3.2 of this Chapter.

[41] Art. 2 ILC Draft Articles, *supra* n. 2; ICJ, 19 December 2005, General List No. 116, Case Concerning Armed Activities on the Territory of the Congo (*Democratic Republic of the Congo* v. *Uganda*), paras. 180 and 245; Eritrea Ethiopia Claims Commission, Partial Award, Civilians Claims, Eritrea's Claims 15, 16, 23 and 27-32, between the State of Eritrea and the Federal Democratic Republic of Ethiopia, The Hague, 17 December 2004, para. 89.

[42] Commentary Art. 2 ILC Draft Articles, *ILC Report 2001*, *supra* n. 2, p. 70.

[43] ICJ, *Diplomatic and Consular Staff*, *supra* n. 25, pp. 31-32, paras. 63, 67.

[44] ICJ, 19 April 1949, *Corfu Channel* case (*United Kingdom of Great Britain and Northern Ireland* v. *Albania*), Merits, *ICJ Reports 1949*, p. 4, pp. 22-23.

[45] *ICRC Commentary Protocol I*, *supra* n. 22, p. 427, para. 1475.

attack, and consideration of the appropriateness of the weapons used to strike them.[46] The necessary preparations also include determining whether there are alternative targets whose destruction would cause less collateral damage while providing the same military advantage.[47] Where danger for the civilian population or other protected persons or objects (e.g., hospitals) is anticipated, a party to an armed conflict is obliged to give advance warning.[48]

2.2 Attribution of the act committed

A state is only responsible for acts attributable to it, without distinction between actions or omissions. Due to its nature as an abstract entity and subject of public international law, a state is unable to act without persons executing the state authority. It is also evident that not every act of an individual person constitutes an act of state. Attribution may follow different rules, as reflected in Articles 4-11 ILC Draft Articles, depending on the nature of the (human) actor and his or her relation with the state. The present investigation will concentrate on attribution of the conduct of state organs, especially a state's military forces;[49] the conduct of persons placed at the disposal of a state by another state;[50] and conduct directed or controlled by a state.[51] The subject of attribution can be an individual state, a group of states, or an international organisation, such as NATO. However, any responsibility of the latter would fall outside the scope of state responsibility,[52] notwithstanding the responsibility of individual NATO Member States.

2.2.1 Attribution of acts of military forces as acts of state

As a general principle of state responsibility, recognised as customary international law[53] and codified in Article 4 ILC Draft Articles, the

> 'conduct of any State organ shall be considered an act of that State under international law, whether the organ exercises legislative, executive, judicial or any other functions, whatever position it holds in the organization of the State, and whatever its character as an organ of the central government or of a territorial unit of the State.'

---

[46] Art. 57(2)(b) Protocol I.

[47] Art. 57(3) Protocol I.

[48] Art. 57(2)(c) Protocol I regarding civilians or civilian objects; Art. 19(1) Geneva Convention IV regarding hospitals; Art. 26 Hague Regulations annexed to Hague Convention IV.

[49] Art. 4 and 7 ILC Draft Articles; Art. 3 sentence 2 Hague Convention IV; Art. 91 sentence 2 Protocol I.

[50] Art. 6 ILC Draft Articles, *supra* n. 2.

[51] Art. 8, ibid.

[52] Art. 57 ILC Draft Articles, *supra* n. 2; For the problem of state responsibility of Member States of international organisations for acts committed by organs of the organisation see Commentary Art. 57 ILC Draft Articles, *ILC Report 2001*, *supra* n. 2, p. 362.

[53] K. Zemanek, 'Responsibility of states: General principles', *EPIL*, *supra* n. 6, pp. 219-229, 223.

The identification of certain conduct as an act of state is hardly open to question when it comes to the conduct of a state's own military forces.[54] They are state organs exercising state authority. This is affirmed in the second sentence of Article 91 AP I, which provides that a state 'shall be responsible for all acts committed by persons forming part of its armed forces'.[55] The term 'responsible' in this sentence refers to the question of attribution.[56] It covers all acts of members of the armed forces who fall within their competence and instructions as well as acts which exceed their competence or contravene their instructions.[57] This interpretation is supported by Article 3 sentence 2 Hague Convention IV and Article 7 ILC Draft Articles, reflecting the general rule of international law[58] that:

> '[t]he conduct of an organ of a State or of a person or entity empowered to exercise elements of the governmental authority shall be considered an act of the State under international law if the organ, person or entity acts in that capacity, even if it exceeds its authority or contravenes instructions.'

Consequently, a state has comprehensive responsibility for members of its armed forces. Attribution relies exclusively on an objective criterion, which is the incorporation of individuals into the armed forces of a state.[59] Even acts *ultra vires* are attributable to the state under international customary law.[60]

2.2.2   *Attribution of acts of joint or incorporated forces*

Under certain circumstances a state is responsible for the conduct of organs of another state. This arises in cases of joint activities of organs of different states under a joint command and control or in cases of conduct of foreign organs placed at the disposal of that state.[61] The attribution of the conduct of organs engaged in joint activities is based on the principle of effectiveness[62] of the state's command and

---

[54] Kalshoven, *supra* n. 29, p. 827.

[55] Similar thereto, Art. 3 sentence 2 Hague Convention IV states that a state 'shall be responsible for all acts committed by persons forming part of its armed forces'.

[56] *ICRC Commentary Protocol I*, *supra* n. 22, p. 1057.

[57] Ibid., p. 1057; Kalshoven, *supra* n. 29, p. 853; Heintschel von Heinegg, *supra* n. 29, p. 40; Commentary Art. 7 ILC Draft Articles, *supra* n. 2, p. 101; *Caire Claim*, French-Mexican Claims Commission, 5 *RIAA* (1929) p. 516, 531; Steinkamm, *supra* n. 29, p. 1360.

[58] Zemanek, *supra* n. 53, p. 224.

[59] Heintschel von Heinegg, *supra* n. 29, p. 28.

[60] Kalshoven, *supra* n. 29, pp. 837 and 853; ICJ, 19 December 2005, General List No. 116, Case Concerning Armed Activities on the Territory of the Congo (*Democratic Republic of the Congo* v. *Uganda*), paras. 213-214; Commentary Art. 7 ILC Draft Articles, *supra* n. 2, p. 99-101 with further references to state practice and judicial decisions.

[61] See Arts. 6 and 8 ILC Draft Articles, *supra* n. 2; Case Concerning Military and Paramilitary Activities in and Against Nicaragua (*Nicaragua* v. *United States of America*), Merits, ICJ Judgment of 27 June 1986, *ICJ Rep.* (1986) pp. 14-546, 64-65, paras. 115-116.

[62] For the principle of effectiveness under public international law see V. Epping, 'Der Staat als die "normalperson" des völkerrechts', in Ipsen, *supra* n. 10, p. 63.

control over the particular force.[63] To attribute the conduct to each of the participating states, the command and control of the force must be exercised by a joint command authority. Equality between the different national command authorities is not necessary as long as every single authority has effective influence over the joint conduct of the armed forces.[64] It is even possible for an authority to delegate its control power to another state authority of the alliance as long as it retains the power to intervene in the decisions and to restore its command and control.[65]

The situation is different in cases where a state places its armed forces at the disposal of another state. Article 6 ILC Draft Articles provides:

> 'The conduct of an organ placed at the disposal of a State by another State shall be considered an act of the former State under international law if the organ is acting in the exercise of elements of the governmental authority of the State at whose disposal it is placed.'

The placement of an organ of one state at the disposal of another state requires the full transmission of effective (full) authority over the organ. The state at whose disposal the organ has been placed remains solely responsible until the moment when the authority over the organ is transferred back from that state. Article 3 Hague Convention IV and Article 91 AP I make no reference to the problem of foreign armed forces under the authority of another state. However, the ICJ in the *Nicaragua case* confirmed the principle that a state can be held responsible for certain conduct committed by forces which cannot be regarded as 'its armed forces' in the sense of Article 91 AP I or as its 'state organ' in the sense of Article 4 ILC Draft Articles if the 'state had effective control of the military or paramilitary operations in the course of which the alleged violations were committed.'[66] However, the criteria for the determination of 'effective control' remain vague.[67] The ICJ provides certain criteria in its Opinion but without clarifying whether they may be regarded as part of a general principle or just appropriate criteria for the particular

---

[63] S.R. Lüder, 'Die völkerrechtliche verantwortlichkeit der Nordatlantikvertrags-Organisation bei militärischen Absicherung der Friedensvereinbarung von Dayton', 43 *Neue Zeitschrift für Wehrrecht* (2001) pp. 107-117, 112.

[64] As stated by the German government in the *Varvarin* case before the German Landgericht Bonn, Germany rejected any responsibility for the air strikes against the Bridge in Varvarin since it was not involved in the combat and had no knowledge about the air strike. The German government also stated that the general principle within the joint command structure of NATO is that only those states that are informed and involved in the preparation of a particular mission will take part in that mission. Landgericht Bonn, Case No. 1 O 361/02, Judgment of 10 December 2003, <http://www.justiz.nrw.de>. See also the decisions of the higher courts, *supra* n. 5.

[65] Commentary Art. 6 ILC Draft Articles, *supra* n. 2, p. 95.

[66] Case Concerning Military and Paramilitary Activities in and Against Nicaragua, *supra* n. 61, p. 65, para. 115.

[67] In addition, a clear distinction between the situations which fall under Arts. 6 and 8 ILC Draft Articles appears to be irrelevant or merely dependent on the fact that the foreign entity has the quality of a state organ of a foreign state.

case investigated by the Court.[68] The criterion of 'effective control' at least corresponds to the wording of Article 6 ILC Draft Articles: 'acting in the exercise of elements of the governmental authority of the State *at whose disposal* it is placed' [emphasis added],[69] or the wording of Article 8 ILC Draft Articles 'if the person or group of persons is in fact acting on the *instructions* of, or under the *direction or control* of that State' [emphasis added]. The decisive difference between Articles 6 and 8 is the fact that, in the case of Article 6, the sending state of an organ consents that this organ shall be controlled exclusively by another state, whereas under Article 8 – which applies not only to foreign state organs, but to every person or group of persons – only factual direction or control is required.

Further indications of the necessary extent of 'overall control' can be drawn from the jurisprudence of the European Court of Human Rights (ECtHR) and the International Criminal Tribunal for the Former Yugoslavia (ICTY). The ECtHR stated in *Ilaşcu and others* v. *Moldova and Russia*:

> 'It is not necessary to determine whether a Contracting Party actually exercises detailed control over the policies and actions of the authorities in the area situated outside its national territory, since even overall control of the area may engage the responsibility of the Contracting Party concerned [...].'[70]

However, participation in a multinational military operation does not of itself suffice to establish the responsibility of a state for the conduct of foreign state organs (foreign armed forces).[71] The quintessence of both decisions is that responsibility may be established along the chain of command in military operations.

---

[68] Case Concerning Military and Paramilitary Activities in and Against Nicaragua, *supra* n. 61, p. 64, para. 115: 'The Court has taken the view (...) that United States participation, even if preponderant or decisive, in the financing, organizing, training, supplying and equipping of the contras, the selection of its military or paramilitary targets, and the planning of the whole of its operation, is still insufficient in itself, on the basis of the evidence in the possession of the Court, for the purpose of attributing to the United States the acts committed by the contras in the course of their military or paramilitary operations in Nicaragua. All the forms of United States participation mentioned above, and even the general control by the respondent State over a force with a high degree of dependency on it, would not in themselves mean, without further evidence, that the United States directed or enforced the perpetration of the acts contrary to human rights and humanitarian law alleged by the applicant State.'

[69] 'Placed at the disposal of' is defined by the ILC as 'the organ is acting with the consent, under the authority of and for the purposes of the receiving State' and '[i]n performing the functions entrusted to it by the beneficiary State, the organ must also act in conjunction with the machinery of that State and under its exclusive direction and control, rather than on instructions from the sending State. Thus article 6 is not concerned with ordinary situations of interstate cooperation or collaboration, pursuant to treaty or otherwise.' Commentary Art. 6 Draft Articles, *supra* n. 2, p. 95.

[70] ECtHR, *Ilaşcu and Others* v. *Moldova and Russia* (Merits), Appl. No. 48787/99, 8 July 2004, para. 315.

[71] ECtHR, *Hussein* v. *Albania and others*, Appl. No. 23276/04, 13 March 2006, second final para.: 'Finally, there is no basis in the Convention's jurisprudence and the applicant has not invoked any established principle of international law which would mean that he fell within the respondent States' jurisdiction on the sole basis that those States allegedly formed part (at varying unspecified levels) of a coalition with the US, when the impugned actions were carried out by the US, when security in the

Still, it may be asked whether the responsibility of the state contributing the commander or that of the states contributing the subordinate personnel would be mutually exclusive. The answer to that question must be no. In a situation of multinational forces with a single command structure, shared responsibility of those states whose personnel are engaged in the relevant conduct along the chain of command (vertical chain of responsibilities) is inherent. As indicated by the *Hussein* judgment,[72] there is no horizontal chain of responsibility. In other words, a state taking part in a multinational mission may be held responsible if its contributed personnel – with respect to the particular conduct – form part of the chain of command, be it at the command level or at the executive level or through assisting conduct.

Critics will argue that the present allocation of the burden of responsibility is unfair and arbitrary since, especially under NATO-led operations, the post of the commander circulates among states contributing to a coalition force. This is a consequence of the political decision of NATO Member States not to have a standing NATO force and from the incomplete integration of national contingents into the NATO command structure.

It does not matter whether a state has difficulties fulfilling its international obligations. Neither the separation of responsibilities under national law nor the factual problems enforcing international obligations exclude the responsibility of a state.[73]

Finally, the ICTY Appeals Chamber pointed out in its *Tadić* decision that the establishment of 'overall control' does not require detailed orders or instructions for each individual operation.

> 'Under international law it is by no means necessary that the controlling authorities should plan all the operations of the units dependent on them, choose their targets, or give specific instructions concerning the conduct of military operations and any alleged violations of international humanitarian law. The control required by international law may be deemed to exist when a State (or, in the context of an armed conflict, the party to the conflict) has a role in organising, co-ordinating or planning the military actions of the military group, in addition

---

zone in which those actions took place was assigned to the US and when the overall command of the coalition was vested in the US. Even if he could have fallen within a State's jurisdiction because of his detention by it, he has not shown that any one of the respondent States had any responsibility for, or any involvement or role in, his arrest and subsequent detention (...). This failure to substantiate any such involvement also constitutes a response to his final submission to the effect that the respondent States were responsible for the acts of their military agents abroad.'

[72] ECtHR, *Hussein* v. *Albania and others*, ibid.

[73] ECtHR, *Assanidze* v. *Georgia*, Appl. No. 71503/01, 8 April 2004, para. 146: 'However, it must be reiterated that, for the purposes of the Convention, the sole issue of relevance is the State's international responsibility, irrespective of the national authority to which the breach of the Convention in the domestic system is imputable. Even though it is not inconceivable that States will encounter difficulties in securing compliance with the rights guaranteed by the Convention in all parts of their territory, each State Party to the Convention nonetheless remains responsible for events occurring anywhere within its national territory.'

to financing, training and equipping or providing operational support to that group.'[74]

Applying these standards to multinational military operations, the use of DU weapons or armour would be attributable to any state forming part of any coalition at a command level. It is irrelevant whether the use of the weapon has been ordered by that state or whether the commander has left the detailed planning of the operation to subordinate entities. The function of a commander not only includes the competence to give orders but also to control their implementation. Of course, this applies only to those operations carried out under a unified command structure.[75]

2.2.3   Responsibility of states for acts of military forces conducted under the authority of international organisations

It is generally recognised under international law that an international organisation can be held responsible for acts violating its international obligations.[76] However, under certain circumstances the Member States of an international organisation remain responsible for the acts committed under the authority of the organisation.

Two conditions must be met before an international organisation can be held responsible. First, the organisation must be a subject of international law. Only then can it be considered to have rights and duties. In particular, the organisation – through its organs – has to have the competence and ability to act as an independent subject. The ability to act depends on the establishment of organs. Their competence, and the competence of the organisation as an independent subject of international law, will be defined by the treaty creating the organisation or derive from the 'implied powers'[77] of the organisation. Third states, which are not State Parties to the statutory treaty of the organisation, are not bound by that treaty or responsible for the organisation's acts. According to customary international law and Article 34

---

[74] ICTY, Appeals Chamber, *Prosecutor* v. *Tadić*, IT-94-1-A, 15 July 1999, para. 137.

[75] Amnesty International goes even further and demands the responsibility of each participating state due to the fact the states collectively agreed to the operation in general. See Amnesty International, NATO/Federal Republic of Yugoslavia, 'Collateral Damage' Or Unlawful Killings? Violations of the Laws of War by NATO during Operation Allied Force', AI Index: EUR 70/18/00, June 2000, p. 13: 'Responsibility of NATO Alliance Members Operation Allied Force was fought by a coalition of NATO member states in the name of the alliance as a whole. The initial decision to resort to force was made collectively, as were subsequent decisions about escalating the air campaign. At no point during the air campaign did any alliance member publicly repudiate any of the attacks carried out by NATO forces. Therefore each NATO member may incur responsibility for the military actions carried out under the NATO aegis.'

[76] K. Ipsen, 'Die völkerrechtliche Verantwortlichkeit Internationaler Organisationen und anderer partieller Völkerrechtssubjekte', in Ipsen, *supra* n. 10, p. 657; Commentary Art. 6 ILC Draft Articles, *supra* n. 2, p. 98.

[77] The ICJ stated in its Advisory Opinion on *Reparation for Injuries Suffered in the Service of the United Nations*: 'Under international law, the Organization must be deemed to have those powers which, though not expressly provided in the Charter, are conferred upon it by necessary implication as being essential to the performance of its duties.' *ICJ Rep.* (1949) pp. 174-220, 182.

Vienna Convention on the Law of Treaties,[78] '[a] treaty does not create either obligations or rights for a third State without its consent'. Nevertheless, third states are free to recognise international organisations as subjects of international law.[79] As long as an international organisation is not recognised, explicitly or through specific conduct, the acts of its organs will be attributed to its Member States as a group.

The second condition is that the act must be attributable to the organisation. While an international organisation is responsible for an act attributable to it,[80] it remains debatable whether the responsibility of an international organisation for an attributable act excludes the responsibility of its Member States.[81] A further problem arises in determining whether a person has acted as an organ of the organisation or as a state official representing the sending state at the international organisation. This is not always easy, especially when the person acts simultaneously as a governmental official and as a representative of the organisation, a practice that remains common in most of the international organisations.[82]

For present purposes, the North Atlantic Treaty Organization[83] is the most significant international organisation. Its status as an international organisation with legal personality[84] has been recognised by the NATO Member States in the Agreement on the Status of the North Atlantic Treaty Organization,[85] especially in Article IV, which states: 'The Organization shall possess juridical personality; it shall have the capacity to conclude contracts, to acquire and dispose of movable and immovable property and to institute legal proceedings.'

Aside from its conventional establishment by way of the North Atlantic Treaty, NATO developed into an international organisation through its effective capacity to enter into relations with other subjects of international law.[86] Article 9 of the NATO Treaty sets up the main organ, the Council, and authorises it to establish further 'subsidiary bodies', in particular a defence committee. By exercising this competence, NATO has established a comprehensive institutional system. Articles 3 and 5 of the NATO Treaty provide it with a wide range of defence and security tasks,

---

[78] Vienna Convention on the Law of Treaties (VCLT) of 23 May 1969, 1155 *UNTS* (1980) p. 331.

[79] V. Epping, 'Internationale organisationen', in Ipsen, *supra* n. 10, p. 461.

[80] Ibid.

[81] The direct responsibility of the Member State may be established in exceptional cases in which the international organisation is designed purely with the purpose to exclude the responsibility of the Member States which otherwise would have been caused by the act. Ibid., p. 463.

[82] K. Ipsen, 'Die völkerrechtliche Verantwortlichkeit Internationaler Organisationen und anderer partieller Völkerrechtssubjekte', in Ipsen, *supra* n. 10, p. 658.

[83] Established by the North Atlantic Treaty, 4 April 1949, 34 *UNTS* (1949) p. 243 (hereinafter, NATO Treaty).

[84] J. Ignarski, 'North Atlantic Treaty Organization', *EPIL*, *supra* n. 6, pp. 646-652, 649.

[85] Agreement on the Status of the NATO, National Representatives and International Staff, 20 September 1951, 200 *UNTS* (1954) p. 3.

[86] Ignarski, *supra* n. 84, p. 647. Gazzini underlines that it is not necessary to declare explicitly the legal personality of an international organisation but refuses to recognise NATO as such. T. Gazzini, 'NATO coercive military activities in the Yugoslav crisis, 1992-1999', 12 *EJIL* (2001) pp. 391-435, 424.

specified by the Washington Declaration[87] and the Strategic Concept of April 1999,[88] updated by the Comprehensive Political Guidance of 2006.[89]

Consequently, as an international organisation, NATO has legal personality under international law in relation to its Member States. It can be held responsible and liable for internationally wrongful acts committed against these states.[90] With regard to third states, any responsibility of NATO would depend on its recognition by these states.[91] As long as these states have not recognised NATO, the only subjects to which military strikes can be attributed are the sending states.

However, even if NATO is regarded as a subject with legal personality, the decisive question in the present context remains whether military strikes executed under the auspices of NATO are attributable to NATO or to each individual Member State. The decisive condition for the change of the responsible subject is the transfer of command and control over the employed forces of the states engaged in military action under the leadership of NATO. The NATO Handbook states:

> 'In assigning forces to NATO, member nations assign operational command or operational control as distinct from full command over all aspects of the operations and administration of those forces. These latter aspects continue to be a national responsibility and remain under national control.
>
> In general, most NATO forces remain under full national command until they are assigned to the Alliance for a specific operation decided upon at the political level. Exceptions to this rule are the integrated staffs in the various NATO military headquarters; parts of the integrated air defence structure, including the Airborne Early Warning and Control Force (AWACS); some communications units; and the Standing Naval Forces as well as other elements of the Alliance's Reaction Forces.'[92]

Operational command will thus be transferred from the national authorities to NATO only in cases of specific operations. Before or after such operations, the full command remains at the national level.[93] Consequently, during armed conflicts in which NATO forces are involved, NATO has operational command. This transfer of power limits the competence of national authorities. However, it would be a misconception to assume that the national authorities no longer retain any responsibility. State

---

[87] The Washington Declaration, signed and issued by the heads of state and government participating in the meeting of the North Atlantic Council in Washington D.C. on 23 and 24 April 1999, Press Release NAC-S(99)63, <http://www.nato.int/docu/pr/1999/p99-063e.htm>.

[88] The Alliance's Strategic Concept, approved by the Heads of State and Government participating in the meeting of the North Atlantic Council in Washington D.C. on 23 and 24 April 1999, Press Release NAC-S(99)65, <http://www.nato.int/docu/pr/1999/p99-065e.htm>.

[89] Comprehensive Political Guidance endorsed by NATO Heads of State and Government, Riga, 29 November 2006.

[90] But see SOFA Agreements as referred to at the end of section 1 of this chapter.

[91] This might be a misconception of Gazzini, *supra* n. 86, when he refuses to accept the legal personality of NATO with a reference to the proceedings before the ICJ.

[92] *NATO Handbook* (Brussels 2001) p. 258 f, <http://www.nato.int>. For the general structure and concept of control see p. 257.

[93] Ignarski, *supra* n. 84, p. 648.

responsibility for acts in combat is not automatically transferred to NATO along with operational command. This conclusion is supported by the military structure of NATO.[94]

There are two different concepts of integrated military cooperation.[95] One concept is to hive off part of the national forces and fully integrate them into another foreign military structure under an independent command. Another concept is to allow a foreign (national or non-national) organ to exercise command over the armed forces, which remain in the national military structure. The latter concept implies that states assigning armed forces recognise the authority of the new power (e.g., NATO organs) to execute the operational command on behalf of the national operational command authorities. A consequence of this construction is that the state which assigns its forces for a specific operation remains responsible for the conduct of its forces executing the orders of the foreign command.

The latter concept applies to NATO forces, which are not under the full command of NATO, which has only operational command over the assigned forces. The competence with regard to the deployed personnel remains at the national level. Disciplinary issues fall within the competence of the national authorities. Consequently, the international obligation to supervise the fulfilment of international humanitarian law obligations,[96] and with it the control of the conduct of the forces in combat, remains at the national level. NATO has no competence to punish those who breach military rules by, for example, instituting disciplinary or criminal proceedings against military personnel of the national forces assigned to it. It has no effective means to implement disciplinary or humanitarian law rules.[97] The assignment of forces depends completely on the will of the national authorities. In case of a disagreement about particular operations, the authorities of the Member States are

---

[94] The combined structure differs from the future integrated structure of NATO. The new structure is described by NATO thus: 'The majority of so-called 'NATO Forces' are forces that remain under national control and only become available to the Alliance in specific circumstances. They are then placed under the responsibility of NATO military commanders. The integrated military command structure is the agreed basis for organising, training and controlling these forces.' NATO Fact sheets: Reform of NATO's Integrated Military Command Structure, 9 August 2000. http://www.nato.int/docu/facts/2000/rf-nimcs.htm. On 12 June 2003 the Alliance Defence Ministers agreed on the design of a new streamlined military command structure. See <http://www.nato.int/issues/military_structure/command/index-e.htm#strategic>.

[95] Integrated military cooperation: 'All nations opting to be members of the military part of NATO contribute forces which together constitute the integrated military structure of the Alliance. In accordance with the fundamental principles which govern the relationship between political and military institutions within democratic states, the integrated military structure remains under political control and guidance at the highest level at all times.' *NATO Handbook, supra* n. 92, p. 249.

[96] Arts. 47 and 49 Geneva Convention I; Arts. 48 and 50 Geneva Convention II; Arts. 127 and 129 Geneva Convention III, Arts. 144 and 146 Geneva Convention IV; Arts. 83, 85 and 87 Protocol I.

[97] This line of argument is supported by the ILC Commentary to Art. 6 ILC Draft Articles which refers to the comparable situation of organs acting under the responsibility of another state: 'Not only must the organ be appointed to perform functions appertaining to the State at whose disposal it is placed. In performing the functions entrusted to it by the beneficiary State, the organ must also act in conjunction with the machinery of that State and under its exclusive direction and control, rather than on instructions from the sending State.' Commentary Art. 6 ILC Draft Articles, *supra* n. 2, p. 95.

in a position to withdraw their forces as a last resort. As long as the national authorities are informed and oversee the operations and supervise compliance with international humanitarian law standards, NATO commanders would not be able to act contrary to the will of the national authorities.

Both arguments combined – the lack of NATO's competence over personnel matters and the remaining influence of the assigning states – lead to the conclusion that the responsibility for actions of assigned forces in NATO operations rests with the Member States.[98]

Finally, the attribution is restricted to a particular mission conducted by the forces of the Member States. As the national authorities of any Member State have no effective influence over the final decisions made by other national authorities, there is no automatic collective responsibility with the effect of attributing all acts which have been committed to every Member State of the alliance.

As a consequence, it is necessary to distinguish between particular actions.[99] With regard to the Kosovo operation, only the US forces employed DU weapons, therefore it is only the United States to which the direct employment of DU is attributable. For the other involved Member States of NATO, only the questions of supply, assistance or participation may arise. The same distinction must be applied to other conflicts, such as in Afghanistan or Iraq.

2.3   Wrongfulness of the act committed

2.3.1   Direct use of depleted uranium

State responsibility requires a wrongful act.[100] A wrongful act exists when a state breaches one of its international legal obligations. According to Article 12 ILC Draft Articles, such a breach exists when 'an act of that State is not in conformity with what is required of it by that obligation, regardless of its origin or character'. The same is valid for Article 3 of the Fourth Hague Convention and Article 91 of AP I.[101] However, the scope of Article 3 of the Fourth Hague Convention is limited to violations of the Regulations respecting the laws and customs of war on land, annexed to the Convention. Article 91 of AP I covers all violations of the four Geneva Conventions or of AP I and is therefore *lex specialis* in case of a violation of the rules of the four Geneva Conventions and AP I.[102]

Not every action by a party to an armed conflict which causes harm to another party to the same conflict or to a third state is prohibited by humanitarian

---

[98] Especially with regard to the failed strikes it is worth noting that a state remains responsible for the acts of its armed forces even if these forces exceed their authority or contravene their instructions. Art. 7 ILC Draft Articles. Therefore, acts based on the personal failure of military personnel remain attributable to the state.

[99] But cf., Amnesty International, which applies a principle of collective responsibility under which only the explicit repudiation by a Member State may exclude its responsibility. *Supra* n. 75, p. 13.

[100] Art. 3 ILC Draft Articles, *supra* n. 2.

[101] *ICRC Commentary Protocol I*, *supra* n. 22, p. 1058, para. 3661.

[102] Heintschel von Heinegg, *supra* n. 29, p. 38.

law. Therefore, state responsibility arises only when a state acts beyond the limits of the legitimate conduct for armed conflicts. These limits may be less restrictive under humanitarian law than under the rules of the law of peace. As to rules valid in peacetime, most of them do not apply in times of armed conflict, especially when they are substituted by rules regulating the rights and obligations of states in times of armed conflict.[103] This general principle also applies to rules protecting the natural environment. Humanitarian law[104] and disarmament law[105] – the latter which includes a prohibition of the use of certain weapons – replace the rules for the protection of the natural environment applicable in times of peace.

Establishing that the use of DU weapons in a particular case was unlawful would turn not only on an ability to show that its use in a particular case actually infringed recognised international legal obligations but also on those international obligations being recognised by the user state at the time when the action was conducted,[106] or having the status of customary international law. This is of particular relevance in the context of NATO-led operations, as the United States and France were not bound by AP I but only by the Fourth Hague Convention during the Kosovo Operation of 1999, and the United States remains a non-State Party.[107] The obligations of Protocol I are not applicable to the acts undertaken by US and French forces to the extent that they do not reflect rules of customary international law. However, Belgium, Canada, Germany, Italy, the Netherlands, Portugal, Spain and the United Kingdom were all parties to Protocol I at the time of the Kosovo operation.[108]

Once a breach of an international legal obligation through the use of DU weapons is proven, it is not necessary to refer to every single act. According to Article 14(2) of the ILC Draft Articles,[109] the breach of an international legal ob-

---

[103] W. Meng, 'War', *EPIL, supra* n. 6, p. 1334; C. Greenwood, 'Scope of application of humanitarian law', in D. Fleck, *The Handbook of Humanitarian Law in Armed Conflict* (Oxford, Oxford University Press 1995) p. 39. See B. Toebes, Chapter 9 of this book. Human rights gain increasing importance even in times of armed conflict. Their influence on the rules regulating lawful conduct in armed conflicts has been underlined by the UN Human Rights Committee (General Comment No. 31 on Art. 2 of the Covenant. The Nature of the General Legal Obligation Imposed on States Parties to the Covenant (adopted at the 2187th meeting on 29 March 2004), CCPR/C/74/CRP.4/Rev.6, para. 11) or the ICJ (Legal Consequences of the Construction of a Wall in the Occupied Palestinian Territory, Advisory Opinion of 9 July 2004, para. 106).

[104] E.g., Arts. 35(3) and 54-56 AP I.

[105] E.g., Convention on the Prohibition of Military or any other Hostile Use of Environment Modification Techniques, 18 May 1977, UN GA Resolution 31/72.

[106] According to Art. 13 ILC Draft Articles: 'An act of a State does not constitute a breach of an international obligation unless the State is bound by the obligation in question at the time the act occurs.'

[107] France became a party to Protocol I on 11 April 2001.

[108] For the Ratification status see <http://www.icrc.org/ihl>.

[109] Art. 14 ILC Draft Articles states: '1. The breach of an international obligation by an act of a State not having a continuing character occurs at the moment when the act is performed, even if its effects continue. 2. The breach of an international obligation by an act of a State having a continuing character extends over the entire period during which the act continues and remains not in conformity

ligation would extend from the first to the last use of DU weapons as long as these actions are seen as a continuing act. A final answer will always depend on the particular circumstances of the case.

2.3.2 Aid and assistance

Article 3 of the Fourth Hague Convention and Article 91 of AP I do not contain any rule about the responsibility of states which provide aid or assistance to any state involved in an armed conflict. Consequently, the decisive rules are once again the general rules of state responsibility as recognised by customary international law. Responsibility for aid and assistance in the commission of wrongful acts does not negate the fact that the main responsibility remains with the assisted state.[110] Nevertheless, an analysis of the use of DU weapons raises the question of the responsibility not only of the participating but also the supporting and assisting states. For those states, the principle applies that:

> 'A State which aids or assists another State in the commission of an internationally wrongful act by the latter is internationally responsible for doing so if:
> (a) That State does so with knowledge of the circumstances of the internationally wrongful act; and
> (b) The act would be internationally wrongful if committed by that State.'[111]

Aside from the question of whether the functions exercised by the allied forces during a military campaign constitute support or assistance, Article 16 of the ILC Draft Articles identifies four conditions under which a state can be held responsible for providing aid or assistance to another state.

First, the state, represented through its competent authorities, to whom the supporting or assisting forces belong, must have been informed about the purpose, method and means of the military strike.[112] The commentary to the Draft Articles states:

> '[T]he relevant State organ or agency providing aid or assistance must be aware of the circumstances making the conduct of the assisted State internationally wrongful …'[113]

Second, the wording of Article 16 of the ILC Draft Articles indirectly expresses the principle of causality. According to the commentary on Article 16 of the ILC Draft

---

with the international obligation. 3. The breach of an international obligation requiring a State to prevent a given event occurs when the event occurs and extends over the entire period during which the event continues and remains not in conformity with that obligation.'

[110] Commentary Art. 16 ILC Draft Articles, *supra* n. 2, p. 155.

[111] Art. 16 ILC Draft Articles.

[112] Sassòli, *supra* n. 3, p. 413.

[113] Commentary Art. 16 ILC Draft Articles, *supra* n. 2.

Articles 'the aid or assistance must be given with a view to facilitating the commission of that act, *and must actually do so.*'[114]

Causality is required between the acts of the state primarily violating international law and the acts of the assisting state. Article 16 of the ILC Draft Articles provides no criteria for distinguishing between relevant and irrelevant assistance. It seems to be clear that direct air support during combat with the purpose of protecting the DU weapons-deploying aircraft from enemy attacks falls within the scope of Article 16. It also seems clear that mere technical ground support for aircraft with the potential to deploy DU weapons would, even under a wide interpretation of Article 16, go beyond its scope. These two examples indicate that it is hardly possible to give a general answer to the question regarding the responsibility of all assisting states.[115] A correct answer would require a detailed analysis of the assistance provided by the individual states.

Third, the use of DU weapons would have to constitute a breach of an international obligation of the assisting state itself if it deployed them in this particular situation. In other words, the completed act must be such that it would have been wrongful if it were committed by the assisting state.[116]

Fourth, responsibility of the aiding or assisting state is only established when the direct employment of DU weapons constitutes a breach of an obligation of the assisted state. The commentary to the Draft Articles states:

> '... it is of the essence of the responsibility of the aiding or assisting State that the aided or assisted State itself committed an internationally wrongful act. The wrongfulness of the aid or assistance given by the former is dependent, *inter alia*, on the wrongfulness of the conduct of the latter.'[117]

The commentary shows clearly the shortcomings of Article 16 of the ILC Draft Articles. Especially in the event of alliances, states could arrange their conduct with regard to their international legal obligations. By choosing the state with a low level of international legal obligations, the rest of the allied states could ignore their own, possibly more far-reaching obligations, by support activities. This could work as long as these support activities would not reach the level of participation, rather than mere assistance or aid.

2.4 Circumstances precluding wrongfulness

After establishing the facts, attributing the acts committed to the particular states and analysing the wrongfulness of the acts, the question arises whether or not the wrongfulness of the act is precluded under international law. But here, the effects of the rules contained in the ILC Draft Articles are fundamentally limited due to the

---

[114] Ibid. Emphasis added.
[115] For the extent of the aid and assistance see ibid., pp. 157-159.
[116] Ibid., p. 157.
[117] Ibid., p. 160.

nature of humanitarian law. The exceptions are based on the nature of humanitarian law. Humanitarian law as a legal system provides for a situation where states are not able or willing to adhere to the law of peace.[118] Preventing a state of lawlessness, humanitarian law establishes a legal system that shall guarantee a certain minimum standard which will allow the parties to the conflict a process of restoration after the conflict. Moreover, it is the fundamental consideration of humanity,[119] or the last bastion of the international legal order,[120] providing the essential protection for human beings affected by an armed conflict. At least rules concerning the protection and the status of protected persons should therefore be regarded as peremptory norms.[121] These considerations restrict the scope for circumstances precluding the wrongfulness of the act of a party to an armed conflict.[122]

Four grounds for precluding wrongfulness appear to be particularly relevant for our analysis, namely self-defence, countermeasures, distress and *force majeure*.

2.4.1   Self-defence

Article 21 of the ILC Draft Articles states: 'The wrongfulness of an act of a State is precluded if the act constitutes a lawful measure of self-defence taken in conformity with the Charter of the United Nations.' However, it is essential to understand that however important or valuable and desirable the intention of the use of force, the legality of the use of force has no influence on the legality of the methods and means of warfare. In other words, the *jus ad bellum* and with it the right to self-defence is without influence on the *jus in bello*.[123] This fundamental distinction found its codification in the preamble of AP I to the four Geneva Conventions, which states

---

[118] W. Heintschel von Heinegg, *Seekriegsrecht und Neutralität im Seekrieg, Schriften zum Völkerrecht Band* (Berlin, Duncker & Humblot 1995) pp. 119 and 125.

[119] See *Legality of the Threat or Use of Nuclear Weapons*, *supra* n. 12, p. 257, para. 79; Hague Convention of 18 October 1907 relative to the Laying of Automatic Submarine Contact Mines (Hague Convention VIII), <http://www.icrc.org>; ICJ, *Corfu Channel* case (Merits), *supra* n. 44, p. 22.

[120] F. Bugnion, 'Just wars, wars of aggression, and international humanitarian law', 84 *IRRC* (2002) pp. 523 at 546.

[121] Sassòli, *supra* n. 3, pp. 414 and 420.

[122] In case of a violation of a peremptory norm of humanitarian law, no preclusion of wrongfulness in accordance with a rule of Chapter V of the ILC Draft Articles would be possible, Art. 26 ILC Draft Articles, *supra* n. 2. See also Art. 53 VCLT.

[123] See for the philosophical conception I. Kant, 'The philosophy of law. An exposition of the fundamental principles of jurisprudence as the science of right', paras. 53 and 57, reprinted in M. Sassòli and A.A. Bouvier, *How Does Law Protect in War* (Geneva, ICRC 1999) p. 86; Inter-American Commission on Human Rights, *Tablada* case, Report No. 55/97, Case No. 11.127: Argentina, OEA/Ser/L/V/II.97, Doc. 38, 18 November 1997, paras. 173 and 174; Trial of Wilhelm List and others, United States Military Tribunal, Nuremberg, 8 July 1947 to 19 February 1948, The United Nations War Crimes Commission, VIII *Law Reports of Trials of War Crimes* (1949) pp. 34-76, section 3 (v); *In re Altstötter and Others* (*Justice* case), United States Military Tribunal, Nuremberg, 4 December 1947, VI *War Crimes Reports* (1948) p. 1 (section [IO]).

'that the provisions of the Geneva Conventions of 12 August 1949 and of this Protocol must be fully applied in all circumstances to all persons who are protected by those instruments, without any adverse distinction based on the nature or origin of the armed conflict or on the causes espoused by or attributed to the Parties to the conflict.'

It also derives from the purpose of humanitarian law, which is to minimise the suffering caused by armed conflicts regardless of their legitimacy.[124] The norms of humanitarian law apply indiscriminately to the aggressor and the defender as well as to the victor and vanquished.[125] The same has been concluded by the ILC. Without stating it in the Draft Articles, the ILC recognises that under certain regimes of law, e.g., humanitarian law, Article 21 of the Draft Article would not apply.[126]

2.4.2   Countermeasures

Within the framework of humanitarian law, the idea of justifying the deployment of DU weapons as a countermeasure can hardly be squared with the ILC Draft Articles. Article 22 states that:

'[t]he wrongfulness of an act of a State not in conformity with an international obligation towards another State is precluded if and to the extent that the act constitutes a countermeasure taken against the latter State in accordance with Chapter II of Part Three.'

The decisive rule is Article 50(1) of the ILC Draft Articles excluding the application of countermeasures under certain condition. Subparagraph (a) prohibits any measure including the threat or use of force. However, this rule derives from the law of peace. When during an armed conflict an injured party to the conflict intends to invoke a countermeasure, the prohibition on use of force is rendered meaningless. The view is supported by the special provisions of humanitarian law excluding reprisals[127] only for actions against certain persons or objects.[128]

---

[124] Heintschel von Heinegg, *supra* n. 29, p. 8.

[125] *ICRC Commentary, supra* n. 22, p. 1055, para. 3652. Restated in the preamble of Addition Protocol I: '*Reaffirming* further that the provisions of the Geneva Conventions of 12 August 1949 and of this Protocol must be fully applied in all circumstances to all persons who are protected by those instruments, without any adverse distinction based on the nature or origin of the armed conflict or on the causes espoused by or attributed to the Parties to the conflict.'

[126] Commentary Art. 21 ILC Draft Articles, *supra* n. 2, p. 178: 'This is not to say that self-defence precludes the wrongfulness of conduct in all cases or with respect to all obligations. Examples relate to international humanitarian law and human rights obligations. The Geneva Conventions of 1949 and Protocol I of 1977 apply equally to all the parties in an international armed conflict, and the same is true of customary international humanitarian law. […] As to obligations under international humanitarian law […], self-defence does not preclude the wrongfulness of conduct.'

[127] The term reprisals is used as a synonym for countermeasures.

[128] See Art. 46 Geneva Convention I, Art. 47 Geneva Convention II, Art. 13(3) Geneva Convention III, Art. 33(3) Geneva Convention IV, Arts. 20, 51(6), 52(1) 53(c), 54(4), 55(2) and 56(4) AP I.

The use of countermeasure is – apart from Article 26 of the ILC Draft Articles – restricted in two ways. First, violations of humanitarian law entitle a party to an armed conflict to use countermeasures only against the state that has breached its obligations under humanitarian law. This limitation derives from the distinction between *jus in bello* and *jus ad bellum*.[129] Second, the scope of possible countermeasure caused by an act violating humanitarian law is limited by the mentioned restrictions flowing from humanitarian law, namely those prohibiting reprisals against protected persons.[130] In conclusion, countermeasures or 'reprisals' can only serve as a very limited ground precluding the wrongfulness of an act of state in breach of international humanitarian law.

2.4.3   *Distress, necessity and force majeure*

Finally – and here only by way of passing reference – the principles of distress[131] and necessity[132] do not apply *vis-à-vis* DU weapons as a matter of *jus in bello*. Again, due to the fundamental distinction of *jus in bello* and *jus ad bellum*,[133] only situations of necessity or distress caused by military actions would provide grounds for the preclusion of the wrongfulness. However, under humanitarian law, necessity is already covered by the concept of 'military necessity'.[134] But military necessity has already been incorporated into the rules regulating the methods and means of warfare.[135] Any additional preclusion of wrongfulness is therefore incompatible with the legal order of humanitarian law.

Something different applies to the exception of *force majeure*. *Force majeure* is defined by Article 23 ILC Draft Articles as 'the occurrence of an irresistible force or of an unforeseen event, beyond the control of the State, making it materially impossible in the circumstances to perform the obligation.' Impossibility of control explains the restriction of Article 91 of AP I, which establishes the responsibility of a party to the conflict 'for all acts'. Therefore, conduct – which cannot, due to the lack of control, be defined as an act – under *force majeure* does not constitute a violation of humanitarian law.[136]

2.4.4   *Restrictions for justification under jus cogens norms*

According to Article 26 ILC Draft Articles the wrongfulness of any act of a state is not precluded if the act is 'not in conformity with an obligation arising under a

---

[129] See section 2.4.1 of this chapter; Sassòli, *supra* n. 3, p. 425.
[130] Art. 50(1)(c) ILC Draft Article; Commentary Art. 50, *supra* n. 2, p. 336.
[131] Art. 24 ILC Draft Articles, *supra* n. 2.
[132] Art. 25, ibid.
[133] See section 2.4.1 of this chapter; Sassòli, *supra* n. 3, pp. 416-417.
[134] Commentary Art. 25 ILC Draft Articles, *supra* n. 2, p. 205.
[135] Ibid., pp. 206-206.
[136] *ICRC Commentary Protocol I*, *supra* n. 22, p. 1058, para. 3661.

peremptory norm of general international law'.[137] Even the primacy of obligations established by a Security Council Resolution under Chapter VII is limited by the norms of *jus cogens*. As stated in the *Kadi* decision, the norms of *jus cogens* will always prevail.[138] The latter aspect is highly relevant when it comes to the question whether actions necessary for the effective implementation of UN Security Council resolutions restoring or maintaining peace could prevail over other international human rights obligations.[139] The difficulty is in identifying those norms that may be regard as *jus cogens*.

3. LEGAL CONSEQUENCES

According to customary international law, a state that is responsible for an internationally wrongful act has to cease that act, if it is continuing;[140] to offer appropriate assurances and guarantees of non-repetition, if circumstances so require;[141] and to make full reparation for the injury caused by the internationally wrongful act.[142] Full reparation shall be given in the form of restitution, compensation and satisfaction.[143] When restitution is not possible[144] (impracticable or unreasonable),[145] compensation is required.[146] However, the injured state is not entitled to punitive or exemplary damages.[147]

---

[137] The ICJ noted in its decision on counterclaims in the case concerning the *Application of the Convention on the Prevention and Punishment of the Crime of Genocide*: 'in no case could one breach of the Convention serve as an excuse for another'. Application of the Convention on the Prevention and Punishment of the Crime of Genocide, Counter-Claims, *ICJ Rep*. 1997, pp. 243 at 258, para. 35.

[138] Court of First Instance, *Yassin Abdullah Kadi* v. *Council and Commission*, T-315/01, 21 September 2005, paras. 226-231.

[139] See the discussion in the *Kadi* case, ibid., but also the still pending case of *Saramati* v. *France and Norway* before the ECtHR, Appl. No. 78166/01.

[140] See Art. 30(a) ILC Draft Articles.

[141] Art. 30(b), ibid.

[142] Art. 31 ibid., see also J. Crawford, Special Rapporteur, Fourth Report on State Responsibility, International Law Commission, 53th session, Geneva, 23 April-1 June and 2 July-10 August 200, A/CN.4/517, Rn. 31; ICJ, Case Concerning Armed Activities on the Territory of the Congo (*Democratic Republic of the Congo* v. *Uganda*), 19 December 2005, General List No. 116, para. 259.

[143] Art. 34 ILC Draft Articles; Steinkamm, *supra* n. 29, p. 1365.

[144] Heintschel von Heinegg, *supra* n. 29, p. 27.

[145] Steinkamm, *supra* n. 29, p. 1365.

[146] Similar to this general obligation, Art. 3 of the Fourth Hague Convention and Art. 91 of AP I state that the belligerent parties to a conflict which are held responsible are obligated to pay compensation. This particular obligation is recognised as customary. See Report on the Protection of War Victims, prepared by the ICRC, Geneva, June 1993, reprinted in Sassòli and Bouvier, *supra* n. 123, pp. 444-458, 457. Compensation based on state responsibility must be distinguished from other forms of payments under humanitarian law, especially for prisoners of war (see Arts. 60-62 and 68 Geneva Convention III), persons compelled to work (Art. 51(3) Geneva Convention IV) or for requisitioned goods in occupied territory (Art. 55(2) Geneva Convention IV) or for internees on release or repatriation (Art. 97(5) Geneva Convention IV). Those payment obligations differ in their purpose from those established by state responsibility.

[147] Heintschel von Heinegg, *supra* n. 29, pp. 31, 39; *ICRC Commentary Protocol I, supra* n. 22, p. 1056, para. 3655.

Deriving from the terms 'which violates the provisions' and 'if the case demands', compensation becomes due only if there is a violation of the Geneva Conventions or AP I which leads to loss or damage of a material or personal nature.[148] Such compensation will be restitution in kind or the full restoration of the situation as it would be without the violation, including compensation in the form of services.[149] However, in case restitution or restoration of a lawful situation is not possible, the responsible state has to pay monetary compensation of a sum corresponding to the value of that which has been damaged.[150] In addition to restitution and compensation, a state is obliged to give satisfaction in the form of acknowledgment of the breach, an expression of regret, a formal apology or another appropriate modality, even if this is not specifically mentioned in Article 91 of AP I and Article 3 of the Fourth Hague Convention.[151]

The obligation to pay compensation in case of a violation of humanitarian law applies equally to each party to the conflict, whether aggressor or defender, victor or vanquished.[152] As such, this obligation is separate from any compensation based on violations of the *jus ad bellum*, which means that an aggressor state is liable to compensation for damages arising from unlawful acts of warfare.[153]

3.1   The beneficiary of the award of compensation

As the matter of principle, the obligation to award compensation can only be owed by one state to another state.[154] Article 3 of the Fourth Hague Convention as well as Article 91 of AP I as such are not meant to establish subjective rights of individuals.[155] This has been approved by national courts, recently the German Landgericht

---

[148] Ibid, p. 1056, para. 3655.

[149] Ibid.

[150] Ibid. See also Arts. 35 and 36 ILC Draft Articles.

[151] Art. 37 ILC Draft Articles; for the general principle of state responsibility see Zemanek, *supra* n. 53, p. 225; E.-C. Gillard, 'Reparation for violations of international humanitarian law', 85 *IRRC* (2003) pp. 529-553, 533.

[152] R. Wolfrum, 'Enforcement of international humanitarian law', in Fleck, *supra* n. 103, p. 543; *ICRC Commentary Protocol I*, *supra* n. 22, p. 1055, para. 3652.

[153] I. Seidel-Hohenveldern, 'Reparations', *EPIL*, *supra* n. 6, p. 179; Heintschel von Heinegg, *supra* n. 29, pp. 24-25.

[154] Report on the Protection of War Victims, *supra* n. 146, p. 457; Heintschel von Heinegg, *supra* n. 29, pp. 31, 39.

[155] So decided by the American Court of Appeals in the Fourth Circuit in the case of *Goldstar S.A. (Panama)* v. *United States*, US 967 F.2d 965, 968 (4th Cir. 16 June 1992) stating: '[...] Goldstar contends that the United States is liable for compensation under Article 3 of the Convention, and that Article 3 must be interpreted as a self-executing waiver of sovereign immunity with regard to such claims. [...] International treaties are not presumed to create rights that are privately enforceable. Courts will only find a treaty to be self-executing if the document, as a whole, evidences an intent to provide a private right of action. The Hague Convention does not explicitly provide for a privately enforceable cause of action. Moreover, we find that a reasonable reading of the treaty as a whole does not lead to the conclusion that the signatories intended to provide such a right. Significantly, the United States Supreme Court has recently construed a provision of the Geneva Convention similar to the presently contested Hague Convention language. The Geneva Convention provides that an illegally

Bonn, and confirmed by the German Supreme Court[156] in a case concerning a bombardment during the NATO Kosovo Campaign.[157] However some authors are of the opinion that even individuals may claim compensation on the basis of Article 3 of the Fourth Hague Convention or Article 91 of AP I.[158] Kalshoven, for instance, refers to the drafting history of the Hague Convention underlining that the private law principle according to which the master is responsible for his subordinates or agents has been the inspiration for Article 3. However, this reference applies only to the question of attribution, not to the issue of the legitimate claimant.[159] To date, no

---

boarded merchant ship "shall be compensated for any loss or damage that may be sustained". [...] Interpreting this language, the Supreme Court found that the "conventions ... only set forth substantive rules of conduct.... They do not create private rights of action...." *Argentine Republic* v. *Amerada Hess Shipping Corp.*, 488 U.S. 428, 442, 109 S.Ct. 683, 692, 102 L.Ed.2d 818 (1989). We can discern no reason why the Hague Convention language, at issue in the present case, should be construed any differently. Furthermore, Article 1 of the Hague Convention states, "[t]he Contracting Powers shall issue instructions to their armed land forces which shall be in conformity with the Regulations...." Hague Convention, art. 1. This language must be taken as further evidence that the Hague Convention is not self-executing, and that, instead, the signatories contemplated that individual nations would take subsequent executory actions to discharge the obligations of the treaty. In sum, we hold that the Hague Convention is not self-executing and, therefore, does not, by itself, create a private right of action for its breach.' [References partly deleted] In the case of *Lindo & Maduro S.A.* v. *U.S.* (506 U.S. 973, 113 S.Ct. 463) the US Supreme Court denied Petition for writ of certiorari which had been supported by a brief of the government of Panama, maintaining that Art. 3 Hague Convention IV does provide for an individual right of persons against a State Party to the Convention. Brief of the Government of Panama, reprinted in Sassòli and Bouvier, *supra* n. 123, pp. 944-948.

[156] German Supreme Court (Bundesgerichtshof), *supra* n. 5.

[157] Landgericht Bonn, *supra* n. 64: 'Die traditionelle Konzeption des Völkerrechts als eines zwischenstaatlichen Rechts versteht den Einzelnen nicht als Völkerrechtssubjekt, sondern gewährt ihm nur mittelbaren internationalen Schutz: Bei völkerrechtlichen Delikten durch Handlungen gegenüber fremden Staatsbürgern steht ein Anspruch nicht dem einzelnen Betroffenen selbst, sondern nur seinem Heimatstaat zu. Der Staat macht im Wege des diplomatischen Schutzes sein eigenes Rechts darauf geltend, dass das Völkerrecht in der Person seines Staatsangehörigen beachtet wird. Das Individuum ist nur über das "Medium" des Staates im dem Völkerrecht verbunden, ohne selbst dessen Subjekt zu sein (vgl. BVerfG, Beschluss vom 13.Mai 1996, Az: 2 BvL 33/93, abgedruckt u.a. in BVerfGE 94, 315, 334 sowie NJW 1996, 2717 f. m.w.N.; Ipsen, Völkerrecht, 4. Auflage, § 7, S.80 f). [...] Die Bestimmungen des Abkommens betreffend die Gesetze und Gebräuche des Landkriegs vom 18.Oktober 1907 (Haager Landkriegsordnung – HLKO) finden "nur zwischen den Vertragsmächten Anwendung" (Art.2 HLKO). Art.3 HLKO sieht allein eine Verpflichtung der "Kriegspartei" (gegenüber der anderen Kriegspartei) zum Schadensersatz vor (vgl. auch BGH, Urteil vom 26.6.2003, AZ: III ZR 245/98, "Distomo"). In dem seitens der Kläger angeführten Genfer Abkommen vom 12.August 1949 über den Schutz von Zivilpersonen in Kriegszeiten (IV.Genfer Abkommen) verpflichten sich in Art.1 gleichfalls allein die "Vertragsparteien" zu dessen Einhaltung und Durchsetzung. Gleiches ergibt sich für das Zusatzprotokoll I zu den Genfer Abkommen vom 12.August 1949 über den Schutz der Opfer internationaler bewaffneter Konflikte, das die Genfer Abkommen zum Schutz der Kriegsopfer ergänzt, Art.1 Abs.3: Auch durch dieses verpflichten sich allein die Vertragsparteien, Art.1 Abs.1; einzelne Zivilpersonen "genießen Schutz" (Art.51), erhalten hingegen keine eigenen Rechte zugesprochen. Auch die in Art.91 normierte Haftungsregelung greift nicht zugunsten des Einzelnen. Im übrigen stellen weder die Genfer Konvention noch deren Zusatzprotokolle ein Verfahren zur Verfügung, das dem Einzelnen die Durchsetzung etwaiger individueller Ansprüche ermöglichen würde.'

[158] Kalshoven identifies Art. 3 Hague Convention IV as a legal basis for individual claims (p. 833) but denies this legal quality for Art. 91 Protocol I (p. 851), *supra* n. 29. For a detailed discussion of further decisions see Gillard, *supra* n. 151, pp. 535-539.

[159] Heintschel von Heinegg, *supra* n. 29, p. 31.

case (international or national) has been reported in which an individual has been awarded compensation exclusively based on Article 3 Hague Convention IV or Article 91 AP I.[160]

3.1.1   Evaluation of the damage

Finally, determining the amount of compensation to be paid by the responsible state depends on the type and quantity of the damage caused by the unlawful act. International law remains silent with regard to any generally applicable guidelines.

> 'In the current state of international law, it would be wrong to presume any specific definition of "injury" or "damage" which is applicable across the board. The many declarations and agreements which lay down primary rules of responsibility do not seem to derogate from any general rule about injury or damage. They do not embody so many special provisions given effect by way of the *lex specialis* principle (article 56). Each is tailored to meet the particular requirements of the context and the balance of a given negotiated settlement.'[161]

The evaluation of damages caused by violations of humanitarian law is particularly complicated. Since the purpose of war is to inflict damage on the enemy,[162] damage which has to be compensated must be distinguished from that which has been caused by lawful acts of warfare or by the war situation as such.[163]

As indicated, not all types of damage or injuries are to be compensated. From the point of view of general principles of state responsibility, there is no doubt that – once its illegality is proven – personal injuries or economic loss caused by the use of DU weapons (munitions or armour) fall under the types of damage that are to be compensated. More interesting is the question whether environmental damage also falls within the scope of damage that can be compensated. In Resolution 687 (1991) the Security Council gave a positive answer to that question stating

> 'that Iraq, ..., is liable under international law for any direct loss, damage – including environmental damages and the depletion of natural resources – or in-

---

[160] Kalshoven had to admit that '[t]he draftsmen stopped far short of providing individual beneficiaries with precise rules for the presentation and settlement of their claims…' (p. 835) and recognised that state practice does not support his view (p. 836), *supra* n. 29.

[161] J. Crawford, Special Rapporteur, Fourth Report on State Responsibility, International Law Commission, 53th session, Geneva, 23 April-1 June and 2 July-10 August 200, A/CN.4/517, Rn. 31.

[162] Steinkamm, *supra* n. 29, p. 1354.

[163] Eritrea Ethiopia Claims Commission, Partial Award, Central Front, Eritrea's Claims 2, 4, 6, 7, 8 and 22, between the State of Eritrea and the Federal Democratic Republic of Ethiopia, The Hague, 28 April 2004, para. 29; Partial Award, Civilians Claims, Eritrea's Claims 15, 16, 23 and 27–32, between the State of Eritrea and the Federal Democratic Republic of Ethiopia, The Hague, December 17, 2004, para. 135: 'Ethiopia is not internationally responsible for losses resulting from sale prices depressed because of general economic circumstances related to the war or other similar factors.'

jury to foreign Governments, nationals and corporations as a result of its unlawful invasion and occupation of Kuwait; ...'[164]

The remaining question concerns the precise amount of damage caused. Paragraph
16 of SC Resolution 687 gives no indication of how to measure or specify the
covered damage.[165] At any rate, in depth expertise would be necessary to be able to
give a reasonable *ex ante* answer to the question of the concrete long-term consequences for the environment of remnants of DU ammunitions or armour. The issues
of sufficient evidence and the burden of proof are immediately raised. As stated by
the ICJ, the injured state has to 'demonstrate and prove the exact injury that was
suffered as a result of specific actions [...] constituting internationally wrongful
acts.'[166]

Faced with the complexity of claims concerning damages caused by military activities in times of war, the Eritrea Ethiopia Claims Commission applied a
less restrictive standard. This was done in three ways. First, the Commission applied a standardised amount of money for compensation for groups of claims.[167]
Second, it lifted[168] or even shifted[169] the burden of proof. Third, it measured the
amount of damage by appraisal of the available facts.[170] The latter method has

---

[164] UN Security Council Resolution 687 (1991) of 3 April 1991, para. 16. One can criticise Res.
687's (1991) reference to violations of the *jus ad bellum*. As to the scope of damages and the obligation
to compensate, this is not relevant since the general rules of state responsibility and the rules concerning the determination of damage apply equally regardless of the source of the violated law.

[165] See C.D. Stone, 'The environment in wartime: An overview', in J.E. Austin and C.E. Bruch,
ed., *The Environmental Consequences of War: Legal, Economic and Scientific Perspectives* (Cambridge, Cambridge University Press 2000) pp. 16-35, 30; see also Decision taken by the Governing
Council of the UN Compensation Commission of 17 March 1992, S/AC.26/1991/7/Rev.1, pp. 7-8,
para. 35: 'These payments are available with respect to direct environmental damage and the depletion
of natural resources as a result of Iraq's unlawful invasion and occupation of Kuwait. This will include
losses or expenses resulting from: (a) Abatement and prevention of environmental damage, including
expenses directly relating to fighting oil fires and stemming the flow of oil in coastal and international
waters; (b) Reasonable measures already taken to clean and restore the environment or future measures
which can be documented as reasonably necessary to clean and restore the environment; (c) Reasonable monitoring and assessment of the environmental damage for the purposes of evaluating and abat-
ing the harm and restoring the environment; (d) Reasonable monitoring of public health and performing
medical screenings for the purposes of investigation and combating increased health risks as a result of
the environmental damage; and (e) Depletion of or damage to natural resources.'

[166] ICJ, 19 December 2005, General List No. 116, Case Concerning Armed Activities on the Territory of the Congo (*Democratic Republic of the Congo* v. *Uganda*), para. 260; see also Art. 14 of the
Rules of Procedure of the Eritrea Ethiopia Claims Commission.

[167] Eritrea Ethiopia Claims Commission, Decision Number 2: Claims Categories, Forms and Procedures and Decision Number 5: Multiple Claims in the Mass Claims Process, Fixed-Sum Compensation at the $500 and $1500 Levels, Multiplier for Household Claims.

[168] Eritrea Ethiopia Claims Commission, Partial Award, Central Front, Ethiopia's Claim 2, between the Federal Democratic Republic of Ethiopia and the State of Eritrea, The Hague, 28 April 2004,
para. 38.

[169] Ibid., para. 73; Partial Award, Western Front, Aerial Bombardment and Related Claims, Eritrea's
Claims 1, 3, 5, 9-13, 14, 21, 25 and 26, between the State of Eritrea and the Federal Democratic
Republic of Ethiopia, The Hague, 19 December 2005, para. 47.

[170] Eritrea Ethiopia Claims Commission, Partial Award, Central Front, Eritrea's Claims 2, 4, 6, 7, 8
and 22, between the State of Eritrea and the Federal Democratic Republic of Ethiopia, The Hague,

considerable advantages. It balances the need for a procedure based on the rule of law with the interest in a less costly procedure of a reasonable duration.

3.1.2   *Conclusion*

If and when the unlawfulness of the use of DU weapons has been established, the obligation to make reparation could cover a wide range of possibilities. It could start with the decontamination of the affected areas (restitution),[171] compensation for possible damage to agriculture arising from the contamination of fields, compensation for costs of medical treatment caused by the use of DU, or compensation for damage to the health of persons and the consequences arising from this damage. The particular nature and amount of compensation differs from case to case depending on the quality and quantity of the damage caused and the contribution to it by the state concerned.[172] But any unlawful conduct of the injured state that provoked the act of the injuring state or increased the damage influences the amount of compensation.[173]

4.   LEGAL REMEDIES UNDER INTERNATIONAL LAW

It is a general principle of international law that states are obliged to resolve their disputes peacefully.[174] Any dispute between states – including those over any responsibility arising out of a violation of humanitarian law – shall be settled 'by negotiation, enquiry, mediation, conciliation, arbitration, judicial settlement, resort to regional agencies or arrangements, or other peaceful means of their own choice.'[175]

It is clear that the commission of the wrongful act, if attributable to a state, gives rise to state responsibility without further actions by the injured state or any other relevant subject of international law being required. Neither its explicit recognition, for example, by an international judicial body, nor the issuing of an appeal or a formal protest are necessary for the establishment of state responsibility. Nev-

---

28 April 2004, para. 29: 'The claims before the Commission involve complex events, some unfolding over many months. In several situations, the Commission has concluded that particular damage resulted from multiple causes operating at different times, including both causes for which there was State responsibility and other causes for which there was not. The evidence does not permit exact apportionment of damage to different causes in these situations. Accordingly, the Commission has indicated the percentage of the loss, damage or injury concerned for which it believes the Respondent is legally responsible, based upon its best assessment of the evidence presented by both Parties.'

[171] See the reference to the obligation to conduct mine-clearing operations after the unlawful laying of mines in Report on the Protection of War Victims, *supra* n. 146, p. 457.

[172] The contribution by the injured state also has to be considered, especially when the conduct of the injured state itself was contrary to international law obligations, e.g., intentional placement of military units or objects close to civilian objects or the civilian population, Art. 39 ILC Draft Articles.

[173] Ibid.

[174] See also Art. 2(3) UN Charter.

[175] Art. 33(1) UN Charter; ICJ, *Case concerning the Legality of the Use of Force* (*Yugoslavia* v. *Belgium*), Request for the Indication of Provisional Measures, Order of 2 June 1999, *ICJ Rep.* (1999) pp. 124-257, 140 para. 48.

ertheless, in many cases the particular facts and the existence of the requirements of state responsibility are disputed by the involved states.

The question whether or not state responsibility has been established is therefore often in question before international courts, tribunals or commissions.

It is the right of every injured state to invoke the responsibility of another state if the obligation breached is owed to the injured state or a group of states that includes the injured state, or even to the international community.[176] The latter only applies when it specially affects that state or the breach of the obligation is of such a character as to radically change the position of all the other states to which the obligation is owed with respect to the further performance of the obligation.[177]

Regarding any claim of individuals, it is still the rule and practice that individuals have no standing under international law to claim any damage against a foreign state based on state responsibility.[178] Such claims of individuals may be brought by their state of nationality against the injuring state by means of diplomatic protection.[179] However, the weakness of this alternative is the lack of an obligation of any state to exercise diplomatic protection. The decision whether a state will claim compensation from another state on the grounds of state responsibility remains at the political discretion of the injured state.[180]

This wide competence to invoke responsibility does not resolve the main practical problem with regard to international legal remedies. The problem is centred on the non-obligatory character of most of the procedures for investigation. From the Fact-Finding Commission[181] established by Article 90 of AP I to the Geneva Conventions, to the ICJ[182] and the *ad hoc* commissions[183] and tribu-

---

[176] Art. 42 ILC Draft Articles. Considering the usual use of the term 'international community' and the statements of governments, the term only refers to the international community of states. Commentary Art. 42 ILC Draft Articles, *supra* n. 2, p. 299.

[177] Ibid. For further possibilities of invocation of responsibility, especially with regard to the groups of states, see Arts. 46 to 48 ILC Draft Articles.

[178] Heintschel von Heinegg, *supra* n. 29, p. 25.

[179] *ICRC Commentary Protocol I, supra* n. 22, p. 1054, para. 3654.

[180] The injured state claims the violation of humanitarian law affecting its citizens as a violation of its own rights.

[181] Established on 25 June 1991. The competence of the Commission has to be recognised expressly by the Member States to the Protocol. As of the end of 2006, 69 states had recognised the Commission. The list of states is available at <http://www.ihffc.org/en/stateparties.html>.

[182] See the Case concerning the Legality of the Use of Force, *supra* n. 175, e.g., Application against the Kingdom of Belgium of 29 April 1999, <http://www.icj-cij.org>. 'The Government of the Federal Republic of Yugoslavia requests the International Court of Justice to adjudge and declare: [...] – by taking part in the use of weapons containing depleted uranium, The Kingdom of Belgium has acted against the Federal Republic of Yugoslavia in breach of its obligation not to use prohibited weapons and not to cause far-reaching health and environmental damage; [...] – by taking part in activities listed above, and in particular by causing enormous environmental damage and by using depleted uranium, The Kingdom of Belgium has acted against the Federal Republic of Yugoslavia in breach of its obligation not to deliberately inflict on a national group conditions of life calculated to bring about its physical destruction, in whole or in part; ....' There has been no decision by the Court on the Merits of case yet.

[183] See United Nations Compensation Commission, established by Security Council Resolution 687 (1991) of 3 April 1991 and 692 (1991) of 20 May 1991; see also Eritrea/Ethiopia Claims Commis-

nals,[184] the establishment, the jurisdiction and the efficiency of these institutions largely depend on the will of the involved states or the political will of third states after a conflict. The future will show the extent to which the International Criminal Court and human rights bodies, such as the ECHR,[185] will have the capacity to strengthen the determination of state responsibility for acts committed in times of armed conflicts.

Fact-finding in general is one of the most important procedural elements when it comes to the determination of responsibility for violations of humanitarian law.[186] But it is confronted with a wide range of obstacles. Even the Fact-Finding Commission established under AP I is of a purely investigative nature. The Commission will not pronounce on the legality of the action or situation investigated. The Commission cannot conduct an investigation *proprio motu*; a request by a state is required to initiate an investigation. Yet, it has never officially been asked to conduct an investigation since its establishment in 1991. Moreover, any investigation depends on the consent of parties to the conflict. This, and the fact that it took more than 20 years after the adoption of Protocol I to establish the Commission, may be symptomatic of the reluctance of states for an independent investigation of the conduct of their military forces in armed conflict.

Other bodies, somewhere between organs of judicial determination and of fact-finding, may assume quasi-judicial functions and make determinations of state responsibility. By Resolutions 687 (1991) of 3 April 1991 and 692 (1991) of 20 May 1991 the UN Security Council established the United Nations Compensation Commission on Iraq to settle claims arising out of Iraq's unlawful invasion and occupation of Kuwait.[187] It is clear from paragraph 19 of Resolution 687 that the Commission is not only a traditional administrative body – as paragraph 18 of Resolution 687 might suggest – but has the authority to evaluate the validity of claims brought before it. In this respect, it exercises a certain judicial power. As stated in a report of the UN Secretary-General:

---

sion, established pursuant to Art. 5 of the Agreement signed in Algiers on 12 December 2000 between the Governments of the State of Eritrea and the Federal Democratic Republic of Ethiopia, 40 *ILM* (2001) p. 260.

[184] Such as the International Criminal Tribunal for the Former Yugoslavia (ICTY). See Final Report to the Prosecutor by the Committee established to review the NATO Bombing Campaign against the Federal Republic of Yugoslavia, in which she evaluated the legality of the use of depleted uranium during the NATO Bombing Campaign against Yugoslavia, reprinted in 39 *ILM* (2000) pp. 1257-1283, 1264.

[185] With reference to NATO combat missions in Yugoslavia: ECHR Decision, *Banković et al* v. *Belgium et al*, Appl. No. 52207/99, 12 December 2001, inadmissibility of the Claim for any violation of the Convention on the Territory of the former Republic of Yugoslavia due to the lack of jurisdiction, <http://www.echr.coe.int/>.

[186] For the International Fact-Finding Commission see J.-M. Henckaerts, 'International legal mechanisms for determining liability for environmental damage under international humanitarian law', in Austin and Bruch, *supra* n. 165, p. 603.

[187] In general see B. Graefrath, 'Iraqi reparations and the Security Council', in 55/1 *ZAöRV* (1995) pp. 1-67.

'the Commission is not a court or an arbitral tribunal before which the parties appear; it is a political organ that performs an essentially fact-finding function of examining claims, verifying their validity, evaluating losses, assessing payments and resolving disputed claims; it is only in this last respect that a quasi-judicial function may be involved.'[188]

The UN Compensation Commissions is open to states as well as individuals.[189]

The strong emphasis on *legal* remedies in the foregoing analysis should, however, not diminish the role of diplomacy. States might be more willing to accept the monetary burden of their unlawful conduct as long as they do not have to formally recognise their responsibility, with the brand of a violator. In some cases, a result reached by diplomacy could be more useful for the injured state and the victims than a long judicial process and the uncertainty of enforcement of its outcome. A few cases have been reported in which compensation has been paid by a state without it recognising any responsibility for a violation of humanitarian law.[190]

5.   STEPS TO BE TAKEN

The following conclusions can be drawn from the preceding analysis. There exists a clear scheme of state responsibility (ready) to be applied to the issue of DU use in armed conflict. The doctrinal structure inherent in the ILC Draft Articles, including the modifications based on the legal nature of humanitarian law, fits the possible scenarios, questions and (hypothetical) answers which have been elaborated. This 'stand-by-system' will be triggered by establishing a breach of international legal obligations through the use of DU weapons in armed conflict. With regard to the possible responsibility of international (governmental) organisations, it has been shown that it is more advantageous to refer to the Member States engaged and not to address the alliance as such, like NATO.

At the present stage, the main problem appears to be the lack of the necessary factual investigation of the particular cases and the uncertainty over effects of the use of DU in combat in general. Any state claiming the responsibility of another state for the illegal use of DU will have to prove as precisely as possible that it has

---

[188] Report of the Secretary-General pursuant to para. 19 of Security Council Resolution 687 (1991) of 2 May 1991, S/22559, section. II. C.

[189] Section II.A.2 *Ratione personae*, Recommendations made by the Panel of Commissioners Concerning individual Claims for Serious personal Injury or death (Category 'B' Claims), UN Doc. S/AC.26/1994/1 of 26 May 1994.

[190] Compensation for the bombing of the Chinese Embassy in Belgrade, see Agreement of the Government of the United States of America and the Government of the People's Republic of China on the settlement of the Chinese claims for property loss and damage suffered by the Chinese side as a result of the US bombing of the Chinese Embassy in the Federal Republic of Yugoslavia on 8 May 1999, Beijing, 16 December 1999, <http://www.state.gov/documents/organization/6599.doc>; Compensation for the bombing of the Richmond Hill Insane Asylum in Grenada by an US military aircraft on 5 November 1983; see the related Case 9213 of the Inter-American Commission on Human Rights (*Disabled Peoples' International* v. *US*), 1 March 1996, Annual Report, reprinted in Sassòli and Bouvier, *supra* n. 123, p. 920.

been afflicted by DU use in violation of rules of international humanitarian law. Therefore, priority should be given to overall/regional fact-finding missions and further scientific research.

The topic of responsibility is also of central importance in a Draft Convention on the prohibition of development, production and use of uranium weapons prepared and published within an anti-DU non-governmental organisation setting.[191] According to Article 16 of that draft, each State Party that uses uranium weapons in a conflict is responsible for clean-up, decontamination and medical care as well as compensation of the victims. It is responsible for all actions committed by persons belonging to its military forces. Moreover, the Draft Convention contains some procedural elements, such as rules on the clarification of questions, fact-finding missions and settlement of disputes. The obligations imposed by Article 16 correspond to the general rules of state responsibility.

---

[191] See the website of the International Coalition to Ban Uranium Weapons (ICBUW), <http://www.bandepleteduranium.org>.

Chapter 11
INTERNATIONAL AND DOMESTIC REMEDIES FOR INDIVIDUALS SUFFERING DAMAGE AS A RESULT OF EXPOSURE TO DEPLETED URANIUM WEAPONS

Avril McDonald[1]

1. INTRODUCTION

If a person develops an illness that he or she believes is linked with exposure to depleted uranium (DU), what are the possibilities for attaining a remedy in international law at the international or domestic levels? What is meant by the term remedy? And who can be considered to be a victim for the purposes of attaining a remedy?

The chapter first identifies who is considered as a victim under international law (section 2). It then elucidates the concept of remedies in international law (section 3) and examines the international legal status and procedural capacity of individuals (section 4) before tackling the subject of individuals and the law on state responsibility (section 5). Thereafter, it focuses on the potential availability of remedies founded in the law of armed conflict (LOAC) (section 6), international human rights law (section 7), international criminal law (section 8) and at the domestic level in the user states (section 9). The chapter rounds off with some final remarks (section 10).

2. WHO IS A VICTIM UNDER INTERNATIONAL LAW?

While the law of armed conflict does not define a victim, it is clear that it uses the term broadly:

> 'It is evident that the [Geneva] Conventions and the Protocols protect all kinds of persons afflicted by *armed conflicts* and not only those suffering as a result of a violation of the laws of armed conflict. ... More specifically, the Protocols use the term "victim" in a context which makes it absolutely clear that the term does not refer, or does not only refer, to persons afflicted by violations of the law of war.'[2]

---

[1] The author is grateful to Brigit Toebes for her comments in general as well as her advice on the question of remedies for economic, social and cultural rights.

[2] M. Sassòli, 'The victim-oriented approach of international humanitarian law and of the International Committee of the Red Cross (ICRC)', in M. Cherif Bassiouni, ed., *International Protection of Victims*, 7 *Nouvelles Etudes Penales* (1988) pp. 147 at 151.

McDonald/Kleffner/Toebes (eds.), Depleted Uranium Weapons and International Law
© 2008, T·M·C·ASSER PRESS, *The Hague, The Netherlands and the Authors*

This does not, however, mean that all victims of armed conflict generally – that is, those who do not actually suffer from a violation of the law – would be entitled to claim legal remedies. Normally, the existence of any legal remedy is contingent on actionable wrongdoing, although legal remedies which are effects- or damage-based are sometimes available.

Generally speaking, a direct or primary victim for the purpose of claiming a remedy is someone who has experienced a violation of international or national law and who has suffered damage as a result. An indirect or secondary victim is someone closely associated with that person who suffers consequentially as a result of the injury to the primary victim.

The Declaration of Basic Principles of Justice for Victims of Crime and Abuse of Power, adopted by the UN General Assembly in 1985, defines victims as

> 'persons who, individually or collectively, have suffered harm, including physical or mental injury, emotional suffering, economic loss or substantial impairment of their fundamental rights, through acts or omissions that are in violation of the criminal laws operative in Member States, including those laws proscribing criminal abuse of power.'[3]

The UN's Basic Principles and Guidelines on the right to restitution, compensation and rehabilitation for victims of gross violations of human rights and fundamental freedoms,[4] which deal with victims of violations of LOAC as well as human rights law, define a victim thus:

> 'For purposes of the present document, victims are persons who individually or collectively suffered harm, including physical or mental injury, emotional suffering, economic loss or substantial impairment of their fundamental rights, through acts or omissions that constitute gross violations of international human rights law, or serious violations of international humanitarian law. Where appropriate, and in accordance with domestic law, the term "victim" also includes the immediate family or dependants of the direct victim and persons who have suffered harm in intervening to assist victims in distress or to prevent victimization.'[5]

The European Court of Human Rights has also recognised that a victim is 'not only the direct victim or victims of the alleged violation, but also any person who would indirectly suffer prejudice as a result of such violation or who would have a valid personal reason in securing the cessation of such violation'.[6] The Inter-American Court has held that both pecuniary and non-pecuniary claims survive and automati-

---

[3] Resolution 40/34 adopted by the General Assembly on 29 November 1985, para. 1.

[4] Resolution 60/147 adopted by the General Assembly on 21 March 2006, Basic Principles and Guidelines on the Right to a Remedy and Reparation for Victims of Gross Violations of International Human Rights Law and Serious Violations of International Humanitarian Law [*on the report of the Third Committee (A/60/509/Add.1)*].

[5] Ibid., para. 8.

[6] *X* v. *Federal Republic of Germany*, Appl. No .4185/69, 35 Eur. Comm. HR Dec.& Rep. 89.

cally pass to a victim's heirs or successors. In *Loayza Tamayo* v. *Peru*, it held that the victim's family members were 'injured parties' within the meaning of Article 63(1) of the American Convention.[7] The Court defined family members broadly to include all persons linked by a close relationship. Rather than using the term 'family', it is preferable to employ the term 'immediate dependents'.[8] This allows family members of non-marital relationships to benefit from any available remedies, avoiding discrimination based on marital status. This is also consistent with the jurisprudence.[9]

Although '[w]ho should be regarded as victims of international crimes was not specifically debated during the elaboration of the Rome Statue',[10] the definition of victim was subject to lengthy discussion during the PrepComs preceding and following the adoption of the Statute of the International Criminal Court (ICC). The Rules of Procedure and Evidence that emerged from the PrepComs define victims to include 'natural persons who have suffered harm as a result of the commission of any crime within the jurisdiction of the Court' (Rule 85(a)) as well as 'organizations or institutions that have sustained direct harm to any of their property, which is dedicated to religion, education, art or science or charitable purposes, and to their historic monuments, hospitals and other places and objects for humanitarian purposes' (Rule 85(b)).

The common elements of all these definitions are the twin requirements of an underlying violation and the necessity to have suffered harm as a result, and it is clear that under international law one cannot claim to be a victim if one cannot show that one has suffered a breach of an identifiable legal obligation and resultant damage.

3.   THE CONCEPT OF REMEDIES IN INTERNATIONAL LAW

Before addressing the question whether individuals who claim to have suffered injury or damage as a result of exposure to DU have any possibility of claiming remedies under international law, it is necessary to briefly explain what is meant and encompassed by the term remedy as a matter of international law. Under national law, the concept may be far broader.

---

[7] *Loayza Tamayo* case, Reparations (Art. 63(1) American Convention on Human Rights), Judgment of 27 November 1998, Inter-Am. Ct. HR (Ser. C) No. 42 (1998), para. 89.

[8] Ibid., para. 92.

[9] See *Aloeboetoe, et al.* case, Reparations (Art. 63(1) American Convention on Human Rights), Judgment of 10 September 1993, Inter-Am. Ct. HR (Ser. C) No. 15 (1993), para. 71 and *Garrido and Baigorria* case, Reparations (Art. 63(1) American Convention on Human Rights), Judgment of 27 August 1998, Inter-Am. Ct. HR (Ser. C) No. 39 (1998), para. 52.

[10] S.A. Fernández de Gurmendi and H. Friman, 'The Rules of Procedure and Evidence of the International Criminal Court', 3 *YIHL* (2000) pp. 289 at 313.

3.1   Primary and secondary rules

Under international law, remedies are secondary rules. Therefore, they do not regulate conduct but are concerned with determining the legal consequences of a breach of a primary rule. There can thus be no international legal remedy in the absence of a pre-existing right/obligation and/or prohibition which can be violated.

International law does not define the meaning of remedy or reparations, although some attempts have been made to indicate what these terms might encompass. The terms 'remedies' or 'reparations' have been loosely if not statutorily defined in the context of state responsibility law.[11]

> '... the basic idea behind all types of legal consequences for breaches of international obligations is to make good the injury caused to persons or property by the State or other subject of international law. To this end, three main types of reparation may be distinguished: restitution, satisfaction, and damages. The award of damages thus presupposes that restitution (restitutio in integrum) or satisfaction are not possible or not sufficient means of reparation. Frequently, however, damages are awarded in conjunction with satisfaction.'[12]

As noted by Parker J. in the *Lusitania* cases, the 'remedy must be commensurate with the injury received. ... The compensation must be adequate and balance as near as may be the injury suffered.'[13]

Is it necessary to suffer injury or some other form of damage in order to claim an international legal remedy? There are different views. According to the definitions offered in the UN Declaration of Basic Principles of Justice for Victims of Crime and Abuse of Power,[14] the Bassiouni Basic Principles and Guidelines,[15] the Rules of Procedure of the International Criminal Court (ICC), and with general principles of international law concerning the law of remedies, it is not sufficient for a prohibited act to have occurred: the victim must also have suffered harm.

---

[11] See I. Brownlie, *System of the Law of Nations: State Responsibility*, Part 1 (Oxford, Oxford University Press 1983) pp. 199 et seq. for a comprehensive discussion of the forms and functions of reparation.

[12] E. Riedel, 'Damages', in 10 *Encyclopedia of Public International Law* (Amsterdam/New York/ Oxford, North-Holland 1987) p. 68. The classical definition of reparations is that given by the Permanent Court of International Justice (PCIJ) in the *Chorzów Factory* case: 'reparation must, as far as possible, wipe out all the consequences of the illegal act and reestablish the situation which would, in all probability, have existed if that act had not been committed. Restitution in kind, or, if this is not possible, payment of a sum corresponding to the value which a restitution in kind would bear; the award, if need be, of damages for loss sustained which would not be covered by restitution in kind or payment in place of it – such are the principles which should serve to determine the amount of compensation due for an act contrary to international law.' PCIJ (Ser. A), No. 17 (1928) p. 47. According to Riedel, 'The Court's dictum is generally accepted as stating a rule of international law, or even a general principle of law in the sense of Art. 38(1)(c) of the Statute of the International Court of Justice.' Ibid.

[13] *Lusitania* cases, 7 *RIAA* pp. 35, 36.

[14] Annex to General Assembly Resolution 40/34, 11 September 1985.

[15] *Supra* n. 4.

Speaking with regard to the law of state responsibility, Judge Jiménez de Aréchaga, in his General Course at The Hague Academy, stated 'that the requirement of damage is derived from a fundamental legal postulate that no one can maintain an action unless he has an interest of a legal nature'.[16] Another view rejects the idea that injury has to be shown as a necessary element of a claim arising from a violation,[17] holding that damage is inherent in any violation.

For practical reasons, however, especially the size of most victim populations, most schemes that have been set up to compensate individuals following the conclusion of armed conflict[18] have not been concerned with determining whether a violation has been committed but have focused on the damage caused.

4. THE INTERNATIONAL LEGAL STATUS AND PROCEDURAL CAPACITY OF INDIVIDUALS

Although, as Gries and Mohr argued in Chapter 10, a majority of jurists continue to believe that individuals can only indirectly claim remedies under the international law of state responsibility, there is a minority view to the contrary. Some jurisprudence and doctrine asserts that individuals may indeed be able to claim remedies directly, both under the law of state responsibility and in particular under LOAC provisions concerning state responsibility for the acts of the armed forces. Individual remedies against states may also be sought in international human rights law. Individuals who have suffered damage linked with exposure to DU weapons may also have a (remote) possibility to seek legal remedies directly against the individual responsible for the violation under international or national law.

---

[16] Quoted in O. Schachter, *International Law in Theory and Practice* (Dordrecht, et al., Martinus Nijhoff Publishers 1991) pp. 205-206.

[17] According to the International Law Commission (ILC), during the drafting of the Articles on State Responsibility, there 'was support for the concept of "objective" responsibility as a fundamental basis for the entire draft, one which rested on solid grounds. The view was expressed that the Commission had taken the truly revolutionary step of detaching State responsibility from the traditional bilateralist approach that had been conditioned upon damage, instead choosing an objective approach based on the transgression of a rule that brought State responsibility closer to the public order system found in modern national law that the idea of criminalizing the State should be abandoned to avoid confusion with domestic law notions that applied solely to individuals and could not be assimilated in international law. ... The notion of objective responsibility was described as an acknowledgement in resounding terms that there was such a thing as international lawfulness, and that States must respect international law even if they did not, in failing to respect it, harm the specific interests of another State, and even if a breach did not inflict a direct injury on another subject of international law.' Report of the ILC on the work of its 15th session, 20 April-12 June and 27 July-14 August 1998, Official Records of the General Assembly, 53rd session, Supplement No. 10.

[18] For example, the Claims Commissions established following the end of various armed conflicts, most recently, the United Nations Claims Commission (established to award claims arising out of Iraqi's unlawful invasion of Kuwait in 1991 by Security Council Resolution 687, adopted under Chapter VII of the UN Charter on 3 April 1990), and the Eritrea-Ethiopia Claims Commission (established by the Agreement Between the Government of the Federal Democratic Republic of Ethiopia and the Government of the State of Eritrea, concluded at Algiers on 12 December 2000).

It is clear that states are no longer considered to be the exclusive international legal actors. While they continue to occupy a unique position in international law, and remain the fundamental elements of the international legal order, it is recognised that there are other international legal persons besides states, albeit with a lesser or different legal status. Both individuals[19] and international organisations,[20] *inter alia*, are recognised to have limited international legal personality.

It has long been recognised that treaties can create direct rights and duties for individuals 'if it was the intention of the parties to do this'.[21] There are a number of treaties that create rights and obligations for individuals, such as the various human rights treaties, the Geneva Conventions and the Statute of the International Criminal Court. Starke has mentioned the Third Geneva Convention dealing with prisoners of war as a particular example of a treaty creating rights and obligations for individuals.[22]

Not only may treaties create rights and obligations for individuals, they may also create procedural capacity for individuals before international or national tribunals.[23] According to the Permanent Court of International Justice in the *Danzig Railway Officials* case:

> '...it cannot be disputed that the very object of international agreement, according to the intention of the contracting parties, may be the adoption by the parties of some definite rule creating individual rights and obligations and enforceable by the national court.'[24]

There are several examples of treaties creating some sort of procedural capacity for individuals who have suffered a violation of international law to seek redress, including the various international and regional human rights bodies which recognise a right of individual petition, such as the European Court of Human Rights (ECtHR) or the Inter-American Commission and Court of Human Rights. Victims of international crimes have some possibility of individual petition to the ICC and of receiving a reparations. On the other hand, there is no possibility of an individual directly bringing a complaint before the International Court of Justice (ICJ), which only has competence to hear contentious cases brought by states against other states.[25]

---

[19] See C. Nørgaard, *The Position of the Individual in International Law* (Copenhagen, Munksgaard 1962); R. McCorquodale, 'The individual and the international legal system', in M.D. Evans, ed., *International Law* (Oxford, Oxford University Press 2003) pp. 325; G. Aldrich, 'Individuals as subjects of international law', in J. Makarczyk, ed., *Theory of International Law at the Threshold of the 21st Century: Essays in Honor of Krzysztof Skubiszewski* (The Hague, Kluwer Law International 1996) p. 851; and M. Shaw, 'The subjects of international law', Chapter 5 of *International Law*, 5th edn. (Cambridge, Cambridge University Press 2003) pp. 175 at 232-241.

[20] Shaw, ibid., p. 241; I. Brownlie, *Principles of Public International Law*, 5th edn. (Oxford, Clarendon Press 1998) p. 677 et seq.

[21] Brownlie, ibid., pp. 585, 559.

[22] J.G. Starke, *Introduction to International Law*, 9th edn. (London, Butterworths 1984) p. 59.

[23] Brownlie, *supra* n. 20, at p. 581.

[24] *Jurisdiction of the Courts of Danzig*, 1928 PCIJ (Ser. B), No. 15 at pp. 17-18.

[25] Art. 34(1) of the Statute of the ICJ provides: 'Only states may be parties in cases before the Court.'

Historically, there are many examples of bilateral treaties concluded at the end of inter-state hostilities that established arbitral tribunals recognising the procedural capacity of individual victims to bring claims for damages sustained in the course of the conflict against the responsible state. Article 5 of the German-Polish Convention of 15 May 1922, dealing with Upper Silesia, authorised private individuals to bring suits before an international court against a state.[26] An agreement of 10 August 1922 between the United States and Germany, entered into pursuant to the 1921 Treaty of Berlin, established a Mixed Claims Commission to hear claims by US citizens against Germany for damage to property, rights and interests in Germany, and other claims for injury to persons or property as a consequence of war.[27] A French-Mexican Claims Commission was established in 1924 to receive claims by French nationals in respect of injuries suffered during the revolutionary activities in Mexico from 1910 to 1920.[28] Indeed, more than 80 mixed arbitral tribunals and claims commissions have been created following armed conflicts and revolutions,[29] with the first one being the US-Great Britain Commission established under the Jay Treaty of 1794[30] and the most recent being the Eritrea-Ethiopia Claims Committee, established pursuant to the Agreement between the Government of the Federal Democratic Republic of Ethiopia and the Government of the State of Eritrea of 12 December 2000.[31] The Eritrea-Ethiopia Claims Commission is mandated 'to decide through binding arbitration all claims for loss, damage or injury by one Government against the other, and by nationals (including both natural and juridical persons) of one party against the Government of the other party or entities owned or controlled by the other party'[32] that, *inter alia*, relate to violations of international humanitarian law or other violations of international law.

Despite the fact that individuals have been able to bring claims against states concerning violations of international law by that state, the general view is that the existence of individual procedural capacity on the international plane in specific cases does not make for a general rule. While individuals do have some procedural capacity under international law, it is not autonomous but relates to the specific creation of competence in a particular case. In other words, individuals have procedural capacity only when states grant it to them in specific cases. The fact that there are many such cases is not indicative of states' intention to create a

---

[26] H. Kelsen, *General Theory of Law and State* (Cambridge, Mass., Harvard University Press 1949) p. 348.

[27] D. Bederman, 'The glorious past and uncertain future of international claims tribunals', in M.W. Janis, ed., *International Courts for the Twenty-First Century* (Dordrecht, Martinus Nijhoff 1992) pp. 189-192.

[28] Ibid.

[29] Ibid.

[30] The Jay Treaty. Treaty of Amity, Commerce, and Navigation, signed at London, 19 November 1794. In H. Miller, ed., *Treaties and Other International Acts of the United States of America*, Vol. Two, Documents 1-4, 1776-1818 (Washington, Government Printing Office 1931).

[31] The Commission sits in The Hague at the Permanent Court of Arbitration. The Agreement can be found at <http://www.pca-cpa.org/upload/files/E-E%20Agreement(1).html>.

[32] Art. 5(1) Agreement between the Government of the Federal Democratic Republic of Ethiopia and the Government of the State of Eritrea of 12 December 2000.

general customary rule relating to individual procedural capacity. Individuals therefore do not have any international procedural capacity in an abstract or general sense: there is no rule of treaty or customary international law which recognises that individuals have the right to a remedy or the right or possibility to bring claims for violations of international law in general or international humanitarian law in particular. Courts and tribunals which do recognise the procedural capacity of individuals to bring claims for violations of international law require individual claimants to prove their *locus standi*, and, since the forum is created by treaty, it recognises only those complaints that are within its competence and jurisdiction. It is for this reason that an individual who has been the victim of a violation cannot forum shop.

5.   INDIVIDUALS AND THE LAW ON STATE RESPONSIBILITY

As a person has no independent procedural capacity except in the specific cases where states grant it to him or her, where such competence has not been created the general view is that individuals cannot operate independently under international law but depend on their state to represent them at the international level. An individual who has suffered harm as a result of an act attributable to a foreign state may be entitled to a legal remedy and may be able to assert a claim,[33] but it is widely believed that he must espouse it through the medium of his own state, and only against another state (only aliens are entitled to assert claims).[34] At its discretion, that state may or may not decide to pursue the matter with the responsible state. As one commentator notes: 'the state of nationality is under no obligation to exercise diplomatic protection on his behalf or even to deliver to him any compensation obtained from the offending state.'[35]

'The theory, then, is that the State is the subject of the claim, whilst the injured individual the mere object of the claim.'[36] In the *Mavrommatis Palestine Concessions* case, the PCIJ stated: 'By taking up the case of one of its subjects and by resorting to diplomatic action or international "judicial" proceedings on his behalf, a State is in reality asserting its own rights – its rights to ensure, in the person of its subjects, respect for the rules of its subject.'[37] In the *Chorzów Factory* case the PCIJ elaborated on the point:

[33] According to Woolsey writing over 100 years ago: 'The right of redress exists in the case of individuals. ... Redress consists of compensation for injury inflicted and its consequences.' T.D. Woolsey, *Introduction to the Study of Introduction Law* (New York, Charles Scribner's Sons 1892) pp. 17-18.

[34] R. Jennings, *Collected Writings of Sir Robert Jennings*, Vol. One (The Hague, Kluwer Law International 1998) pp. 143 at 144; Brownlie, *International Law*, supra n. 20, p. 585.

[35] M. Kamminga, 'Legal consequences of an internationally wrongful act against an individual', in T. Barkhuysen, M. van Emmerik and P.H. van Kempen, *The Execution of Strasbourg and Geneva Human Rights Decisions in the International Legal Order* (The Hague, et al., Martinus Nijhoff Publishers 1999) pp. 65 at 68.

[36] Jennings, *supra* n. 34, pp. 143 at 144.

[37] *Mavrommatis Palestine Concessions* case, PCIJ (Ser. A), No. 2 (1924).

> 'For the purposing of asserting a claim, the injury, if there be any, is deemed to be an injury to the state of the alien's nationality. ... Hence, if the alien is a stateless person, has the nationality of the state alleged to have acted wrongfully, or, although formally a national of the espousing state, his/her nationality is not entitled to recognition on the international plain, the claim will have to be dismissed.'[38]

On the other hand, the damage done to individuals is often the basis for calculating the amount of reparations to be paid.

6.   REMEDIES ARISING UNDER THE LAW OF ARMED CONFLICT

While an individual victim could not directly claim a remedy against a state based on the general principles of state responsibility, might he or she be able to directly claim a remedy against a state based on state responsibility for violations of LOAC?

Before coming to the question of whether LOAC may be directly relied on by individuals, a preliminary matter that must be addressed is whether LOAC recognises, in the first instance, primary rights of individuals, and in the second instance, the possibility of claiming remedies for breaches of those rights.

6.1   Does the law of armed conflict recognise rights of individuals?

The traditional view was that while individuals may benefit from certain protections arising from the law, these could not be considered as rights, and a failure to uphold these protections did not give right to any remedy. Writing in 1971, Draper said:

> 'in our own time jurists are not yet in agreement that the individual enjoys legal rights under the modern law of armed conflict. States may be enjoined by that law to ensure certain humanitarian standards of treatment to war victims such as POWs, civilians in occupied territory, and the sick and wounded in the armed forces, but that is not the same thing as conferring rights to such treatment flowing directly from international law.'[39]

Is this view still correct? Or is it possible to maintain that individuals do have primary rights to protection under LOAC and secondary rights to remedies when the former are breached? If states and combatants are the subject of obligations under the law, does this not create corresponding primary rights to protection for the civilian population, *inter alia*? Can one meaningfully speak of obligations of states and

---

[38] *Chorzów Factory* case. 13 September 1928. PCIJ (Ser. A), No. 17, p. 29.

[39] G.I.A.D. Draper, 'The relationship between the human rights regime and the law of armed conflict', in M.A. Meyer and H. McCoubrey, eds., *Reflections on Law and Armed Conflicts: The Selected Works on the Laws of War by the late Professor Colonel G.I.A.D. Draper, OBE* (The Hague, et al., Kluwer Law International 1998) pp. 125-126.

combatants to desist from specified acts without there being any concomitant right to protection from such acts being vested in the object of such protection?

It seems clear that rights arise under international law other than under human rights treaties, *stricto sensu*. The Geneva Conventions employ the terminology of rights. For example, Article 5 of Geneva IV refers to the '*rights and privileges*' arising under the present Convention, and again to the '*full rights and privileges* of a protected person under the present Convention'. [emphasis added.] Common Articles 7/7/7/8 state that protected persons may in no circumstances validly renounce 'in part or in entirety the *rights* secured to them by the present Convention'. The rights holder is the individual while the duty holder in principle can be both the state and the individual, and possibly also other actors, including, in principle, armed groups.[40] Common Articles 6/6/6/7 provide that 'No special agreement shall adversely affect the situation of [protected persons] nor restrict the *rights* which it confers upon them.' 'As the state is not the sole holder of the rights, it cannot validly waive them for the individual: the same is true for the individual vis-à-vis the state.'[41]

According to Meron:

> 'Although traditional humanitarian law originally governed almost exclusively interstate rights and obligations, contemporary humanitarian law, just like human rights law, contains a large component of rights of individuals vis-a-vis public authorities of their own countries (Article 75 of the First Geneva Protocol).'[42]

The Genocide Convention is couched in the language of prohibitions rather than rights. Yet, in the *Case Concerning Application of the Convention on the Prevention and Punishment of the Crime of Genocide (Bosnia-Herzegovina v. Yugoslavia), Preliminary Objections* the ICJ recognised rights arising under the Genocide Convention.[43] Even if the Genocide Convention does not recognise secondary remedies in the sense of an explicit right to a remedy for breaches of the prohibitions, a remedy under the Genocide Convention for breach of those rights could be considered to consist of the obligation on states to prosecute persons believed to have committed genocide.[44] The same could be said for victims of grave breaches of the

---

[40] See L. Zegveld, *Armed Opposition Groups in International Law: The Quest for Accountability* (Cambridge, Cambridge University Press 2002).

[41] R. Provost, *International Human Rights and Humanitarian Law* (Cambridge, Cambridge University Press 2002) p. 28.

[42] T. Meron, 'The convergence between human rights and humanitarian law', in *Human Rights and Humanitarian Law: The Quest for Universality* (The Hague, Martinus Nijhoff Publishers 1997) pp. 97 at 101. But cf., Provost, ibid., pp. 32, 45.

[43] The Court found that: '... the rights and obligations enshrined by the Convention are rights and obligations *erga omnes*'. 11 July 1996, para. 31.

[44] During the Diplomatic Conference to draft the Genocide Convention: 'A proposal to require reparations for genocide was defeated because delegates feared it might detract from the Convention's emphasis on criminal punishment.' N. Roht-Arriaza, ed., *Impunity and Human Rights in International Law and Practice* (New York, Oxford, Oxford University Press 1995) p. 26. See also W.A. Schabas, *Genocide in International Law* (Cambridge, Cambridge University Press 2000) pp. 400-401.

Geneva Conventions. Therefore, it is clear that the existence within a treaty of a complaints mechanism by which individuals can claim remedies in respects of rights breached is not determinative for not considering it a human rights treaty or a treaty which recognises rights. For example, the International Covenant on Economic, Social and Cultural Rights contains no such mechanism but no one would argue that it is not a convention that recognises primary human rights.

Another example concerns crimes against humanity. While under customary international law, crimes against humanity are framed in terms of prohibition rather than rights to protection from their commission, Lauterpacht argued that the acknowledgment by the international community that crimes against humanity existed in customary international law necessarily implied the recognition of corresponding fundamental human rights for the individual.[45]

It is thus clear that the term 'right' covers a variety of relationships: 'an entitlement to something from the bearer of a corresponding obligation, an immunity against encroachment of certain fundamental interests, a privilege to do something, a power to create a legal relationship, and several other variations'.[46] One could therefore conclude that primary rights exist outside of human rights law *stricto sensu*, including in LOAC, and that the existence of a primary right should be considered apart from any secondary means to enforce it; in other words, the primary right should not be considered as less than complete because it lacks an enforcement mechanism.

6.2   Does the law of armed conflict recognise secondary rights of victims of violations of the law?

The law of armed conflict does not explicitly recognise the right of an individual victim of a violation to a legal remedy, that is, it does not recognise any secondary rules relating to individual victims,[47] although it does recognise state responsibility for acts of the armed forces.[48] Nor does LOAC provide for any procedural mechanism through which individuals who have been the object of a breach of the law can claim a remedy.

Thus, even if one could make the argument that the use of DU in a particular case breached a rule of LOAC, such as the principle prohibiting superfluous injury or unnecessary suffering or the principles of distinction or proportionality, and there was a legal basis for a claim (to the effect that LOAC does recognise the right to a remedy of individual victims), there is no forum before which an individual victim can bring a complaint; no International or European or American

---

[45] H. Lauterpacht, *International Law and Human Rights* (London, Stevens & Sons 1950) pp. 35-37.

[46] Provost, *supra* n. 41, pp. 17-18.

[47] It does, however, address secondary rules to individuals fighting in the conflict through the grave breaches regime.

[48] Art. 3 Hague Regulations Respecting the Laws and Customs of War on Land annexed to the Hague Convention (IV) on the Laws and Customs of War on Land and Art. 91 AP I.

Court of LOAC. There is not even anything akin to the weak but useful international human rights bodies, such as the Human Rights Committee. The only mechanism for conducting investigations of possible violations of LOAC that is provided
for in the law – the Fact-Finding Commission established by Article 90 of Protocol
I – has no power to receive and examine complaints by individuals or to order states
to compensate individual victims.[49]

However, it cannot be excluded that individuals could seek a remedy against
states at the national level based on violations of LOAC.

6.3   **Can the law of armed conflict be directly enforced by individuals
        viz. states in domestic jurisdictions?**

As Gries and Mohr observed in Chapter 10, it would appear that the same limitations applicable in the case of state responsibility generally apply *vis-à-vis* state
responsibility for violations of the laws and customs of war. That is, it is generally
assumed that individual victims can bring claims based on state responsibility for
violations of the laws of war only through the medium of their states. Almost all the
cases that have dealt with the question have found that Article 3 of the Hague Convention does not create direct rights for individuals *vis-à-vis* enemy or their own
states.[50] This is also the view of the International Committee of the Red Cross.[51]

On the other hand, there is emerging doctrine and jurisprudence to support
the view that LOAC creates directly enforceable rights for individuals and that it
can have direct effect in national jurisdictions.[52] This was the view put forward by

---

[49] See F. Kalshoven, 'The International Humanitarian Fact-Finding Commission established by
the First Additional Protocol to the Geneva Conventions', in F. Kalshoven, *Reflections on the Law of
War: Collected Essays* (Leiden/Boston, Martinus Nijhoff Publishers 2007) pp. 843 at 849-853.

[50] See *Soc. Timber, Soc. Zeta, Soc. Obla* v. *Ministeri Esteri e Tesoro*, Italian Court of Cassation,
18 *ILR* (1951) p. 621; *Filipina ex-'comfort women'* v. *Japanese Government*, Tokyo District Court,
9 October 1998; *Filipino English ex-prisoners of war et al.* v. *Japanese Government*, Tokyo District
Court, 26 November 1998 and *Dutch ex-prisoners of war et al.* v. *Japanese Government*, Tokyo District Court, 30 November 1998. For commentary, see H. Kastutani, 2 *YIHL* (1999) pp. 389-390; *Shinto*
case, Tokyo District Court, 1 October 1999; *Tel Oren* v. *Libyan Arab Republic*, 726 F2d 774, 810. (D.C.
Cir. 1984); *Leo Handel et al.* v. *Andrija Artukovic*, 601 F. Supp. 1421 (United States District Court for
the Central District of California 1985); *Goldstar (Panama) SA* v. *US*, 967 F.2d 965, 970 (4th Cir. 1992)
pp. 58-59; *Hugo Princz* v. *FRG*, 26 F.3d 1166 (D.C. Cir. 1994); *Fishel* v. *BASF Group et al.*, Civil No.
4-96-CV-10449, Lexis 21230 (S.D. Iowa 1998); *Burger-Fischer et al.* v. *Degussa AG*, 65 F.Supp.2d
248, 273 (D. New Jersey 1999).

[51] The ICRC Commentary to the 1949 Fourth Geneva Convention states: 'The Convention does
not give individual men and women the right to claim compensation. The State is answerable to another contracting State and not to the individual ... It is inconceivable, at least as the law stands today,
that claimants should be able to bring a direct action for damages against the State in whose service the
person committing the breach was working. Only a state can make such claims on another State, and
they form part, in general, of what is called "war reparations".' J.S Pictet, *Commentary IV Geneva
Convention Relative to the Protection of Civilian Persons in Time of War* (Geneva, ICRC 1958) at
pp. 211 and 603.

[52] 'An international treaty is self-executing or directly applicable if it can be directly applied by
national courts and national authorities, if it establishes subjective rights and duties for the individual,

the expert witnesses in the Prisoner of War cases (Professors Frits Kalshoven, Christopher Greenwood and Eric David) heard by several Japanese courts.[53]

One must distinguish between the questions whether international law recognises the right to individuals to bring claims against states and whether states do. As Greenwood points out, the right of individuals to raise claims against third states arises under international law rather than national law, whereas the US cases denying such a right were really concerned with the jurisdiction of US courts to hear such claims, which is a question of US domestic law rather than international law.[54] This was clearly stated to be the case in *Leo Handel*.[55] Therefore, one can argue that while LOAC may recognise the right of individuals to raise claims, and there is nothing in LOAC to prevent them from presenting claims, the enforceability of those claims before particular national courts will depend on the jurisdiction of those courts to hear those claims. Thus claims might be made in one jurisdiction, whose national law recognises the competence of its courts to hear such claims, but not in others, which do not.

It should also be pointed out that most of the cases that rejected the direct enforceability of LOAC did so on policy rather than on strictly legal grounds. In the case of *Leo Handel, et al.* v. *Andrija Artuković*,[56] the District Court for the Central District of California found that

> 'Recognition of a private remedy under the Convention would create insurmountable problems for the legal system that attempted it; would potentially interfere with foreign relations; and would pose serious problems of fairness in enforcement. [...] The code of behaviour the Conventions set out create perhaps hundreds of thousands or millions of lawsuits by the many individuals, including prisoners of war, who might think their rights under the Hague Convention violated in the course of any large-scale war. Those lawsuits might be far beyond

---

and if the individual can rely on it before national courts and national authorities. (...).' A. Bleckmann, 'Self-executing treaty provisions', in 7 *Encyclopedia of Public International Law* (Amsterdam/New York/Oxford, North-Holland 1987) pp. 414 at 414-415. See also M. Janis, R. Kay and A. Bradley, *European Human Rights Law: Text and Materials*, 2nd edn. (Oxford, Oxford University Press 2000) p. 468; T. Buergenthal, 'Self-executing and non-self-executing treaties in national and international law', Collected Courses of the Hague Academy of International Law, 35 *Recueil des Cours* (1992) p. 303.

[53] See Claims for compensation from Japan arising from injuries suffered by former POWs and civilian internees of the ex-Allied Powers, Decision rendered by the Civil Division No. 31 of the Tokyo District Court, 26 November 1998; Claims for compensation from Japan arising from injuries suffered by former POWs and civilian internees of the Netherlands, Decision rendered by the Civil Division No. 6 of the Tokyo District Court, 30 November 1998. Both decisions, and all the expert opinions, are partially reproduced in English translation, in *War and the Rights of Individuals: Renaissance of Individual Compensation*, H. Fujita, I. Suzuki and K. Nagano, eds. (Tokyo, Nippon Hyoron-sha Co., Ltds. Publishers 1999).

[54] See also Expert Opinion by C. Greenwood, in *War and the Rights of Individuals*, ibid., p. 68.

[55] *Leo Handel et al.* v. *Andrija Artukovic*, 31 January 1985 [United States District Court for the Central District of California, 601 F. Supp. 1421 (1985).

[56] Ibid.

the capacity of any legal system to resolve at all, much less accurately and fairly; and the courts of a victorious nation might well be less hospitable to such suits against that nation or the members of its armed forces than the courts of the defeated nation might, perforce, have to be. Finally, the prospect of innumerable private suits at the end of the war might be an obstacle to the negotiation of peace and the resumption of normal relations between nations.'

In the Japanese cases concerning POWs and the so-called 'comfort women', the courts' decisions have turned not only on the fact that the courts considered that Article 3 does not create a right of action for individuals but crucially the fact that individual claims against states were considered as barred by virtue of the peace treaties conducted by Japan at the end of the Second World War which provided for lump sum payments and by which the States Parties waived their rights and their nationals' rights to bring further claims.[57]

Kalshoven has argued that the drafting history of Article 3 of the Fourth Hague Convention clearly reveals the intention to create rights that are directly enforceable by individuals. His position has been set out in a number of publications[58] and was expressed by him in his expert evidence to the Tokyo District Court.[59] He maintains that '[a] prime purpose of the introduction of the Article was from the outset to provide individual persons with a right to claim compensation for damages they suffer as a result of acts in violation of the Regulations.'[60]

---

[57] See the citations at *supra* n. 50.

[58] F. Kalshoven, 'State responsibility for warlike acts of the armed forces', 40 *ICLQ* (1991) p. 827.

[59] Expert Opinion by Frits Kalshoven, in *War and the Rights of Individuals*, *supra* n. 53, pp. 31 et seq.

[60] Ibid., p. 37. Eric David adds that 'the German representative at the Conference declared that the object of the proposed provision was to allow individuals injured to obtain compensation for an act by an officer or a soldier by addressing himself directly to the responsible Government ... the Swiss representative observed that ... the proposal conferred a "right of compensation" in favour of individuals.' The Swiss representative added that '[t]he principle that it lays down is applicable to each injured individual, whether nationals of neutral States or nationals of enemy States. The only distinction established between these two categories of victims and of rightful claimants concerns the regulation of compensation, and the distinction between them, on this point, is in the very nature of things.' Whereas nationals of neutral states could immediately claim compensation, those of enemy states had to wait until the conclusion of hostilities. Finally, '[t]he British delegation concluded that (...): "I do not contest the obligation which exists for a belligerent Power to compensate those who have been the victims of violations of the laws and customs of war and Great Britain does not wish in any way to avoid its obligations". Even if Great Britain expressed certain reservations about the practical difficulties concerning the establishment of damage caused, it did not question the basic principles of the German draft proposal. At the end of the conference Great Britain withdrew its reservations and accepted "the principle of compensation for breach of the Hague Regulations" without restriction. Only one State made and maintained a reservation with regard to this principle: Turkey, without providing any explanation. Apart from Turkey no other State which voiced an opinion, challenged the right of victims of a breach of the Regulations, to obtain compensation from the Government responsible. Some of them even emphasised this right (...).' ... 'No State questioned the right of compensation to people injured, a right which was expressed by those intervening in the discussions on the draft German proposal. The preparatory works therefore show clearly that Article 3 of the 1907 Convention was intended to and does give a direct right of compensation to individuals.' Expert Opinion by E. David in *War and the Rights of Individuals*, *supra* n. 53, pp. 49 et seq.

Although Article 3 itself does not give any clue as to whether it is a self-executing provision, Kalshoven argues that 'both on the basis of its drafting history and on systematic grounds of interpretation ... Article 3 must be recognized as having the implicit character of a self-executory rule and hence must be applied as such on the domestic level.'[61] This is so, he claims, because the provision was contained in the Convention rather than in the annexed Regulations, which, according to Article 1 of the Fourth Convention, need to be implemented into national law. On the other hand, there is no such stipulation regarding the Convention itself. This distinction indicates that the former was intended to be self-executing.

Eric David asserts that the private law principle that a person who 'by an act contrary to law damages, with intent or negligence, the right of another, is obliged to provide redress for the resulting damage' is 'equally applicable to international law'.[62]

A few national judgments have upheld direct enforceability of Article 3. For example, in 1997 a Greek court found that it was competent to hear a claim brought by a large number of individual plaintiffs and the prefecture of Boetia (Voetia) against Germany for material and moral damage in respect of torts (summary executions and the complete obliteration of the village of Distomo on 10 June 1944 by German forces, including wilful murders and property damage carried out by the Germany Occupying Forces),[63] notwithstanding the fact that a Treaty between Germany and Greece had supposedly settled the matter. Significantly, the Court noted:

> 'a) that the clause contained in the said article 3 of the 1907 Hague IV Convention "if the case demands" does not constitute a built-in flexibility clause but a particular emphasis on the fact that pecuniary damages must be sustained . . . b) that the contested claims are admissibly presented in their individual capacity and not by the State whose citizens they are since that is not precluded from an rule of international law ... c) that the judicial pursuit of the plaintiffs' claims is not precluded by either the express reservation of Law 2023 of 10/13 March 1952 on the suspension of the state of war with Germany according to which "the existing state of war between Greece and Germany is suspended from 10 June 1951, notwithstanding the settlement by the Peace Treaty to be concluded of outstanding issues and existing disputes arising from the war", or the suspensive clause of article 5 para. 2 of the London Agreement of 27 February 1953 concerning German Foreign Debts, to which Greece became a party on 21 April 1956, having previously ratified the said Agreement by Law 3480/1956 ....'[64]

Gavouneli states the court was 'at pains in establishing that there is no rule in international law preventing private parties from seeking redress in their individual

---

[61] Expert Opinion by Frits Kalshoven, in *War and the Rights of Individuals*, *supra* n. 53, pp. 44 et seq.

[62] Expert Opinion by Eric David in *War and the Rights of Individuals*, *supra* n. 53, p. 50.

[63] *Multi-member Court of Levadia 137/1997*, 30 November 1997. For commentary see M. Gavouneli, 'War reparation claims and state immunity', 50 *RHDI* (1997) pp. 595 at 600.

[64] Quoted in Gavouneli, ibid., pp. 601-602.

capacity rather than as citizens protected by their state of nationality. In this respect it found propitious ground on the cited decision 33/93 of the German Bundesverfassungsgericht, which made the same point under similar circumstances.'[65]

However, these occasional instances of judicial willingness to recognise that individuals can directly invoke Hague and Geneva Law must be set against the many cases where this view has been rejected.

One could conclude that even if Article 3 of the Fourth Hague Convention of 1907 established a direct right of individuals to compensation from third states, this treaty-based right has, through subsequent state practice and *opinio juris* to the contrary, fallen into abeyance or been replaced by a modified right, which may amount to this: international law establishes the right of individuals to compensation in respect of violations of the laws and customs of war, but that right is generally only enforceable through states although its direct effect in some national jurisdictions cannot be excluded.[66] The right to an individual remedy for violations of LOAC before national courts is therefore dependent on whether a particular state recognises the direct effect of international law in the national legal system and the competence of individuals to bring complaints founded on breaches of the law.

7.   REMEDIES ARISING UNDER HUMAN RIGHTS LAW

7.1   **The right to a remedy under human rights law**

The concept of a legal remedy for violations of one's human rights is enshrined in almost all of the major human rights treaties. It is contained in the general human rights conventions, including the 1966 International Covenant on Civil and Political Rights (ICCPR),[67] in conventions addressing discrete rights, including the 1966 International Convention on the Elimination of All Forms of Racial Discrimination;[68] the 1980 Convention on the Rights of the Child,[69] the 1984 Convention

---

[65] Ibid., pp. 603-604.

[66] Provost states: 'Despite some indications in the *travaux préparatoires* of the 1907 Hague Convention IV that the obligation to compensate imposed by Article 3 could have been intended to benefit individuals as well as states, subsequent practice indicates that the corresponding customary rule has evolved to envisage claims from states only.' *Supra* n. 43 at p. 45.

[67] Art. 2(3) of the ICCPR provides: 'Each State Party undertakes: (a) To ensure that any person whose rights or freedoms as herein recognized are violated shall have an effective remedy, notwithstanding that the violations has been committed by persons acting in an official capacity; (b) To ensure that any person claiming such a remedy shall have his right thereto determined by competent judicial, administrative or legislative authorities of the State, and to develop the possibilities of judicial remedy; (c) To ensure that the competent authorities shall enforce such remedies when granted.' Art. 9(5) provides that anyone who has been the victim of unlawful arrest or detention shall have an enforceable right to compensation, while Art. 14(6) provides for compensation, when a person has suffered punishment as a result of a miscarriage of justice.

[68] Art. 6.

[69] Art. 39.

against Torture and Other Cruel, Inhuman or Degrading Treatment or Punishment (Torture Convention)[70] and the 1989 International Labour Organisation (ILO) Convention,[71] and in regional human rights treaties, including the 1950 European Convention on Human Rights,[72] the 1969 American Convention on Human Rights[73] and the 1981 African Convention on Human and Peoples' Rights.[74] Sometimes the right to a legal remedy is expressed in general terms; at other times it is expressed as a right of access to justice when the guaranteed rights are violated. In all cases, the right pertains to an effective remedy at a national level; international channels of redress only become available when domestic remedies have been exhausted.[75]

The right of individuals to a remedy in respect of breaches of their human rights, and to have access to a complaints mechanism, was upheld in broad terms in the Declaration on the Right and Responsibility of Individuals, Groups and Organs of Society to Promote and Protect Universally Recognized Human Rights and Fundamental Freedoms, adopted by the UN GA in 1998.[76] Moreover, Article 47 of the Charter of Fundamental Rights of the European Union,[77] adopted in 2000, recognises that everyone whose fundamental rights and freedoms guaranteed by the law of the Union are violated has 'the right to an effective remedy ...'. This includes the right to a fair and public hearing, within a reasonable time.

In contrast with the situation under LOAC, almost all of the international and regional human rights treaties establish at least some sort of individual complaints mechanism, by virtue of which victims can directly petition the bodies or bodies established to hear complaints, although, in most cases, they are still optional. There is a trend towards establishing individual complaint mechanisms where they do not already exist and strengthening existing mechanisms. Victims' chances

---

[70] Art. 13.

[71] Art. 15(2) provides for the right to 'fair compensation for damages'; Art. 16(4) provides for 'compensation in money'; and Art. 16(5) provides for 'full compensation for any loss or injury'.

[72] Art. 13 of the ECHR provides: 'Everyone whose rights and freedoms set forth in this Convention are violated shall have an effective remedy before national authorities notwithstanding that the violation has been committed by persons acting in an official capacity.'

[73] The American Convention on Human Rights provides does not include the general right to a remedy as a substantive civil and political right, but as a matter of the jurisdiction and functions of the Court. Art. 63(1) provides: 'If the Court finds that there has been a violation of a right or freedom protected by this Convention, the Court shall rule that the injured party be ensured the enjoyment of his right or freedom that was violated. It shall also rule, if appropriate, that the consequences of the measure or situation that constituted the breach of such right or freedom be remedied and that fair compensation be paid to the injured party.' However, Art. 10 gives as a civil and political right a narrower right to compensation: 'Every person has the right to be compensated in accordance with the law in the event he has been sentenced by a final judgment through a miscarriage of justice.'

[74] Art. 7 of the African Charter on Human and Peoples' Rights provides: 'Every individual shall have the right to have his cause heard. This comprises: (a) the right to an appeal to competent national organs against acts violating his fundamental rights as recognized and guaranteed by conventions, laws, regulations and customs in force ...'

[75] Art. 26 ECHR; See A.A. Cançado Trindade, *The Application of the Rule of Exhaustion of Local Remedies in International Law* (Cambridge, Cambridge University Press 1983).

[76] GA Res. 53/144, 9 December 1998, Art. 9.

[77] 2000/C 364/01.

of success in actually attaining remedies through these mechanisms vary considerably under each treaty regime.

The most sophisticated system for the protection and enforcement of human rights is established by the European Convention on Human Rights, which operates under the auspices of the Council of Europe.[78] Since the implementation of Protocol No. 11 and the institution of the new ECtHR on 1 November 1998, the European Commission of Human Rights has been abolished and its functions have been absorbed by the Court. Access by individuals to the Court has been broadened with the replacement of the Optional Protocol by a provision making mandatory the right of individual petition.[79] In practice, this will not affect existing States Parties, most of which were already party to the Optional Protocol.[80]

Unlike the European Convention, the American Convention does not explicitly recognise a general right to a remedy but rather a right to 'judicial protection' (Art. 25(1)).[81] Further, Article 10 of the American Convention provides for the right of every person 'to be compensated in accordance with the law in the event he has been sentenced by a final judgment through a miscarriage of justice'.[82]

Under Article 44 of the American Convention and Article 26 of the Inter-American Commission's Regulations, '[a]ny person, or group of persons or any non-governmental entity legally recognized in one or more member states of the Organization [OAS] can submit a complaint before the Commission.'[83] Article 26 of the Regulations of the Commission states that any person can submit individual

---

[78] See generally C. Kilpatrick, T. Novitz and P. Skidmore, eds., *The Future of Remedies in Europe* (Oxford, Hart Publishing 2000); C. Ovey and R.C.A. White, *The European Convention on Human Rights*, 3rd edn. (Oxford, Oxford University Press 2002).

[79] Under Art. 34: 'The Court may receive applications from any person, non-governmental organization or group of individuals claiming to be the victim of a violation by one of the High Contracting Parties of the rights set forth in the Convention or the protocols thereto. The High Contracting Parties undertake not to hinder in any way the effective exercise of this right.' For analysis, see H. Steiner and P. Alston, *International Human Rights In Context: Law, Politics, Morals, Text and Materials*, 2nd edn. (Oxford, Oxford University Press 2000) pp. 797-804.

[80] 'By 28 June 1994, all 30 of the states then party to the Convention had agreed to recognize the admissibility of private petitions, ordinarily called 'applications' in the practice of the Commission.' M. Janis, R. Kay and Anthony Bradley, *European Human Rights Law: Text and Materials*, 2nd edn. (Oxford, Oxford University Press 2000) p. 29.

[81] Art. 25(1) guarantees everyone 'the right to simple and prompt recourse, or any other effective recourse, to a competent court or tribunal for protection against acts that violate his fundamental rights recognized by the constitution or laws of the state concerned or by this convention, even though such violation may have been committed by persons acting in the course of their official duties'. In consequence, under subpara. 2, the States Parties undertake '(a) to ensure that any person claiming such remedy shall have his rights determined by the competent authority provided for by the legal system of the state; (b) to develop the possibility of a judicial remedy; and (c) to ensure that the competent authorities shall enforce such remedies when granted'.

[82] C.M. Cerna, 'The Inter-American Court of Human Rights', in Janis, *supra* n. 27, pp. 117 at 120; C.M Cerna, 'The structure and functioning of the Inter-American Court of Human Rights, 1979-1992', 63 *BYIL* (1992) pp. 135 at 142.

[83] V. Krsticević, 'The development and implementation of legal standards relating to impunity in the Inter-American system of human rights protection', 10 *Interrights Bulletin* (1996) p. 91.

petitions 'on one's own behalf or on behalf of third persons'.[84] Moreover, in a further departure from the European system, the regulations allow the Commission to initiate a case *sua sponte* (of its own accord) without receiving a complaint.[85] Article 45 of the Convention allows States Parties, upon ratification, to recognise the competence of the Commission to receive and examine communications submitted by other States Parties alleging a violation of one of the human rights guaranteed in the Convention. Article 63(1) guarantees that victims of violations of the Convention shall be awarded reparations.[86] Article 68(2) provides: 'That part of a judgment that stipulates compensatory damages may be executed in the country concerned in accordance with domestic procedure governing the execution of judgments against the state.'

The ICCPR established a Human Rights Committee to monitor compliance with its terms.[87] The Committee can hear individual complaints from persons resident in states that have signed Optional Protocol I.[88] However, the Committee's decisions are not legally binding in the sense of being enforceable judicial decisions, but are referred to as 'views' in Article 5(3) of the Optional Protocol. 'After the Committee has made a finding of a violation of one or more provisions of the Covenant, it usually proceeds to ask the State Party to take appropriate steps to remedy the violation.'[89] Although States Parties that have signed up to the Optional Protocol are legally obliged to afford victims effective remedies, if a State Party refuses to implement the view, there is little that the Committee, lacking enforcement machinery, can do to persuade it otherwise.

7.2 Remedies for victims of depleted uranium-related illnesses under human rights law

In order to claim a remedy before a human rights body, a victim must show that there has been a violation of a primary human right, and that he or she has suffered damage as a result. This means that the individual victim must show an actual violation of a human right that is protected under the particular Convention pursu-

---

[84] Art. 26(1) of the Regulations of the Inter-American Commission on Human Rights, reprinted in *Basic Documents Pertaining to Human Rights in the Inter-American System*, OEA/Ser.L.V/II.82 doc.6 rev.1 at 103 (1992).

[85] Ibid.

[86] 'If the Court finds that there has been a violation of a right or freedom protected by this Convention, the Court shall rule that the injured party be ensured the enjoyment of his right or freedom that was violated. It shall also rule, if appropriate, that the consequences of the measure or situation that constituted the breach of such right or freedom be remedied and that fair compensation be paid to the injured party.'

[87] Art. 28(1).

[88] For general commentary see H. Steiner, 'Individual claims in a world of massive violations: What role for the Human Rights Committee?', in P. Alston and J. Crawford, eds., *The Future of Human Rights Treaty Monitoring* (Cambridge, Cambridge University Press 2000) p. 38.

[89] T. van Boven, 'The right to restitution, compensation and rehabilitation', in *International Meeting on Impunity of Perpetrators of Gross Human Rights Violations, Geneva, 2-5 November 1992* (Dijon-Chenôve, Imprimerie Abrax 1992) pp. 301 at 306.

ant to which the individual is making the application. A complaints mechanism set up under a human rights treaty is only competent to hear complaints related to breaches of the treaty establishing it; for example, the ECtHR cannot entertain claims relating to breaches of the American Convention on Human Rights. There are also limits on all of these courts' personal jurisdiction; usually the petitioner has to be within territory or jurisdiction of the state against which he or she is bringing the complaint.

As shown in this book, in particular – limited – cases, the use of DU weapons could potentially implicate violations of several legal rules, including rules of LOAC and rules of international human rights law. But an individual victim wishing to pursue a claim before a human rights complaints mechanism arising out of DU use can only assert a violation of human rights law. International human rights bodies are not competent to hear complaints of breaches of other bodies of law, for example, LOAC. In the *Los Palmeras* case,[90] the Inter-American Court rejected the argument of the Inter-American Commission in the *Tablada* case[91] that the Commission and Court are competent to apply LOAC directly, specifically, common Article 3 of the 1949 Geneva Conventions. However, it found that it is competent to refer to the Geneva Conventions whenever necessary to interpret a provision of the American Convention.[92] While the Court did not expressly address whether it would be competent to find a state guilty of violations of LOAC and order it to pay compensation to the victims, it made clear that the American Convention only confers competence on the Inter-American Court or Commission to hear violations of the rights protected by that Convention.[93] Of course, a particular act could be at the same time a violation of human rights law and LOAC; in that case, the complaint would have to be based on the human rights violation in order to be admissible before a human rights body.

As Toebes noted in Chapter 9, the right to life would not offer any general protection to a combatant during an armed conflict, it being lawful to kill him under LOAC as the *lex specialis*. Thus, for such a person, there would be no possibility of a remedy for a denial of the right to life. For a civilian, human rights still offer protection during an armed conflict, and the right to life remains absolute for them, although civilian deaths that are not disproportionate are permitted under the LOAC,

---

[90] Judgment on Preliminary Objections of 4 February 2000, Inter-Am. Ct. H.R. (Ser. C) No. 67 (2000).

[91] *IACHR Report No. 55/97*, Case No. 11.137, Argentina, Oea/Ser.L/V/II.97, Doc. 38 (30 October 1997). The Commission reached the same conclusion in its 1999 Report on Colombia, See IACHR, Third Report on the Human Rights Situation in Colombia, Oea/Ser.L/V/II.102, Doc. 9 Rev. 1 (26 February 1999), Chapter IV, B.2, para. 13, where it stated that it was required to utilise humanitarian law, along with human rights law, in general reports, 'for the purpose of analyzing a State's international responsibility relating to violence, where much of that violence occurs in the context of an armed conflict. ... Humanitarian law ... may provide the Commission ... with a set of accepted legal standards that enable it to examine the conduct not only of the State's security forces, but of its armed opponents as well, and to note and classify the acts of violence that infringe these standards.'

[92] *Supra* n. 90, paras. 32 and 33.

[93] Ibid., para. 34.

which takes precedence over human rights law during situations of armed conflicts. If, however, civilian deaths during armed conflict as a result of the use of DU could be considered as disproportionate under LOAC, the right to life under human rights law might also be infringed. Such civilians – or rather their next of kin – might have grounds to bring a case before a human rights body alleging a violation of the right to life with respect to the use of DU.

The human rights bodies may be of particular service in trying to make public information relating to the use of DU. Toebes has mentioned the right to information in Chapter 9. People living in countries where DU has been used could have the possibility to bring a case demanding that states make full disclosure regarding the sites where DU has been used and in what amounts. There is a theoretical, if remote, possibility of using the human rights mechanisms as a way to force states to undertake widespread testing of persons exposed to DU. Persons living in areas where DU has been used could argue that unless testing and monitoring are carried out, their right to information regarding the full effects of DU is being breached.

The problem is that human rights bodies hear claims brought by victims against their own state or at least one to whose jurisdiction they are subject. This would make it impossible for the ECtHR, for example, to hear a case brought by a Serbian national against the United Kingdom, following the finding of the ECtHR in *Banković* that for a state to have extra-territorial jurisdiction it must exercise effective control over another territory.[94] However, such jurisdiction would arise in cases of occupation. It is not fanciful, therefore, to imagine that a case could be brought by an Iraqi or Afghan national before the ECtHR against the United Kingdom relating to DU use by the United Kingdom in Iraq or Afghanistan or that similar suits be brought against the United States before the Inter-American Commission.

While the ICCPR has the Human Rights Committee,[95] an equivalent complaints mechanism does not exist for the ICCPR's sister convention, the 1966 International Covenant on Economic, Social and Cultural Rights, although there has been some discussion about adding an optional protocol to the ICESCR.[96] Indeed, the latter convention is one of the only human rights conventions not to expressly recognise a right to a remedy for breaches of its provisions. This means that there is no forum before which victims can bring a claim alleging a violation of the right to health or to food, for example.

---

[94] ECtHR, *Banković and Others* v. *Belgium and 16 Other Contracting States* (Appl. No. 52207/99), 12 September 2001.

[95] For general commentary see Steiner, *supra* n. 88, p. 38.

[96] Following the adoption in 1996 by the Committee on Economic, Social and Cultural Rights of a draft Optional Protocol (E/CN.4/1997/105), an expert meeting on an Optional Protocol to the ICESCR was held on 30 November 2001 in Leuven, Belgium. That meeting urged the UN Commission on Human Rights to give high priority to the consideration of a draft Optional Protocol to the ICESCR and to the establishment of a Committee competent to hear individual complaints. See W. Vandenhold, 'An Optional Protocol to the International Covenant on Economic, Social and Cultural Rights', prepared by the Institute for Human Rights, KU Leuven, for the Expert Seminar on an Optional Protocol to the International Covenant on Economic, Social and Cultural Rights (30 November 2001).

One should note the limitations of the remedies that human rights bodies can grant. The precise remedy that will be offered is at the discretion of the particular complaints mechanism. The remedy can include material compensation, and ordering the offending party to desist from the illegal activity that gives rise to the claim for a remedy. However, even in the event that a human rights body found the use of DU in a particular case to constitute a violation of human rights, it seems only a remote possibility that it would order the offending state to desist from all uses of DU, or that a state would comply with such an order.

8.   REMEDIES UNDER INTERNATIONAL CRIMINAL LAW

It would be difficult to imagine an individual who has suffered ill effects as a possible result of exposure to DU being able to bring a successful case before the ICC or any other international court or tribunal. The use of DU in a particular case would have to amount to a war crime, a crime against humanity or genocide. It would be tenuous, to say the least, to argue that the use of DU *per se* represents a widespread or systematic attack on a civilian population, or a deliberate attempt to destroy a protected group in whole or in part.[97] Its effects on the civilian population are collateral to its primary use as an anti-*matériel* weapon against military objectives, and it is not used with the deliberate or criminal intention of causing harm to civilians (or indeed to combatants).

In order for DU use to constitute a war crime, its use would have to be considered as a grave breach of the Geneva Conventions or AP I by or as a violation of common Article 3 or AP II, the breach of whose provisions is now generally considered to give rise to individual criminal responsibility.[98] It is difficult to see how the use of DU as an anti-*matériel* weapon could constitute a grave breach of the Geneva Conventions. Grave breaches are violations committed against persons or property protected by the Conventions. Military objectives are not protected objects and soldiers are not protected persons, so using DU against tanks could hardly be considered as a grave breach. If personnel experience superfluous injury or unnecessary suffering as a result of their exposure to DU, this may be unlawful where a breach of the principle was specifically intended, but it is not a war crime. Particularly where the results of the use of DU were secondary to its intended purpose, its use cannot be considered to give rise to criminal culpability.

---

[97] See J. Kleffner and T. Boutruche, Chapter 6 of this book, at pp. 125 et seq.

[98] As demonstrated by their inclusion in Art. 8(2)(c) and (e) of the Statute of the International Criminal Court. Violations of LOAC applicable in non-international armed conflict have been prosecuted by the International Criminal Tribunals for the Former Yugoslavia and Rwanda and by the Special Court for Sierra Leone. Moreover, the ICRC Customary International Humanitarian Law study found that individual criminal responsibility arises for war crimes committed in both international and non-international armed conflicts as a matter of customary international law. J.-M. Henckaerts and L. Doswald Beck, *Customary International Humanitarian Law*, Vol. 1: Rules (Cambridge, Cambridge University Press 2005) p. 551.

The only possibility would be if, knowing DU to have deadly effects, it was deliberately used with the specific intention of killing or causing inhuman treatment, or wilfully causing great suffering or serious injury to body or health of civilians. This scenario seems unlikely, however. Even if DU can be considered to have indiscriminate effects, and to have possible negative impacts on the civilian population which is exposed to the secondary effects of its use against hard military targets, the possible breach of a principle or rule of LOAC through the use of DU weapons does not mean that a war crime has been committed. The criminal intent is missing.

The International Criminal Tribunal for the Former Yugoslavia (ICTY), in the report of the Office of the Prosecutor on the Decision not to Prosecute NATO in relation to its bombing of the Federal Republic of Yugoslavia (FRY), considered the use of DU by the alliance during the conflict. The investigating panel found that the use of DU was not an action giving rise to the possible criminal responsibility of individuals.[99]

Therefore, it seems unlikely that any court would bring charges against an individual with respect to the use of DU weapons, even in the event that it would be able to exercise jurisdiction over nationals of the main user state, the United States.

It is impossible to consider any circumstance in which the use of DU of itself could contravene the Genocide Convention, as some have alleged.[100]

Finally, it should be noted that most persons suffering from so-called Gulf War or Balkan War Syndromes or any illness which is anecdotally linked with exposure to DU are not actually calling for the criminal prosecution of those responsible for sanctioning its use or those who fired it in battle. The significance of criminal prosecution as a remedy in relation to DU use is therefore negligible to non-existent.

9. REMEDIES AVAILABLE AT THE DOMESTIC LEVEL FOR PERSONS AFFECTED BY EXPOSURE TO DEPLETED URANIUM

We have concluded that even if individuals have rights under LOAC, that law does not specifically recognise the right to a remedy of individual victims of violations

---

[99] The Final Report to the Prosecutor by the Committee Established to Review the NATO Bombing Campaign Against the Federal Republic of Yugoslavia referred, *inter alia*, to the question of the legality of the use of DU weapons during the NATO bombing campaign and whether it might *prima facie* constitute a war crime. In declining to initiate an investigation into the use of DU projectiles by NATO during the campaign against the FRY, the report stated: 'There is no specific treaty ban on the use of DU projectiles. There is a developing scientific debate and concern expressed regarding the impact of the use of such projectiles and it is possible that, in future, there will be a consensus view in international legal circles that use of such projectiles violate general principles of the law applicable to the use of weapons in armed conflict. No such consensus exists at present.' ICTY OTP, *Final Report to the Prosecutor by the Committee Established to Review the NATO Bombing Campaign Against the Federal Republic of Yugoslavia* (8 June 2000).

[100] R.C. Koehler, 'Depleted Uranium (DU) – Silent Genocide', Tribune Media Services (25 March 2004), <http://www.mindfully.org/Nucs/2004/DU-Silent-Genocide25mar04.htm>; M. Daud Miraki, 'The Silent Genocide from America', <http://www.rense.com/general37/InvisibleGenocid.html>.

of those rights and there is in any event no forum at the international level where they could take such a complaint. Human rights law offers some possibilities of a remedy for persons suffering harm as a result of exposure to DU. The only other possibility of attaining a remedy is available at the national level. This could potentially include an array of possibilities arising under LOAC and human rights law as incorporated into domestic law, national civil law, military law, labour law, administrative law and contract law, etc. An individual might try and bring a case against his or her own or another state or state organ or agent before their own courts or in another jurisdiction claiming a violation of LOAC. As noted, the chances of such as suit succeeding will depend not only on the applicant proving his or her *locus standi* but also that the damage caused is attributable to DU use by the respondent party and on the willingness of the domestic courts to recognise the direct enforceability of LOAC by individuals against states within the legal system. An individual could alternatively seek whatever non-adversarial possibilities exist at the national level for having his or her condition recognised and treated and for attaining compensation or disability payments. Except in cases where treatment or compensation does not depend on proving the cause of the disability, the main obstacle will be proving that the illness has resulted from exposure to DU.

So far, there have only been a few instances where a domestic judicial body was willing to recognise that exposure to DU was responsible in whole or in part for the pathology alleged. In early 2004 a UK pensions tribunal was the first judicial body anywhere to make an award specifically related to DU poisoning. The Scottish Pensions Appeal Tribunal on 2 February 2004 found that Mr Kenny Duncan had suffered exposure to DU during his service in the Gulf during the 1991 war. According to a press report: 'Mr Duncan has suffered increasing breathlessness and aching joints which he has linked to DU'. All three of his children born post-1991 have suffered physical problems. His exposure to DU was attributable to his service in the Gulf.[101]

On 10 July 2004 a Rome Court ordered the Italian Defence Ministry to pay 500,000 euros in compensation to the family of Stefano Melone, an Italian soldier who served in the Balkans and who died of cancer in 2001. The court stated that Melone had died 'due to exposure to radioactive and carcinogenic substances', and listed DU amongst those substances.[102]

9.1   Non-adversarial remedies available in the United States for members of the US Armed Forces

If an individual member of the US Armed Forces suffers a war related injury, he or she is entitled to compensation. The legal authority is chapter 61 of title 10, United

---

[101] M. Williams, 'First award for depleted uranium poisoning claim', *The Herald* (Glasgow) (4 February 2004); National Gulf Veterans & Families Association. Press Release, 3 February 2004.

[102] E. Tinjak, F. Boric and H. Griffiths, 'Bosnians say NATO bombs brought "angel of death"', IWPR's Balkan Crisis Report, No. 526, 15 November 2004.

States Code.[103] The general rule is that any disability discovered during military service is presumed to have been incurred during military service for purposes of eligibility for compensation, unless medical evidence proves to the contrary. The compensation is not fault but results-based (there is no admission of liability). Furthermore, the compensation does not relate to the use of a particular method or means of warfare but to the particular manifestation of disease. A veteran would receive his disability payment because he is suffering from cancer *per se*, not because he is suffering from cancer associated with the use of DU. There is thus no linkage between the cause of the effects manifested and the disease itself. Thus, while many Gulf War veterans have suffered illnesses and received disability payments, these payments do not recognise that the veteran has suffered from Gulf War Syndrome or DU-related illnesses.

The Department of Defense rules with respect to the disability compensation system for military members apply to current members of the military. For former members of the military seeking compensation for military injuries and illnesses, the rules of the Department of Veterans Affairs[104] take over.[105]

9.2 Non-adversarial remedies available in the United Kingdom for members of the UK Armed Forces

A report by the UK Government Select Committee on Defence Eleventh Special Report,[106] addressed the subject of Gulf Veterans' Illnesses.[107] On the question of financial assistance it noted that the House of Commons Defence Select Committee had stated that 'it is possible that we may never find the answer to the causes of Gulf veterans' illnesses. What is important, though, is that those who have served their country feel that they are adequately compensated if they have suffered illnesses as a result of their service.'[108] It said that it accepted that 'those who have served their country should be adequately compensated for ill health arising from their Service. This is currently provided for by War Pensions, the Armed Forces Pension Scheme attributable benefits and preferential rates of pensions and supplementary allowances compared with most Social Security equivalents.'

The House of Commons Report had argued in favour of waiving the seven-year rule in relation to Gulf veterans because 'the nature of their illnesses means that it may be some years before symptoms, and their severity, are apparent'.[109]

---

[103] The primary Department of Defense regulation is DoD Instruction 1332.38, <http://www.dtic.mil/whs/directives/corres/html/133238.htm>.

[104] <http://www.va.gov/>.

[105] The author is grateful to Mr John Casciotti, US Department of Defense Office of General Counsel for providing this information.

[106] House of Commons Defence Select Committee, *Seventh Report of Session 1999-2000, on Gulf Veterans' Illnesses* (HC 125) (11 May 2000) (hereinafter, HCDSC), <http://www.publications.parliament.uk/pa/cm199900/cmselect/cmdfence/753/75302.htm>.

[107] Annex A contains the government's response to the HCDSC's Seventh Report on Gulf Veterans' Illnesses.

[108] HCDSC, *supra* n. 106, para. 74.

[109] Ibid., para. 77.

The government, in its response, stated:

> 'The Government believes that the legislation covering claims made outside the seven year rule is sufficiently generous to allow war pensions to succeed. For deaths arising, or disablement claims lodged within seven years of termination of service, the onus lies with the Secretary of State to show beyond reasonable doubt that the disablement or death is not due to service. There is no onus on the claimant to show any link between disablement and service. These provisions are in Article 4 of the Naval, Military and Air Forces (Disablement and Death) Service Pensions Order 1983, as amended. Even where a claim for disablement is made more than seven years after termination of service, or where death occurs more than seven years after service, the onus of proof is still more generous than the burden of proof in civil tort which rests on a balance of probabilities. Article 5 of the Naval, Military and Air Forces (Disablement and Death) Service Pensions Order 1983, as amended provides that it is necessary for the claimant only to raise reasonable doubt, based on reliable evidence, that the death or disablement is due to service. The benefit of any reasonable doubt is always given to the claimant. As these provisions have applied to claims for war pensions for half a century it would not be feasible to restrict any concession on the seven year rule to Gulf veterans.'[110]

The government denied that it was 'using its position of strength to the detriment of the ill veterans' and stated that the

> 'MOD has assisted veterans through its policy of openness, making available the information it possesses which is relevant to Gulf veterans' illnesses. MOD has also agreed not to rely on the defence of limitations under the Limitation Act 1980 in respect of negligence claims concerning Gulf veterans' illnesses. This means that such claims can in effect be launched at any point of the veterans' choosing instead of before the normal cut off point of 3 years, unless and until the Department gives notice to Gulf veterans to issue and serve such proceedings and they fail to do so within the notice period given. MOD has also agreed to extend the standard provisions of the Access to Medical Records Act of 1990 which normally entitles individuals only to their medical records complied after 1 November 1991. In the case of Gulf veterans we have arranged access to records back to 1 August 1990 to ensure their whole period of Gulf service is covered. In short MOD has done nothing to deprive Gulf veterans of their right to take legal action against MOD in accordance with the provisions of the Crown Proceedings (Armed Forces) Act 1987.'[111]

Veterans are also entitled to all of the services offered by the National Health Service, while the health needs of serving personnel are met by the Defence Medical Services. 'The Government believes that the current financial provisions are the most appropriate way of compensating Gulf War veterans; they are able to apply for no fault compensation through the War Pensions Agency, and for an Armed

---

[110] At p. 9 of Annex A.
[111] At pp. 10-11 of Annex A.

Forces Pension or an attributable Benefit for Reservists. (The scope of the latter scheme has just been extended).'[112]

10. CONCLUDING REMARKS

While the term remedies obviously has a legal connotation – international law recognises the concept of reparations for wrongs suffered – it can also be understood in a far wider sense. The discussion about what is meant by remedies in the context of victims of damage associated with exposure to DU, and what remedies are and should be available to any victim, cannot be separated from an analysis of what these victims actually need and want.

Where an individual has suffered an injury that he or she attributes to exposure to DU, he or she will have several needs and will require various forms of help. Most immediately, such an individual will need medical attention, and medical care will likely be an ongoing, long-term requirement. Medical care should address not only the victim's physical but also his or her psychological needs. In case the individual cannot work, he or she will obviously require some sort of disability allowance. The individual might be able to perform other work, in which case, he or she will require retraining. The individual might also feel entitled to some sort of punitive compensation for the illness that he attributes to DU. If a primary victim dies as a result of an illness which is linked to exposure to DU, the greatest need of any existing secondary victims will probably be financial. If an individual has been exposed, even if no ill effects have manifested themselves, he or she will likely need access to information regarding the level of exposure and the likely consequences, and the precautions to take. Persons living in areas that have been exposed will need information regarding where the hot spots are located and how best to minimise any personal risk.

Individuals with illnesses which they associate with exposure to DU are of course entitled to pursue any available remedy – legal or non-legal – even to the extent of trying to sue their own or another state for breaches of international law through the use of DU weapons. But the most effective responses to the needs of such persons will likely be either non-legal or non-adversarial in nature. Most victims do not want to have to jump through hoops – legal ones, in particular –in order to be recognised and to receive what they need.

Since the most pressing needs of persons exposed to DU are for information, medical care and disability payments or financial compensation, and responses should be based on effects and needs presenting, these persons above need assurance that these basic needs will be addressed. A lottery-based system, where the outcome is neither guaranteed nor fairly certain, would not be the most appropriate one. Therefore, litigation should be considered as a remedy of last resort. As seen, the United States and the United Kingdom at least provide for a continually func-

---

[112] HCDSC, *supra* n. 107, p. 12.

tioning system of disability and compensation payments and access to medical treatment for their serving personnel and veterans.

As far as trying to bring a case based on a violation of LOAC is concerned, a great limitation in LOAC as a body of rules is the absence of any mechanism for bringing claims based on breaches of the primary rules. As for bringing cases in national fora based on breaches of LOAC rules, the law of state responsibility, including the *lex specialis* rules in LOAC, recognises only inter-state claims, meaning that individuals cannot bring claims directly against states. The state of the affected individual's nationality could pursue a claim against the user state on the basis of state responsibility, but the success of such a claim would by no means be guaranteed, and there is moreover no legal obligation on the claimant state to distribute any reparations received amongst the victims of the violation.[113]

There is a minority view that argues that international law in general and LOAC in particular can have direct effect, and some jurisprudence to support it, and there is actually nothing in international law to prevent this. Generally, the lack of standing of individuals to bring claims against states before domestic fora is a limitation of national rather than international law. On the other hand, even if the possibility of doing so cannot be theoretically excluded, it is a remote one, and, as such, of limited practical value to persons suffering illnesses linked with DU exposure.

This means that even if individuals do have rights under the law, there is practically speaking no forum at the international or domestic level before which they can exercise them.

Even to the extent that such fora may exist, the legal obstacles to proving that a violation of international law has been committed in a particular case as a result of the use of DU weapons are such that an approach which does not depend on a finding of a violation of international law but that is effects- or damage-based would be more practically useful to persons claiming to have suffered ill effects as a result of exposure to DU.

For individuals suffering illnesses which they associate with their exposure to DU, international law may be of little practical help. Even if the possibility of a legal remedy can be shown to exist and there is a forum before which an individual has standing to bring a claim, the bar to attaining relief is simply too high. International law's most useful contribution is in establishing that there is a legal basis for remedies for persons affected by exposure to DU. Of most practical assistance to veterans and civilians will be remedies available at the national level that are not faulted-based but only depend on showing that damage has been caused. However, the problems of relying too much on remedies available at the national level must also be recognised. In particular, they are only available to persons within the state's territory or subject to its jurisdiction, and, therefore, largely inaccessible to nationals of other states affected by DU.

---

[113] *Mavrommatis Palestine Concessions* case, *supra* n. 37; *Civilian War Claimant* v. *The King* [1932] A.C. 14; *Rustomjee* v. *The Queen* (1876) 1 QBD 487; *German Property* v. *Knopp* [1933] 1 Ch. 439; J.G. Gardam and M. Jarvis, *Women, Armed Conflict and International Law* (The Hague, Kluwer Law International 2001) p. 230.

Part Four
CONCLUSIONS AND THE WAY FORWARD

Chapter 12
AVERTING FORESEEABLE AND UNEXPECTED DAMAGE: THE CASE FOR A PRECAUTIONARY APPROACH *VIS-À-VIS* DEPLETED URANIUM WEAPONS

Avril McDonald[1]

1.   THE DEBATE SURROUNDING THE EFFECTS AND LEGALITY OF THE USE
     OF DEPLETED URANIUM WEAPONS

The debate regarding the effects on human health of exposure to depleted uranium (DU) is polarised to an extent that makes assessment of the subject a challenging task. While some of its opponents allege that DU is responsible for life-threatening pathologies in persons exposed to it and horrific birth defects in their offspring,[2] some of its supporters claim that DU is so safe that you can eat it.[3] Such diametrically opposed views are by no means to be found only at the margins; even in the scientific mainstream, the safety of DU is both affirmed and disputed. Amongst medical and scientific professionals, both civilian and military, opinions on the effects of DU on human health swing from those which assert that exposure to DU particles or dust can cause cancers and other serious illnesses[4] to

[1] The author is very grateful to her colleagues who read earlier drafts of this chapter and offered valuable critical remarks: Dan Fahey, William Fenrick, Dieter Fleck, Charles Garraway, Françoise Hampson, Eric Jensen, Frits Kalshoven, Jann Kleffner, A.P.V. Rogers, Michael Schmitt, Steven Solomon, Brigit Toebes, Kenneth Watkin, and other reviewers who wish to remain anonymous. Those who provided critical comments do not necessarily concur with the views expressed in the chapter. The author alone is responsible for any errors.

[2] See for example, S. Horst-Gunth, *Uranium Projectiles – Severely Maimed Soldiers, Deformed Babies, Dying Children, Documentation of the aftermaths of the Gulf War 1993-1995* (Freiburg, Ahriman Publishing House 1996).

[3] According to a Reuters report, a 'U.S. Army briefer assured reporters it was safe enough to eat'. D. Hamilton 'Bid to bury plutonium factor dismays NATO', Reuters (20 January 2001). See also D. Hastings, 'Sickened Iraq war veterans cite 'depleted uranium'', Associated Press (12 August 2006); 'It's something that we eat, and drink and breathe everyday', said Dr Michael Kilpatrick, Deputy Director, Deployment Health Support Directorate, Office of the Deputy Assistant Secretary of Defense for Force Health Protection and Readiness; D. Sample, 'Pentagon officials say depleted uranium powerful, safe', *American Forces Press Service* (14 March 2003), <http://www.defenselink.mil/news/newsarticle.aspx?id=29292>.

[4] Depleted uranium (DU) has been described as a 'low level alpha radiation emitter which is linked to cancer when exposures are internal, chemical toxicity causing kidney damage. . . aerosol DU exposure to soldiers on the battlefield could be significant with potential radiological and toxicological

McDonald/Kleffner/Toebes (eds.), Depleted Uranium Weapons and International Law
© 2008, T·M·C·ASSER PRESS, *The Hague, The Netherlands and the Authors*

those which claim that no negative health effects are associated with its use.[5]

Given that there is disagreement within the scientific community as to the effects of DU, it should come as no surprise that the legality of DU *per se* and of its use as a weapon during armed conflict is also disputed.

The rest of section 1 sets out the views of various interested parties as to the legality of the use of DU: the conclusions of the authors of this study; the user and other states; and relevant international and non-governmental organisations.

Based on our legal findings and the concerns raised by other parties, section 2 sets out an approach for managing the use of DU that is aimed at (1) avoiding harmful (and potentially unlawful) levels of exposure to DU of the civilian population and of enemy and friendly combatants and consequential liability of states for the use of DU based on foreseeable risks; (2) forestalling unknown but possible risks (long-term effects which cannot yet be predicted). The term used to describe the package of proposals is a precautionary approach, although, as shown later, not too much should be read into that phraseology.

1.1   The conclusions of this study as to the legality of the use of depleted uranium weapons

Den Dekker found in Chapter 3 that DU, whether in the form of ammunition or armour, is not subject to any ban or restriction under the law of arms control, whether

---

effects.' Report prepared by the US Army Production Base Modernisation Activity, Picatinny Arsenal, New Jersey, July 1990; '… prolonged exposure could cause illness', Memo from the Department of the Navy outlining the need to make the Saudi Arabians aware of the effects of DU weapons, dated 8 September 1990; 'When soldiers inhale or ingest DU dust, they incur a potential increase in cancer risk. … Expected physiological effects from exposure to DU dust include possible increased risk of cancer (lung or bone) and kidney damage.' Memo from US Army Chemical Medical School on Depleted Uranium Safety Training, written by Col. Robert G. Claypool, director of Professional Services, 18 August 1993; 'Strong evidence exists to support [a] detailed study of potential DU carcinogenicity.' Dr David McClain, the U.S. military's top depleted uranium researcher, speaking to a presidential committee investigating Gulf War illnesses', quoted in D. Fahey, 'The final word on depleted uranium', 25 *Fletcher Forum of World Affairs* (2001) pp. 189 at 197.

[5] Dr. Michael Kilpatrick from the Pentagon's Office of the Special Assistant (incorrectly) told a NATO press briefing that: 'We have seen no cancers or leukaemia in this group, which has been followed since 1993. There has been no evidence of any subsequent medical problems that can be attributed to DU exposure.' NATO HQ Brussels, Briefing by Mr Mark Laity, NATO Acting Spokesman, Lt. Col Scott Bethal, Dr Michael Kilpatrick and Col Eric Daxon, 10 January 2001, <http://www.nato.int/docu/speech/2001/s010110b.htm>. See further D. Fahey, Chapter 2. More recently, a report performed by Batelle analysed the health hazards from DU dust inhalation for surviving crew members of an Abrams tank (with conventional and/or DU armour) and a Bradley Fighting Vehicle hit by DU and conventional penetrators. The properties of the dust were determined in penetrator impact tests, and the health hazards were calculated using ICRP models, based on various exposure scenarios. The report concluded that: 'For all vehicle configurations and exposure times modeled (up to 2 h), predicted radiation doses are not likely to cause adverse health effects' and 'For all vehicle configurations and exposure times, except for the unventilated Abrams tank perforated through conventional armor, predicted uranium concentrations in the kidney are not likely to cause adverse chemically-induced health effects.' Depleted Uranium Aerosol Doses and Risks: Summary of U.S. Assessments (October 2004), <http://www.deploymentlink.osd.mil/du_library/du_capstone/index.pdf>.

in customary or treaty form. As it is not a nuclear, chemical or biological weapon, it cannot be considered as a weapon of mass destruction. As a type of conventional weapon, it is not currently expressly prohibited by any international treaty, including the Certain Conventional Weapons (CCW) Convention. In particular, it cannot be considered as an incendiary weapon within the meaning of Protocol III to the CCW Convention.

Den Dekker did conclude that international regulations concerning the transfer of and trade in fissile material, such as the International Atomic Energy Agency's safeguards system, apply to DU. Although the transfer of and trade in DU weapons are not prohibited by international law, they are subject to export controls and licenses. But these types of restrictions aim at controlling the trade in and transfer of DU for security reasons (non-proliferation), considering that it is a dual-use material which could be converted into a nuclear weapon, rather than at restricting the use of the weapon during a situation of armed conflict.

The question of the legality of a weapon as such – which should, in principle, be determined at the stage of a weapon's review, before it is ever fielded – and the lawfulness of its use in a particular military operation during a situation of armed conflict must be distinguished. Even if DU is not specifically prohibited or restricted under the law of arms control, its use during armed conflicts, as with any weapon, is subject to the law of armed conflict (LOAC). All weapons must be used in a way that comports with that law, in particular, the principles of superfluous injury and unnecessary suffering to combatants and distinction between military objectives and civilians and civilian objects. However, LOAC does not impose any prohibitions on particular categories of weapons nor does it prohibit the use of any weapon in all circumstances (other than poison weapons and expanding bullets).

As Carnahan explained in Chapter 4, in deciding at the strategic level which weapons provide the greatest military utility, states can take into account not only a weapon's military effectiveness but also logistical factors, such as its transportability, its cost, its effects on enemy morale, whether it increases (or reduces) troop safety, as well as the availability of alternatives.[6] At the point of deciding which weapons to introduce into their arsenals, states should assess whether, notwithstanding a weapon's utility, it will inevitably cause superfluous injury or unnecessary suffering to combatants, because it has been designed to, for example, or whether it is *per se* indiscriminate or poisonous. If so, the weapon should not be fielded at all. In battle, the use of weapons is largely governed by the principles of unnecessary suffering and proportionality.

The contributors to this book have concluded that the use of DU weapons *per se* is currently not in violation of LOAC, although there may be situations where breaches of the law could potentially be implicated.

As far as the protection of combatants from certain methods and means of warfare is concerned, weapons of a nature (that have the effect) to cause superfluous injury and unnecessary suffering to combatants should not be used.[7] The ICJ,

---

[6] B. Carnahan, Chapter 3, pp. 75 et seq.

[7] Art. 35(2) AP I.

in its Advisory Opinion on *The Legality of the Threat or Use of Nuclear Weapons*, stated that unnecessary suffering is 'a harm greater than that unavoidable to achieve legitimate military objectives'.[8] Zwanenburg argued in Chapter 5 that '[t]he principle calls for a balancing of humanitarian concerns (the suffering and injury inflicted) and military necessity. The test is not whether there is a vast amount of injury or suffering but whether the suffering is needless, superfluous or manifestly disproportionate to the military advantage reasonably expected from the use of the weapon.'[9]

This author disagrees that what is required is a balancing test that measures whether the suffering or injury is (manifestly) disproportionate to the military necessity, in the sense of the proportionality test used to measure whether injury to civilians is excessive. It is not as though, as the military necessity of the use of a weapon decreases, the suffering to combatants which is considered as unnecessary increases in inverse proportion, even if the question of the weapon's military necessity is highly relevant in assessing whether suffering and injury will ultimately be considered as unnecessary and superfluous. Dinstein and others have rightly criticised such a proportionality approach to determining what is superfluous injury and unnecessary suffering.[10]

The correct test is whether in targeting one uses weapons that have the effect of causing superfluous injury or unnecessary suffering to combatants. This has two aspects: first, a threshold of suffering or injury must be met. The test for this is high, given that it is, after all, permitted to kill and injure combatants, using both anti-personnel and anti-*matériel* weapons. The St. Petersburg Declaration still provides some guidance as to what is required: the objective of defeating the greatest number of men 'would be exceeded by the employment of arms which uselessly aggravate the sufferings of disabled men, or render their death inevitable.'[11] According to the ICRC Customary International Humanitarian Law Study, '[t]he causing of permanent disability or the rendering of death inevitable are two factors that some states consider relevant for finding that a weapon causes superfluous injury or unnecessary suffering.'[12] Second, even if the suffering and injury meets the required threshold, it may not be considered as unnecessary or superfluous if there is a military advantage to using that weapon. However, if an alternative weapon is

---

[8] *Legality of the Threat or Use of Nuclear Weapons*, Advisory Opinion of 8 July 1996, para. 238, ICJ General List No. 95, 35 *ILM* 809 & 1343 (1996).

[9] M. Zwanenburg, Chapter 5, pp. 111 et seq.

[10] See Y. Dinstein, *The Conduct of Hostilities Under the Law of International Armed Conflict* (Cambridge, Cambridge University Press 2004) p. 59: 'The reference to proportionality is linked to the issue of collateral damage to civilians and has nothing to do with injury or suffering sustained by combatants.' See also H. Meyrowitz, 'The principle of superfluous injury or unnecessary suffering from the Declaration of St. Petersburg of 1868 to Additional Protocol I of 1977', 34 *IRRC* (1994) pp. 98 at 109-110.

[11] 1868 St. Petersburg Declaration Renouncing the Use, in Time of War, of Explosive Projectiles Under 400 Grammes Weight, reprinted in A. Roberts and R. Guelff, *Documents on the Laws of War*, 2nd edn. (Oxford, Clarendon Press 1989) p. 29.

[12] J.-M. Henckaerts and L. Doswald-Beck, eds., *Customary International Humanitarian Law*, Vol. 1: Rules (Cambridge, Cambridge University Press 2005) p. 241 (hereinafter, CIHL)

available which provides the same advantage, the suffering and injury would be unnecessary and superfluous.

Zwanenburg showed in Chapter 5 that there is a presumption that the principle prohibiting superfluous injury and unnecessary suffering only applies to (enemy) combatants and not to civilians. He found that it does not prohibit particular categories of weapons but only particular uses of a weapon, with the possible exception of weapons expressly designed to cause superfluous injury or unnecessary suffering. He found no state practice to support the view that the principle could absolutely prohibit the use of a weapon which causes superfluous injury or unnecessary suffering in a majority of cases where it is not designed to have this effect. Still, if at the review stage a weapon was found to cause superfluous injury or unnecessary suffering in most cases, it would be unlikely to pass legal muster.

As regards the fundamental question (for the use of DU weapons and the enemy combatants affected by their use) whether the principle applies only to anti-personnel weapons or also to anti-*matériel* weapons, Zwanenburg said that there was strong evidence that states take the restrictive view that it applies only to anti-personnel weapons. He argued that there are good reasons for taking a more progressive approach and applying the principle to anti-*matériel* weapons as well, particularly given the nature of modern warfare, but found that state practice did not support this position. Other writers argue that it is actually impossible to apply the principle to anti-*matériel* weapons since their use against *matériel* will inevitably involve the projection of more force than would be necessary to render combatants inside it *hors de combat*.

This author disagrees with Zwanenburg's conclusions in this respect and, given that the principle is the only one protecting enemy combatants from some of the worst effects of warfare, it is worth explaining the legal basis for the disagreement.

It is true that when states have been moved to prohibit by treaty the use of weapons based on the principle which prohibits the causing of superfluous injury and unnecessary suffering to combatants it has been in relation to anti-personnel weapons. But this does not imply that they consider the rule to cover only these weapons and not anti-*matériel* weapons as well. In the case of anti-*matériel* weapons which are directed against *matériel*, there will not usually be any breach of the principle if a combatant is killed or injured. It has not been necessary to ban by treaty anti-*matériel* weapons on the basis that they cause superfluous injury or unnecessary suffering because such injury and suffering is usually – although not always – necessary and therefore not unlawful. But states apply the principle to all weapons, particularly at the point of a legal review of the weapon, in order to determine whether their intended use would breach it. Indeed, it is the application of the principle that will influence whether a weapon is designated only for use against *matériel* (if its military utility can be shown) and not for anti-personnel use. The prevalent view amongst weapons reviewers, which is shared by the US Department of Defense, is that the principle not only applies during the actual use of a weapon in an attack; it is a standard to be applied during weapons reviews as well, and to all types of weapons.[13]

In conducting its legal review of DU ammunition, both the United States Army and Air Force reviewers considered its legality according to the principle prohibiting superfluous injury and unnecessary suffering, as discussed below in section 1.2.1. It influenced their decision to recommend its use solely for anti-*matériel* purposes, unless other weapons were not available.

The British Military Manual does not directly answer the question whether the principle applies both to anti-personnel and anti-*matériel* weapons, but it does admit that it must be applied in a very different reality from that in which it was conceived, and implicitly recognises that it applies to anti-*matériel* weapons that can have incidental effects on combatants.[14]

It is therefore not true that state practice indicates that the principle prohibiting causing superfluous injury or unnecessary suffering to combatants only applies to anti-personnel weapons. There is practice and *opinio juris*, most importantly that of the only two acknowledged DU-user states, to show that states do consider the principle in relation to anti-*matériel* weapons in such a way as to restrict the use of such weapons to hard targets and to prohibit or limit their use against personnel, where alternatives are available.

Turning to the argument that the principle cannot be applied to anti-*matériel* weapons, Dinstein, referring to Solf, claims that:

> 'In essence, the injunction between "superfluous injury or unnecessary suffering" hangs on a distinction between injury/suffering that is avoidable and unavoidable. This requires a comparison between the weapon in question and other options. Two issues arise in particular: (a) whether an alternative weapon is available, causing less injury or suffering and, shifting the focus, (b) whether the effects produced by the alternative weapon are sufficiently effective in neutralizing enemy personnel. Inescapably, the "test is valid only for weapons designed exclusively for anti-personnel purposes", inasmuch as (for instance) artillery explosives designed to pulverize military fortifications "may be expected to cause injuries to personnel in the vicinity of the target which would be more severe than necessary to render these combatants *hors de combat*".'[15]

It is submitted that Dinstein and Solf are applying the wrong test.

It should be noted, first, that it can be somewhat formalistic to speak of two separate and entirely distinct categories of anti-*matériel* and anti-personnel weapons. Anti-*matériel* weapons directed against *matériel* are usually used with the in-

---

[13] This was the view expressed to the author by several military reviewers of this chapter who wish to remain anonymous.

[14] UK Ministry of Defence, *The Manual of the Law of Armed Conflict* (Oxford, Oxford University Press 2004) p. 102, para. 6.1.2.

[15] Dinstein, *supra* n. 10, p. 60, referring to W. Solf, 'Article 35', *New Rules for Victims of Armed Conflicts*, M. Bothe, K.J. Partsch and W.A. Solf, eds. (The Hague, Martinus Nijhoff Publishers 1982) p. 196.

tention not only to disable the hard target but also the combatants inside it. The anti-*matériel* and anti-personnel effects of the weapon cannot be separated. Although it would be admittedly impossible or at least redundant to apply the principle of superfluous injury or unnecessary suffering to the use of an anti-*matériel* weapon against *matériel* where no combatants would be affected, the principle can be applied to the use of anti-*matériel* weapons where those weapons are intended to disable not only *matériel* but also personnel. Anti-*matériel* weapons are also sometimes used directly against personnel.

When assessing the legality of the use of an anti-*matériel* weapon against *matériel* (and personnel inside it) according to the principle prohibiting superfluous injury or unnecessary suffering, the correct test to be applied is the following: (a) whether an alternative weapon is available, causing less injury or suffering and (b) whether the effects produced by the alternative weapon are sufficiently effective in disabling enemy *matériel* (and personnel). If there is a choice of weapons, and the alternative is equally effective in disabling *matériel* (and personnel), all other things being equal, provided that the suffering or injury meets the requisite threshold, any additional injury or suffering caused by one relative to the other could potentially be considered as superfluous and unnecessary according to the principle.[16]

Where an anti-*matériel* weapon's military utility is proven, and there is no equally or more effective weapon, or there is but there are other military advantages arising from the first weapon's use, such as the fact that it is cheaper or increases troops' safety, it can be used, and any suffering or injury caused to combatants inside the target would be considered to be necessary even if reached a high level. States are entitled to use weapons that are necessary to destroy hard military objectives and to deliberately or incidentally disable any combatants inside them, including by killing or injuring them. But the military necessity of defeating a hard military objective and of disabling, including by killing or injuring, any combatants inside it does not relieve parties from the obligation to take into account the effects of the use of anti-*matériel* weapons on combatants and determine if these exceed what is needed to destroy the *matériel* and disable the greatest number of men. These include both intended and incidental primary and secondary effects. If an anti-*matériel* weapon, such as DU, were to uselessly aggravate the suffering of disabled men, or render their death inevitable, its use would violate the principle if an alternative offering an equal military advantage existed.

The question whether the intended and incidental primary effects of exposure to DU – death or serious injury as a result of the impact of the strike and the burning fragments released but also the spalling effect[17] produced inside the tank, which is designed to incapacitate combatants – causes superfluous injury and un-

---

[16] The ICRC Study on customary international humanitarian law states: 'Some States also refer to the availability of alternative means as an element that has to go into the assessment of whether a weapon causes unnecessary suffering or superfluous injury.' The two states it cites as including such a provision in their military manuals are the United States and the United Kingdom. CIHL, *supra* n. 12, p. 240.

[17] 'Spall', Wikipedia, <http://en.wikipedia.org/wiki/Spall>.

necessary suffering can only be answered by reference to the question whether such suffering and injury is in itself great enough to be considered as unnecessary and superfluous and whether there is an alternative weapon. It is doubtful that this is the case as far as the primary effects are concerned, given that death, where it occurs as the result of the strike and spalling effect, is almost instantaneous, and in fact this is usually the goal of the strike. Even if an alternative weapon was used, it would have much the same primary effect[18] and the intention would be to have this effect, given that this is exactly what armoured piercing projectiles are designed to achieve.

It might be assumed that a soldier inside a tank stands a greater chance of being directly killed by the strike and the resultant fires and spalling (the primary effect) than by temporally removed after-effects of any exposure to DU, such as the development of cancer or another serious illness years later (the secondary effect). If the primary effect is not unlawful, it might seem impossible for any secondary effects to violate the principle prohibiting superfluous injury or unnecessary suffering. The question would be moot if the results of the primary effects were generally successful. But, as Fahey noted in Chapter 1, in the Gulf War 89 percent of the crewmen in US tanks hit by DU rounds in friendly fire incidents survived the strikes. Those personnel would be more affected by any potentially detrimental secondary effects than by the primary effects of the strikes. While the principle prohibiting superfluous injury and unnecessary suffering would not apply to these friendly combatants, this statistic indicates that it is relevant to consider the secondary effects of DU on enemy combatants.

As the application of the principle of superfluous injury or unnecessary suffering is based on the effects of the weapon,[19] it can be argued that all effects, incidental secondary as well as intended and incidental primary, should be considered in applying it, where they can be expected, from the fact that the combatant will inevitably die, but in a way in which his suffering is needlessly aggravated, to the fact that he might develop a life-threatening illness if he survives. Even effects which are remote in time should be considered, if they are foreseeable. As a practical matter, however, it is particularly difficult to convince the military of the need for concern about the secondary effects of a weapon on combatants' health and well-being when it is not illegal to kill or injure those combatants as the primary effect of the use of a particular weapon.

Where a designated anti-*matériel* weapon is used as an anti-personnel weapon, the principle prohibiting unnecessary suffering will also come into consideration, and here the question of the availability of alternatives is again paramount. As Carnahan noted, as far as air warfare is concerned, the impossibility of changing weapons mid-flight means that effectively there is no alternative as regards targets of opportunity. The military necessity (and the possibility of infringement of the principle prohibiting superfluous injury and unnecessary suffering) is different in

---

[18] Tungsten, for example, also produces a spalling effect.

[19] The ICRC in its study on customary international humanitarian law states: 'The prohibition by means of warfare which are of a nature to cause superfluous injury or unnecessary suffering refers to the effects of a weapon on combatants.' CIHL, *supra* n. 12, at p. 240.

this case than where DU weapons are used as anti-personnel weapons where alternatives are available. Given that alternative weapons, causing less suffering and injury to combatants, are usually available to disable combatants, there is generally no military necessity for using DU weapons against personnel, and such use could potentially involve a breach of the unnecessary suffering principle.

In conclusion, it is this author's view that the principle prohibiting causing superfluous injury or unnecessary suffering could be relevant to the use of DU ammunition even if it is considered and mainly used as an anti-*matériel* weapon. The expected primary and secondary effects of the use of DU weapons should be considered in deciding whether its use might involve a breach of the principle. Whether DU weapons could be said to actually cause superfluous injury or unnecessary suffering when used in an anti-*matériel* capacity is difficult to say. It is not possible to come to a definitive conclusion on the question whether the suffering and injury caused is, objectively speaking, serious enough to be considered as unnecessary or superfluous, but there is a good case to be made that it is, given the serious concerns that have been raised about DU's secondary effects on combatants exposed to it. The other difficulty lies in assessing DU's military utility. We will come to that question in section 1.2.1. As for DU use directly against personnel, again, it is difficult to know if the suffering and injury caused is sufficiently serious to meet the standard. If it is, DU use would only be justified against personnel when no other alternatives were practically available (during air warfare). There would certainly seem to be no necessity to tank-fire DU rounds against personnel.

As far as the principle of distinction/prohibition against indiscriminate attacks is concerned, Kleffner and Boutruche determined that DU use could potentially violate only one provision of Article 51 of Additional Protocol I (AP I), namely, paragraph 4(c), which prohibits using methods or means of combat which cannot be limited as required by the Protocol. As for the other provisions, either the use of DU weaponry does not raise specific issues in relation to them, or it could not be considered to violate them. The authors concluded that DU weapons are not inherently indiscriminate, and therefore 'whether and to what extent the use of DU ammunition violates the prohibition of indiscriminate attacks under Article 51(4)(c) and the corresponding rule of customary international humanitarian law will have to be determined on a case-by-case basis. While it does not do so under all circumstances, it may do so in some.' The fact that the effects of DU may be removed in time and space from its initial use may make it difficult to prove that they were caused by it; however, if such a link could be established, effects which are remote in time and space are not excluded by Article 51(4)(c) AP I.

The use of DU could also potentially violate the principle of proportionality in particular cases, where the incidental loss of civilian life, injury to civilians or damage to civilian property is excessive in relation to the military advantage anticipated from its use. This assessment will have to be made on a case-by-case basis. The use of DU in urban areas could have more adverse effects than its use in an isolated rural area. While long-term effects can be considered in making the proportionality assessment, provided they are foreseeable, weighing those effects in making the proportionality calculation is particularly difficult in relation to DU.

Finally, DU use could only be considered to violate the prohibition to render useless objects indispensable to the survival of the civilian population if carried out with that purpose. During all military operations, parties to armed conflicts are required to ensure that precautions are taken.

As to the question whether DU weapons can be considered as poison or poisonous weapons, Kleffner found in Chapter 7 that, even if an effects rather than a premeditation approach is applied, it is very difficult to consider DU as such. He was unable to conclude that DU weapons violate the prohibition on poison *per se*. As to the use of DU weapons in specific situations, he found that the current level of scientific knowledge is such that it is not possible to show that the toxic properties of DU residues cause death or serious damage to health in the ordinary course of events.

Koppe concluded in Chapter 8 that damage to the natural environment through the use of DU weaponry is foreseeable to a certain extent. However, it is unlikely that such damage will pass the high threshold – widespread, long-term and severe – of Articles 35(3) and 55 of AP I. He also found that it is unlikely that the use of DU would violate the customary prohibition to cause wanton or excessive collateral damage to the environment.

Human rights law obviously does not provide any legal basis for prohibiting the use of a weapon *per se*, but Toebes found in Chapter 9 that there is some possibility to argue, at least at a theoretical level, that DU use and states' behaviour in the aftermath of such use could in certain circumstances constitute a violation of several human rights, including the right to life, the right to health, the right to food and the right to information. Human rights law might also serve as a legal basis for claiming certain rights and remedies in the aftermath of DU use, as shown by McDonald in Chapter 11. LOAC, on the other hand, is less useful as the basis for claiming a remedy. In fact, the most practically useful remedies may be those found at the national level, which do not depend on having to establish that DU was the cause of the injury. Gries and Mohr found in Chapter 10 that, in principle, international legal rules concerning state responsibility are relevant to the use of DU weapons, but, as these are secondary rules, they will depend on showing the breach of a primary rule.

In sum, then, DU is not *per se* prohibited by treaty nor does LOAC prohibit its use *per se*. The only rules of LOAC that the use of DU could potentially violate are those prohibiting causing superfluous injury or unnecessary suffering to combatants and launching indiscriminate, including disproportionate, attacks against civilians.

Let us now turn to the views of states as to the legality of the use of DU weapons.

1.2   The views of states as to the legality of the use of depleted uranium weapons

1.2.1   *The user states*

The two main user states – the United Kingdom and the United States – assert that the use of DU weapons is in conformity with international law. Britain has stated:

'Their use is not prohibited under any international agreements, including the Geneva Conventions.'[20]

The United States is the only country to have acknowledged undertaking a legal review of DU weapons, even though it is not a party to AP I.[21] The US Air Force and Army legal reviews of munitions incorporating DU penetrators, already alluded to by Carnahan in Chapter 4, were, however, undertaken at a time when less was known about the substance's effects. Moreover, the reviewers had in mind a very different military justification for its use than that which has ensued.

The US Air Force conducted its first legal review of munitions incorporating DU in 1975,[22] many years before the first reported use by US Armed Forces of DU on the battlefield during the 1991 Gulf War and well before any ill effects of DU in combatants or civilians had been observed. This was, in fact, the first official legal review of DU ever conducted by any state.[23]

The review examined the legality of DU under three applicable principles of LOAC: the prohibition of unnecessary suffering, the prohibition of poison and the principle of proportionality/prohibition of indiscriminate effects. The DU-containing armour-piercing incendiaries in question passed the test of legality under all three principles. The review found that 'the use of depleted uranium, described as radioactive, while presenting international and national political issues, does not violate existing international law.'

The review found that the use of DU did not contravene the prohibition against poison, as 'DU's toxic radiological and chemical properties are an inherent

---

[20] Then-UK Prime Minister, Tony Blair, in response to a question by Mr Tam Dalyell, M.P., during Parliamentary Questions, 10 June 1999, PQ85351. This view was repeated in a statement by the Secretary of State for Defence, Mr Doug Henderson, in response to a parliamentary question by Dr Lynne Jones: 'The use of depleted uranium ammunition is not prohibited under any international agreement, including the Geneva Convention, and there are no circumstances under which such ammunition may be defined as a chemical weapon. (...).' PQ 91763, 20th June 1999. The UK Minister of State and the Ministry of Defence have both stated that depleted uranium ammunition is not proscribed by any of the international agreements to which the United Kingdom is a party, and that the Ministry of Defence reserves the right to issue depleted uranium-based weapons if the safety of British troops requires a capability against modern armour, see House of Commons – Foreign Affairs – Fourth Report, para. 51, <www.parliament.the-stationary-off.../pa/cm199900/cmselect/cmfaff/28/2814.htm>.

[21] Art. 36 of AP I requires States Parties to undertake legal reviews of new weapons. See further section 2.2.1 of this chapter. The legal basis for the US review was the US Department of Defense Instruction 5500.15 (16 October 1994), Army Regulation 27-53 (1 January 1979) and Department of Defense Directive 5000.1 (15 March 1996), according to which legal review of the intended acquisition of a potential weapon shall be carried out, to ensure compliance with the international legal obligations of the United States, including the laws of war. Additionally, legal review of new, advanced or emerging technologies which may lead to the development of weapons or weapon systems, including ammunition, is encouraged.

[22] Department of the Air Force, HQUSAF (JA) Legal Memorandum concerning review by Harold R. Vague, Major General, USAF, The Judge Advocate-General United States Air Force, of the High Explosive Incendiary and Armour Piercing Incendiary munitions, 14 March 1975.

[23] It specifically concerned 30 mm ammunition fired by a GAU-8 gun installed in the A-10 close air support aircraft. The gun fired several types of rounds, including an armour-piercing incendiary incorporating a DU penetrator, which is used against hard targets, such as heavy tanks, armoured personnel carriers, etc.

characteristic of the substance and not a designed, added in, characteristic.' It found that '[u]ranium does not appear to be any more chemically toxic than lead.' Anyway, any long-range toxicity effects were not considered significant when set beside the military necessity of its use.

As for any possible incompatibility with the prohibition against indiscrimination, the review noted the potential risk of DU at a localised level and that the risk varied according to the amount and the area where it was used and could be 'locally significant'. Yet, the review considered these risks to be not disproportionate considering the weapon's military utility.

The 1994 US Army review of DU[24] examined the legality under international law of the M829A2 tank ammunition, which utilises a DU penetrator. The review considered the ammunition's legality by reference to the principle prohibiting unnecessary suffering and superfluous injury, the principle prohibiting the use of poison weapons and the prohibition in AP I of methods or means of warfare that are intended or may be expected to cause widespread, long-term and severe damage to the natural environment. The review concluded that the ammunition was fully consistent with the law of war obligations of the United States.

It is interesting to note that allusions in the 1975 review to the toxic effects of DU and the fact that it could have locally significant impact were replaced in the 1994 review with the confident assertion that 'Post-Desert Storm studies by environmental and health officials conclude that the health risk in the use of DU is negligible'. The review stated that: 'Radiological exposure to external sources of DU occurs through the proximity of personnel to munitions, armor and contaminated equipment. These are low-level, low-dose-rate exposures that are within current safety and health standards of the US Nuclear Regulatory Commission.' The reviewer's concern seemed to extend only to the US's own troops (to whom the prohibition against superfluous injury or unnecessary suffering does not even apply) – and even then his assertion has been proven to be no longer factually correct – as the dose of radiation that would be received by an enemy – or indeed any – combatant sitting inside a tank hit by a DU penetrator (internal exposure) could hardly be described as low level, given that it has been shown to be 1,400 times the daily normal dosage.[25]

Finding DU use against *matériel* not incompatible with the principle of superfluous injury or unnecessary suffering, the reviewer noted that 'DU is not employed to increase the suffering of individual enemy combatants'[26] and 'there is no evidence to suggest that DU increases combatant suffering'. As to the first statement, it should be remembered that the principle prohibiting superfluous injury or unnecessary suffering to combatants is concerned with the effects of weapons. It

---

[24] Memorandum for US Army Armament Research, Development and Engineering Center (SMCAR-GCP [Mr Parise]). Subject: M829A2 Cartridge, 120 mm, APFSDS-T (Depleted Uranium Tank Round); Law of War Review, 27 December 1994.

[25] See D. Fahey, Chapter 2, pp. 29 et seq.

[26] *Supra* n. 24.

applies to weapons whose effects cause superfluous injury or unnecessary suffering, even if they are not deployed with that purpose. As to the second, it is no longer true to say that there is no evidence that DU increases combatant suffering; rather, there is evidence that it can do so due to the secondary effects of the weapons. As stated, these effects should be considered due to the high percentage of combatants who may survive the strike on a tank using DU weapons.

The review also referred to the view of the Office of the Surgeon-General of the Army that the long-term health effects of embedded DU-fragments, though not well defined, are minimal. The passage of time has shown that this conclusion is not longer accurate. In fact, the embedding of DU fragments is one of the most potentially hazardous forms of exposure for a combatant, and serious health effects have been reported in veterans exposed in this way.[27] In such cases, it is not difficult to link the pathologies presenting in these veterans with their exposure to DU, as they are consistent with the expected effects of internal exposure to high levels of DU.

Both the United States[28] and the United Kingdom[29] continue to maintain that the main military utility for the use of DU ammunition is its superior penetrating ability. However, it is no longer true to say, as the US Army reviewer did in 1994, that the performance advantage DU offers remains substantive and 'well above' that of competing materials such as tungsten steel. Matt Kagan, a former munitions analyst with *Jane's Defence Weekly,* was quoted as saying that the latest developments in tungsten technology have made it 'almost as effective as D.U.'[30]

So long as DU provides even a marginal performance advantage, this suffices to establish the military necessity for its use as a matter of the law of armed conflict – provided its use does not breach an absolute rule, such as distinction, in which case no military necessity would justify its use. But, as DU's performance advantage diminishes relative to that provided by alternatives,[31] it will be more difficult to justify its use on this ground, unless another reason for its military utility can be found which overrides the humanitarian concerns it gives rise to. As other munitions – at least for some applications – match or even outperform DU, there will be greater room to argue that, in particular cases where the alternative is suffi-

---

[27] See D. Fahey, Chapter 2, pp. 29 et seq.

[28] See B. Carnahan, Chapter 4, pp. 99 et seq.

[29] The UK Ministry of Defence has stated that 'no satisfactory alternative material to DU exists to achieve the level of penetration necessary to defeat Main Battle Tanks'. Ministry of Defence, *Memorandum: Gulf War Illnesses* (London, MoD 1991), presented to the House of Commons Defence Select Committee on 26 April 2001 (London, HMSO 2001), p. 37.

[30] In B. Mesler, 'The Pentagon's radioactive bullet', *The Nation* (21 October 1996), <http://www. thenation.com/doc/19961021/19961021mesler/4>.

[31] And not only Tungsten. 'For example, DARPA's Structural Amorphous Metals (SAM) program is advancing a new class of bulk materials with amphorous or "glassy" microstructures that have previously unobtainable combinations of hardness, strength, damage tolerance and corrosion resistance. Possible uses of SAM alloys include … self-sharpening penetrators that could replaced depleted uranium'. In Bridging the Gap: Defense Advanced Research Projects Agency, February 2005. <http:// www.darpa.mil/body/news/2005/BridgingTheGap_Feb_05.pdf >

cient to the job, DU use is not necessary as the advantage it offers is not actually required for the task at hand. This much has tacitly been admitted by the United States and the United Kingdom, both of which have taken the decision in recent years to phase out DU use for some weapons systems, variously based on the belief that tungsten is sufficient in the circumstances and on the grounds of humanitarian concerns.[32]

1.2.2   *The views of other states*

The United States and United Kingdom, although they are the only avowed user states, are almost alone in their view that the use of DU to destroy hard armour is either required or justified, given the humanitarian concerns its use gives rise to. As shown in Chapter 1, most armed forces use tungsten rather than DU for armour piercing ammunition, *inter alia*, on the grounds of the possible health and environmental impacts of DU. Fahey observed that many states, including Belgium, Canada, the Czech Republic, Germany, the Netherlands, Norway and Switzerland, shy away from DU use as a matter of policy. In March 2007 Belgium became the first nation to ban the use of DU on its territory. Such states are increasingly taking the view that, particularly when fighting an enemy lacking DU-reinforced armour, the currently available alternative substance, tungsten, is perfectly adequate.[33]

On the other hand, Canada, while not a user state, echoes the US and UK view that the use of DU is not banned by arms control law or LOAC.[34] Some other states have expressed views as to the legality of DU use in international fora (see below section 1.3).

1.3   **The United Nations**

DU has been on the active agenda of the United Nations Sub-Commission on Prevention of Discrimination and Protection of Minorities for over a decade. In 1996 the Sub-Commission passed a resolution calling for a moratorium on the use of DU and some other weapons (15 yes, 1 no, 8 abstentions).[35] It delegated to the Secretary-General (SG) responsibility to prepare a report on information gathered on the use of a number of weapons, including DU.[36] The SG did not pursue an indepen-

---

[32] See D. Fahey, Chapter 1, pp. 3 et seq.

[33] See D. Fahey, Chapter 2, pp. 29 et seq.

[34] Ms Aileen Carroll, Parliamentary Secretary to the Minister of Foreign Affairs of Canada, stated that '[t]he use of depleted uranium munitions is not prohibited or restricted under the 1980 UN Convention on prohibitions or restrictions on the use of certain conventional weapons and related protocols, nor is it otherwise prohibited by international humanitarian law. This is because it is not deemed to be excessively injurious or to have indiscriminate effects.' Canada, 37th Parliament, 2nd Session, edited Hansard Number 114, 9 June 2003, <www.parl.gc.ca/37/2/parlbus/chambus/house/debates/114_2003-06-09/han114_1845-E.htm>.

[35] 'International peace and security as an essential condition for the enjoyment of human rights, above all the right to life' E/CN.4/Sub.2/Res/1996/16, <http://ap.ohchr.org/documents/E/SUBCOM/resolutions/E-CN_4-SUB_2-RES-1996-16.doc>.

[36] By Resolution 1996/16 of 29 August 1996.

dent investigation into the legality of the use of any of these weapons, but solicited the views of states on the matter. Just two responded, Croatia and Nigeria, and only Nigeria specifically addressed the use of DU weapons. It stated that such weapons, *inter alia*, 'violate the rights to life, health, physical security and other human rights, such as economic, social and cultural rights and the right to development'.[37] It encouraged the international community to ban the production and use of such weapons, in conformity with international humanitarian law. Such language indicates a certain lack of appreciation of LOAC, which does not forbid the production of any weapon and does not contain any blanket prohibition on the use of any particular weapon.

The Secretary-General issued an addendum to his report on 28 July 1997,[38] which incorporated comments subsequently received by Cuba, Trinidad and Tobago and the Philippines.[39] Only the latter made specific reference to DU weapons: '… the Philippines adheres to peaceful, non-military approaches to conflict and renounces the use of nuclear and chemical weapons, fuel-air bombs, napalm, cluster bombs and biological weaponry containing depleted uranium.' The turn of phrase also suggests a lack of understanding of the nature of DU weapons. As neither Nigeria nor the Philippines are DU possessor or user states only limited significance can be attached to their comments. The failure of other states to offer an opinion on the matter indicates that they did not regard it as being of pressing importance at that time.

By its Resolution 1997/36 of 28 August 1997 the Sub-Commission on Prevention of Discrimination and Protection of Minorities[40] welcomed the report of the Secretary-General.[41] It urged 'all States to be guided in their national policies by the need to curb the testing, the production and the spread of weapons of mass destruction, or with indiscriminate effect, or of a nature to cause superfluous injury or unnecessary suffering', and decided to authorise Ms Clemencies Forero Ucros to prepare a working paper, 'in the context of human rights and humanitarian norms, assessing the utility, scope and structure of a study on weapons of mass destruction or with indiscriminate effect, or of a nature to cause superfluous injury or unnecessary suffering.' However, it did not assert that DU was a weapon to have such effects. The working paper should have been presented in 1998, but Ms Forero Ucros did not deliver it. It was rescheduled for delivery in 1999, 2000 and 2001, but never materialised.

---

[37] E/CN.4/Sub.2/1997/27 of 24 June 1997, para. 12.

[38] E/CN.4/Sub.2/1997/27/Add.1 of 28 July 1997.

[39] UN Sub-Commission on Prevention of Discrimination and Protection of Minorities Resolution 1997/27, UN Doc. E/CN.4/Sub.2/RES/1997/27/Add.1 of 28 July 1997, para. 9.

[40] In 1999 the Economic and Social Council changed its name from Sub-Commission on Prevention of Discrimination and Protection of Minorities to Sub-Commission on the Promotion and Protection of Human Rights.

[41] UN Sub-Commission on Prevention of Discrimination and Protection of Minorities Resolution 1997/36 of 28 August 1997, UN Doc. E/CN.4/Sub.2/RES/1997/36.

By Sub-Commission Decision 2001/119, Mr Yeung Sik Yuen was mandated to prepare the working paper in lieu of Ms Forero Ucros.[42] It was submitted to the Sub-Commission at its 54th session.[43] The subsequent discussion on the report especially singled out DU for focus. By decision 2002/113 the Sub-Commission requested Mr Sik Yuen to deliver an undated working paper.[44]

The updated paper was submitted to the Sub-Commission during its 55th session.[45] While Mr Yuen's ultimate conclusion that DU weapons can be considered as prohibited, whether or not there is a specific treaty regime banning them,[46] is not shared by this author, anymore than his assertion that: 'Weapons which are the subject of a specific treaty should also be considered universally banned for all States, regardless of whether the State is a signatory',[47] his analysis of the legality of the use of DU weapons remains the first and only one ever conducted under UN auspices, so for that reason is of some interest.

1.4   The European Parliament

The European Parliament has adopted four resolutions in which it called for a moratorium on the use of DU. The first, adopted on 17 January 2001,[48] at the urging of Italy, Belgium, Finland, Portugal and Norway, following the deaths from leukaemia of several Italian peacekeepers who had served in Kosovo, concerned the so-called 'Balkan syndrome'.[49] It requested NATO to 'consider other types of munitions until the results of investigations on depleted uranium are known.' It also called for 'a thorough investigation … into all the effects of the military operations conducted during the conflicts in Bosnia and Kosovo on military personnel, the civilian population and the environment.'[50] While conceding NATO's official position that 'there

---

[42] UN Sub-Commission on Prevention of Discrimination and Protection of Minorities Decision 2001/119 of 16 August 2001, UN Doc. E/CN.4/Sub.2/DEC/2001/119.

[43] Human Rights and weapons of mass destruction, or with indiscriminate effect, or of a nature to cause superfluous injury or unnecessary suffering, UN Doc. E/CN.4/Sub.2/2002/38, Working paper submitted by Y.K.J. Yeung Sik Yuen in accordance with Sub-Commission Resolution 2001/36.

[44] UN Sub-Commission on Prevention of Discrimination and Protection of Minorities, Decision 2002/113, UN Doc. E/CN.4/Sub.2/DEC/2002/113.

[45] Human Rights and weapons of mass destruction, or with indiscriminate effect, or of a nature to cause superfluous injury or unnecessary suffering, E/CN.4/Sub.2/2003/35, 2 June 2003. Working paper submitted by Y.K.J. Yeung Sik Yuen in accordance with Sub-Commission Resolution 2002/113.

[46] 'In particular, from the information he has studied, it is clear to the author that these weapons must necessarily be considered banned as causing superfluous injury or undue suffering or because of a real threat to the environment. These weapons could also be viewed as poisonous.' Ibid., at p. 9, para. 12. Further: 'The main legal conclusion reached by the author is that all the weapons under review in his two papers should be considered banned, whether or not there is a specific treaty banning them.' At p. 16, para. 55.

[47] Ibid., p. 15, para. 55.

[48] European Parliament, Use of Depleted Uranium in Bosnia and Kosovo (Balkan syndrome) (17 January 2001).

[49] See 'Euro parliament calls for moratorium on use of depleted uranium rounds', AFP (17 January 2001); S. Castle, 'Euro MPs defy Nato and seek ban on DU', *The Independent* (18 January 2001).

[50] See 'Depleted uranium concerns boost nonradioactive bullet: The European Parliament yesterday called for a suspension of DU use pending trial', *The Christian Science Monitor* (18 January 2001).

is now neither medical nor statistical proof clearly establishing the existence of a link between the use of depleted uranium and the appearance of leukaemia and other forms of cancer or other sicknesses among soldiers and policemen'[51] who served in the Balkans, it called for the establishment of an 'independent European medical work group' to study the medical complaints that were being reported.

The second resolution of the European Parliament, adopted on 13 February 2003[52] and concerning the harmful effects of unexploded ordnance (landmines and cluster submunitions) and DU ammunition, noted the harmful effects of these weapons and called on EU Member States to 'immediately implement a moratorium on the further use of cluster ammunition and depleted uranium ammunition (and other uranium warheads), pending the conclusion of a comprehensive study of the requirements of international humanitarian law.'

On 17 November 2005, the European Parliament reiterated 'its call for a moratorium – with a view to the introduction of a total ban – on the use of so-called "depleted uranium munitions".'[53] A fourth resolution was adopted on 15 November 2006.

1.5 The International Committee of the Red Cross

The International Committee of the Red Cross (ICRC) published some comments on DU weapons in 2001. While not pronouncing on the legality of the use of DU as such, the ICRC, referring to the obligation on states pursuant to Article 36 of AP I to carry out legal reviews of new weapons, 'strongly urges all States which study, develop, acquire or adopt munitions containing depleted uranium to carry out such legal reviews if they have not already done so, and would welcome an exchange of views and information on these reviews.'[54] As to the possible effects on health of the use of DU weapons, the ICRC stated:

> 'Currently available scientific information provides evidence that the increase in levels of uranium is marginal in areas where depleted uranium munitions have been used, except at the point of impact of depleted uranium penetrators. Nevertheless, the ICRC welcomes the additional studies which are being carried out by various international organizations, in particular field studies in Kosovo and other regions where munitions containing depleted uranium weapons have been used. Hopefully these studies will not only concentrate on international staff but will also include the local population and in particular children.'[55]

---

[51] See 'Euro parliament calls for moratorium on use of depleted uranium rounds', Agence France-Press (17 January 2001).

[52] European Parliament resolution on the harmful effects of unexploded ordnance (landmines and cluster submunitions) and depleted uranium munitions, P5_TA(2003)0062, <http://www.europarl.europa.eu/sides/getDoc.do?language=EN&pubRef=-//EP//TEXT+TA+P5-TA-2003-0062+0+DOC+XML+V0//EN>.

[53] European Parliament resolution on non-proliferation of weapons of mass destruction: A role for the European Parliament (2005/2139(INI)), P6_TA(2005)0439.

[54] 'Depleted uranium munitions: Comments of the International Committee of the Red Cross', 83 IRRC (2001) pp. 543 at 545.

[55] Ibid., p. 543.

1.6   **Human rights organisations**

In a 2000 report concerning the 1999 NATO bombardment of the Federal Republic of Yugoslavia, Amnesty International referred to the use of DU by NATO forces, specifically the United States and United Kingdom. The report noted: 'The use of depleted uranium munitions is not prohibited by international law and Amnesty International does not oppose their use *per se.*' While Amnesty did not characterise the use of DU as a war crime or even as an unlawful act, it did recommend further research into its effects: 'NATO and its member states should also investigate and cooperate fully with independent investigations of the possible long-term health and environmental risks posed by the use of depleted uranium weapons. They should also consider suspending the use of these weapons pending the outcome of such investigations.'[56] Other major human rights NGOs have not weighed into the DU debate. In its comprehensive report on the conduct of the Iraq War, titled, *Off Target*, Human Rights Watch, while pronouncing on the lawfulness of the use of weapons such as cluster munitions, shied away from dealing with DU.[57]

1.7   **Other non-governmental organisations**

A number of NGOs have sprung up specifically to rally the anti-DU cause[58] and some existing NGOs have expanded their mandates to include DU. Although many of the NGOs and individuals campaigning against the use of DU are well-intentioned, the information they disseminate about DU is not always accurate and at times provides a misleading picture as to both the effects and lawfulness of the use of DU weapons. Those organisations campaigning for a ban by definition do not believe that DU use in general is already illegal. However, some commentators have asserted that, even in the absence of any ban, DU use is already unlawful under LOAC not only in particular instances of use but in all cases.[59] At the more extreme fringe of the anti-DU lobby are persons who go so far as to call the use of DU in all and any cases 'genocidal'[60] or a 'war crime'.[61] For reasons explained in

---

[56] Amnesty International, *Collateral Damage or Unlawful Killings? Violations of the Laws of War by NATO during Operation Allied Force, 1999.* AI Report EUR 70/18/00 (June 2000), <www.amnesty.org/ailib/intcam/kosovo/docs/nato_all.pdf>.

[57] *Human Rights Watch, Off Target: The Conduct of the War and Civilian Casualties in Iraq* (December 2003), <http://www.hrw.org/reports/2003/usa1203/>.

[58] In particular, the International Coalition to Ban Depleted Uranium, a coalition of groups whose principal aim it is to lobby for a ban on DU, <http://www.bandepleteduranium.org>. According to its website, ICBUW's membership includes 90 groups in 25 countries worldwide.

[59] See for example, K. Parker, 'The illegality of DU weaponry', available at <http://www.webcom.com/hrin/parker.html>; P. Bein and K. Parker, 'Uranium Weapons Cover-ups – a Crime against Humankind' (January 2003), <www.ratical.org/radiation/DU/Uweps-CAH.pdf>.

[60] See R.C. Koehler, 'Depleted uranium (DU), silent genocide', <http://www.mindfully.org/Nucs/2004/DU-Silent-Genocide25mar04.htm>; A. Lambremont Webre, 'Canada's role in depleted uranium weapons worldwide', July 2007, <http://commonground.ca/iss/0707192/cg192_du.shtml>.

[61] See Parker, *supra* n. 59.

Chapter 11, the use of DU *per se* or in the normal circumstances of its use cannot be considered as a war crime nor can it be considered as genocidal.

2.   A PRECAUTIONARY APPROACH TO THE USE OF DEPLETED URANIUM
        WEAPONS

2.1   **The need for a precautionary approach and clarification of its legal basis**

Although some effects of DU may remain unexpected or unknown, this does not imply that all its effects are unforeseeable. Notwithstanding the range of views as to the safety and legality of DU, and the lack of a complete understanding of its effects, arguably there is already sufficient evidence of the harm DU can cause to humans and the environment on which to base a reasoned and reasonable opinion as to its foreseeable effects and consequent safety and legality.

While we do not have sufficient evidence to conclude that DU is unlawful *per se* under LOAC, we have not been able to exclude cases where the use of DU could infringe the law, even given today's only partial knowledge regarding its effects. We do know that DU is a chemically toxic and radioactive substance,[62] and that certain uses of DU pose a higher chance of negatively impacting on the civilian population than others. In such cases, there is a greater likelihood of an infringement of a legal rule.

Hundreds of studies of DU have been conducted, although Fahey described their limitations. DU has been shown to cause cancers and neurological diseases in laboratory animals, and there are grounds for concluding that it can cause illness in certain cases of human exposures. DU has been shown to cause Hodgkin's lymphoma in some exposed service-members. Thousands of US veterans who have been exposed to DU have developed serious, sometimes life-threatening, illnesses, and patterns of pathologies are presenting. Especially where combatants have suffered internal exposure to DU, it would be more than negligent to exclude DU as a possible causal factor of illnesses presenting.

What is needed is a framework for approaching the use of DU and its consequences which takes into account that this is a weapon about whose effects something but not everything is known. Such an approach would focus on minimising any current (expected or foreseeable) harm and potential liability and forestalling possible future (currently unforeseeable) damage. Such an approach could be described as 'a precautionary approach'.

The so-called 'precautionary approach' advocated herein has a number of facets. Each of these has in common that it is founded on the concept of taking

---

[62] 'DU is radioactive and poisonous. Exposure to sufficiently high levels might be expected to increase the incidence of some cancers, notably lung cancer, and possibly leukaemia, and may damage the kidneys.' *The Health Hazards of Depleted Uranium Munitions, Part 1* (London, The Royal Society May 2001) p. 21.

precautions, but it does not mean that each facet has the same legal basis or, above all, that the legal basis for each or any facet is what has come to be known as the 'precautionary principle'. The precautions concerning the use of DU advocated herein have a varied legal basis, some of which are founded in LOAC, some human rights law. Other aspects have no conventional or customary legal basis but are merely rooted in common sense and advised as a matter of prudence, to minimise future problems in line with the military's own credo of expecting the best but planning for the worst. Before elaborating on the legal bases of this precautionary approach, a few words are necessary about why neither the precautionary principle or the principle of precaution of themselves can provide a comprehensive legal basis for the approach suggested here.

International environmental law recognises the existence of the so-called 'precautionary principle'.[63] In general terms, where it appears that an action is likely to cause unacceptable harm to the environment, the precautionary principle enjoins states to desist from that activity, even in the absence of a specific legal prohibition. The International Court of Justice in the *Nuclear Tests* case enunciated the precautionary principle in the following terms: 'when there is sufficient evidence that an activity is likely to cause unacceptable harm to the environment, ... responsible public and private powerholders prevent or terminate the activity'.[64] In their dissenting opinions in the *Nuclear Tests* case, Judges Weeramantry and Palmer both referred to the precautionary principle. The former said that the precautionary principle 'is gaining increasing support as part of the international law of the environment',[65] while the latter stated that 'as the law now stands it is a matter of legal duty to first establish before undertaking an activity that the activity does not involve any unacceptable risk to the environment.'[66] He also stated that: 'the norm involved in the precautionary principle has developed rapidly and may now be a principle of customary international law relating to the environment.'[67]

The 1992 Rio Declaration on Environment and Development, adopted by the United Nations Conference on Environment and Development, contains a widely

---

[63] The literature on the precautionary principle is vast and growing. See, *inter alia*, T. O'Riordan, J. Cameron and A. Jordan, eds., *Reinterpreting the Precautionary Principle* (London, Cameron May 2001); C. Raffensberger and J. Tickner, eds., *Protecting Public Health and the Environment: Implementing the Precautionary Principle* (Island Press, Washington, DC. 1999); E. Fisher, J. Jones and R. von Schomberg, eds, *Implementing the Precautionary Principle: Perspectives and Prospects* (Cheltenham, UK and Northampton, MA, USA, Edward Elgar 2006); C.R. Sunstein, *Laws of Fear: Beyond the Precautionary Principle* (New York, Cambridge University Press 2005); P. Harremoës, D. Gee, M. MacGarvin, A. Stirling, J. Keys, B. Wynne and S. Guedes Vaz, *The Precautionary Principle in the 20th Century: Late Lessons from Early Warnings* (Earthscan, 2002); K.J. Arrow and A.C. Fischer, 'Environmental preservation, uncertainty and irreversibility', 88 *Quarterly Journal of Economics* (1974) pp. 312-319.

[64] *Request for an Examination of the Situation in Accordance with Paragraph 63 of the Court's Judgement of 20 December 1974 in the Nuclear Test Cases* (*New Zealand* v. *France*), ICJ, Order of 22 September 1995 (*ICJ Rep.* 1995) para. 64.

[65] Ibid., Dissenting Opinion of Judge Weeramantry, p. 342.

[66] Ibid., Dissenting Opinion of Judge Palmer, para. 87.

[67] Ibid., para. 91.

accepted formulation of the precautionary principle. It states in Principle 15: 'In order to protect the environment, the precautionary approach shall be widely applied by States according to their capabilities. Where there are threats of serious or irreversible damage, lack of full scientific certainty shall not be used as a reason for postponing cost-effective measures to prevent environmental degradation.'[68] In cases of doubt, the burden of proving that something is legal falls to the state wishing to take the action.

The Rio Declaration represented the first codification of the precautionary principle; it has been included in numerous subsequent international environmental law agreements.[69] The 1992 Convention on Biodiversity states in its preamble: 'where there is a threat of significant reduction or loss of biological diversity, lack of full scientific certainty should not be used as a reason for postponing measures to avoid or minimize such a threat' (para. 9).[70] A precautionary approach is also incorporated into the international environmental law of many states.[71]

The precautionary principle was formulated not only because the prevention of long-term or irreversible damage to the environment is a good in itself but also because human health and development depends on care for the environment in the short- and long-term. The link between damage to the environment and human health has been made in human rights law.[72]

Still, notwithstanding the view of Judge Weeramantry in the *Nuclear Tests* case that the precautionary principle may have become a rule of customary law, it is submitted that the precautionary principle is an emergent – although widely recognised – rule in environmental law rather than a generally recognised principle of international law considered as binding on all states. Moreover, it is a principle that has been developed in relation to protection of the environment rather than the use of weapons during armed conflict. According to the IRCR's Customary International Humanitarian Law study, there does not appear to be any national practice

---

[68] Available at <http://www.unep.org/Documents.multilingual/Default.asp?DocumentID=78&ArticleID=1163>.

[69] Including Arts. 6 and 5(c).1995 Agreement on Fish Stocks; the Preamble to the Convention on Biological Diversity; Art. 3(3) of the 1992 Convention on Climate Change; Annex II, Art. 3(3)(c) of the 1992 Convention for the Protection of the Marine Environment of the North-East Atlantic; Art. 3(1) of the 1996 Protocol to the 1972 London Dumping Convention; the 1994 Code of Practice on the Introduction and Transfer of Marine Organisms adopted by the International Council for the Exploration of the Seas; the IMO's 1997 Guidelines for Preventing the Introduction of Unwanted Aquatic Organisms and Pathogens from Ships' Ballast Water and Sediment Discharges; and the FAO's 1995 Guidelines on the Precautionary Approach to Capture Fisheries and Species Introduction.

[70] Available at <http://www.biodiv.org/convention/articles.asp>. See also Bergen UN Economic Commission for Europe Ministerial Declaration on Sustainable Development, Art. 7: 'In order to achieve sustainable development, policies must be based on the precautionary principle. Environmental measures must anticipate, prevent and attack the causes of environmental degradation. Where there are threats of serious or irreversible damage, lack of full scientific certainty should not be used as a reason for postponing measures to prevent environmental degradation.'

[71] For example, Israel's Water Law of 1959 and Planning Law and the Czech Republic's Environmental Protection Act; French Law No. 95-101 of 1995.

[72] See B. Toebes, Chapter 9, pp. 187 et seq.

(legislation, military manuals or case law) concerning the application of the precautionary principle in the context of armed conflict or in relation to LOAC.[73]

While the precautionary principle, as a guiding principle and framework for regulation, could be applied *vis-à-vis* DU, and arguably could provide at least a policy basis for approaching the use of weapons suspected but not yet conclusively proven to have negative effects for the environment – and consequently human health – it needs to be emphasised that, of itself, the precautionary principle does not provide the legal basis for the precautionary approach advocated herein nor could it serve as the legal basis for preventing the use of DU weapons even in particular cases.

Neither does the principle of precaution recognised by LOAC – as distinct from the precautionary principle just mentioned– provide a complete or even primary legal basis for all the precautions outlined herein, although it could serve as a legal basis for certain aspects.

The rule that those conducting hostilities must take all feasible precautions in attack has been part of conventional LOAC since 1907,[74] albeit in a narrower form than the rule set out in Article 57(1) of AP I. The rule to take constant care to spare the civilian population and civilian objects during the conduct of military operations has been found by the ICRC in its study of Customary International Humanitarian Law to be customary in nature, during both international and non-international armed conflict.[75]

Beckett asserts that Article 57 of AP I imposes on states a duty to prove the legality of a weapon. Applied to DU, it would mean 'that those states wishing to use DU weapons must, in order to have taken all feasible precautions, prove their legality. This would involve refuting any and all serious allegations of illegality. In other words, the burden of proof would shift to those wishing to deploy weapons, and factual inconclusiveness would not lead to legality, but to interim illegality.'[76] It is submitted that this is a misreading of Article 57 and of what it requires.

As mentioned, states to have a duty to ensure that the weapons they use are *per se* lawful (that is, not unlawful) and that the generally envisaged circumstances of their use will not contravene the relevant provisions of LOAC. However, its legal basis is Article 36 of AP I and the general duty to comply with LOAC in the

---

[73] The ICRC study on customary IHL found no state practice or national legislation concerning the precautionary principle, although it was not clear in precisely what context they were looking for evidence of any practice concerning the precautionary principle. J.-M. Henckaerts and L. Doswald-Beck, eds, *Customary International Humanitarian Law*, Vol. II: Practice (Cambridge, Cambridge University Press 2005) pp. 871-872.

[74] Para. 3 of Art. 3 of 1907 Hague Convention IX Concerning Bombardment by Naval Forces in Time of War, concerning bombardment of undefended ports, towns, villages, dwellings, or buildings states: 'If for military reasons immediate action is necessary, and no delay can be allowed the enemy, it is understood that the prohibition to bombard the undefended town holds good, as in the case given in paragraph 1, and that the commander shall take all due measures in order that the town may suffer as little harm as possible.' (Emphasis added.)

[75] CIHL, *supra* n. 12, p. 51.

[76] J.A. Beckett, 'Interim legality: A mistaken assumption? – An analysis of depleted uranium munitions under contemporary international humanitarian law', 3 *Chinese JIL* (2004) pp. 43 at 82-83.

use of weapons, not the principle of precaution in Articles 57 and 58 of AP I. Once weapons are shown to be *per se* not unlawful and where their use in a particular case is not obviously unlawful, states are free to use them, even in the midst of allegations as to their unlawfulness. States are certainly under no general obligation to refute each and every allegation of illegality of the use of a weapon, including DU, although if a credible case of the breach of international law through the use of DU weapons were to be alleged, it should be investigated. Factual inconclusiveness does not lead to interim illegality; quite the contrary. As noted, in determining whether the use of a weapon is unlawful in a particular situation, a commander can only take into consideration expected effects. If these are not unlawful, then the weapon can be used.

2.2   The elements of the precautionary approach and their legal bases

As applied to DU weapons, a precautionary approach necessarily implies precaution in their use during the conduct of hostilities but it is much broader and emphasises precaution, and therefore risk minimisation, at various stages, as well as prescribing action for dealing with the aftermath of their use.

The precautionary approach advocated here has four main facets:

1. Legal reviews of DU weapons by states.
2. Precaution in targeting: restricting the deployment of DU weapons in civilian areas.
3. Precautions in the aftermath of DU use:
   - Remedial and risk reduction measures;
   - Testing of exposed individuals and populations and the conduct of further medical and scientific research by military and civilian bodies.
4. The voluntary adherence by user states to a moratorium on the use of DU weapons.

While the precautionary approach is presented as a package of proposals, it is not intended to suggest that all states should or must adopt them all. For most states, the issue is simply irrelevant, but it is important *vis-à-vis* user or possessor states or those considering acquiring or developing DU weapons, that is at least 18 states.[77] While all of the elements are addressed to user states, some elements (such as undertaking legal reviews of DU weapons) are also addressed to non-user but possessing or developing or acquiring states. Ideally, user states would adopt all or most aspects, although the adherence by states to a moratorium on use would obviously make redundant the need for precautions in targeting, for example. The adoption of a moratorium to allow for a period of further research and testing fits squarely within a precautionary approach, although it is not argued here that states are any legal duty to adopt a moratorium, as Beckett suggested above, for the reasons

---

[77] See D. Fahey, Chapter 1, pp. 3 et seq.

already given. However, even if states were to adopt only one or more of the steps recommended, it would reduce risk.

The precautionary approach outlined herein is hardly original. Most of its constituent elements have already been advocated by one or several international organisations, national and international non-governmental organisations and members of civil society, including many people within the scientific and medical communities. There is a groundswell of opinion behind all of the steps advocated here. As shown earlier, many sound minds have called for taking some or all of these steps, either because they are already legally required or because they are the prudent thing to do, given the potential risks and resultant liability based on what is already known about DU and given the even greater risks if DU should transpire to be more hazardous than currently assumed. There is thus nothing radical about this legal advice or these policy recommendations.

Why should user states adopt a precautionary approach? The several benefits of adhering to the various elements of the precautionary approach will be discussed below in more detail. Suffice it here to say that the principal justification is not founded in law but pragmatism: states stand to lose little (tungsten is almost as effective and usually sufficient to the task) and may forestall harm (even if the probability, nature and extent of that future harm remain somewhat speculative). To an extent, user states are already following a precautionary approach by switching to other munitions for several platforms where they can without losing a military advantage. Even where DU use remains a matter of military necessity, states are not always required to use the best weapons they have. Even if one weapon provides an advantage relative to another, and may be necessary in legal terms, this advantage may not be necessary in military terms to do the job. If states took the precaution of restricting DU use to those occasions where it was absolutely required, because nothing else would do, they would go a long way towards reducing any potential risks while losing little in terms of their military advantage.

This so-called precautionary approach aims not only to protect civilian populations from the dangers of military operations and at avoiding superfluous or unnecessary injury to combatants, but also to, at least partially, address the fact that international law does not cover the relationship between a state and its own forces. States are recommended to apply this approach not only because it will minimise their potential legal liability and financial burden *vis-à-vis* enemy civilians or personnel or states but also because it will benefit their own personnel. In the end, the greatest motivation for states to adopt a precautionary approach is the fact that DU may be a cause of illness in their own forces. There is nothing to LOAC to address this problem. It may be more prudent to use alternatives where possible even against *matériel,'* even where not legally required, because it may not simply be worth the risk to friendly forces and contractors involved in clean-up operations afterwards.

The adoption of a precautionary approach is without prejudice to any attempts to secure a treaty ban or restriction on DU use under the law of arms control. Such moves could solve many of the problems associated with DU use, and limit states' potential liability, but they would not address problems concerning those persons or environments already exposed. Even if a treaty was introduced to regulate the use of DU weapons, depending on its content it might still be necessary for states to follow some of the precautions outlined herein, particularly as regards remediation of sites that have already been contaminated and testing, treatment and possible compensation of already exposed individuals. No treaty regime can address all the needs arising out of the use of DU. Given the unlikelihood of a treaty to deal with DU being adopted in anything like the short-term, a precautionary approach is all the more necessary.

2.2.1   Legal reviews of depleted uranium weapons by states

2.2.1.1 The obligation to carry out a legal review

An important legal element of a precautionary approach is that all states possessing or using, or intending to possess or use, DU weapons carry out legal reviews to ensure their compatibility with LOAC, as the ICRC and others have advocated. Pursuant to Article 36 (New Weapons) of 1977 AP I, states wishing to acquire, develop or use new weapons are legally obliged to undertake legal reviews. The fact that the legality of a weapon must be determined not only by the rules established in Protocol I but also by customary rules[78] means that norms which have emerged following as well as those preceding the adoption of the Protocol must be considered.[79]

The antecedent of Article 36 is Article 23(e) of the 1907 Hague Regulations. Article 36 also arises out of and logically follows from the basic rules concerning methods and means of warfare in Article 35 of AP I, as well as the other general principles of LOAC,[80] including, in particular, the principle of precaution in Article 57. 'Although new to international humanitarian law [in 1977], such a requirement is included in the existing internal legislation or regulations of some States.'[81]

While Article 36 cannot of itself serve as the legal basis for prohibiting a weapon such as DU, it offers a potentially useful mechanism for regulating and perhaps forestalling the use of DU weapons, for drawing greater attention to the question of their legality by user states, and for discouraging their acquisition by non-possessor states and their use by those already possessing them but which have

---

[78] *Commentary on the Additional Protocols of 8 June 1977 to the Geneva Conventions of 12 August 1949* (Geneva, International Committee of the Red Cross/Martinus Nijhoff Publishers 1987) para. 1472.

[79] See I. Daoust, R.M. Coupland and R. Ishoey, 'New wars, new weapons? The obligation of states to assess the legality of means and methods of warfare', 84 *IRRC* (2002) p. 352.

[80] According to Greenwood: the requirement in Art. 36 'is a novelty but one which is uncontroversial and, in any event, follows logically from the obligations of Article [sic] 35(1).' C. Greenwood, 'Customary law status of the 1977 Geneva Protocols', in *Humanitarian Law of Armed Conflict: Challenges Ahead*, A.J.M. Delissen and G.J. Tanja, eds. (The Hague, Martinus Nijhoff 1991) pp. 93 at 105.

[81] Bothe, Partsch and Solf, *supra* n. 15, p. 199, para. 2.1.

not yet used them. Given that the last legal review of DU was conducted in 1994 (by the United States), and a considerable body of evidence has accumulated since then which indicates that DU has more serious implications for human health and the environment than known at that time, it is possible that reviews of it undertaken now would reach a different conclusion as to its lawfulness, at least in certain circumstances of its use. If several states were to undertake legal reviews which concluded that DU use was either unlawful or should be restricted, it would contribute to the building of support for a moratorium or ban on DU use.

In assessing whether a weapon is lawful, '[t]he determination is to be made on the basis of normal use of the weapon as anticipated at the time of the evaluation.'[82] This is why the US reviews of DU ammunition assessed its legality primarily in relation to its use as an anti-*matériel* weapon. The reference in Article 36 to 'some or all circumstances' is not meant to imply an obligation 'to foresee or analyze all possible misuse of a weapon, for any weapon can be misused in a way that would be prohibited.'[83] The meaning of the phrase is to require a determination whether the employment *for its normal or expected use* would be prohibited under some or all circumstances.

Article 36 seems to cover only weapons acquired or used after states become party to AP I and not those it already possesses. 'It cannot be expected that states will introduce specific prohibitions on the basis of general principles, when such prohibitions could be considered as *a posteriori* condemnation of prior use of such weapons.'[84] This is particularly apposite in relation to DU weapons. The United Kingdom, which became party to AP I in 1998[85] following its development and use of DU weapons, takes the view that Article 36 does not operate retroactively. Britain has not undertaken a legal review of DU weapons, as DU was introduced in the United Kingdom prior to it becoming a party to AP I.[86] While possessing states may seem disinclined to undertake the legal review of a weapon they acquired before becoming a party to AP I, they are legally obliged to do so at least prior to its use if that postdates their becoming a party to the AP, and a failure to do so does not protect them from any legal liability arising out of its use.

Given that new information about the effects of a weapon can come to light after it has passed an initial legal review, and the military utility of the use of the weapon can change – as may well be happening in relation to DU weapons – a progressive interpretation of Article 36 is that it imposes a continuing obligation, which does not end with any initial review at the time of the weapon's development, acquisition or use. The stated UK policy is to keep the legality of weapons systems under regular consideration, particularly in light of changes in interna-

---

[82] *Commentary, supra* n. 78, para. 1466.

[83] Official Records of the Diplomatic Conference on the Reaffirmation and Development of International Humanitarian Law applicable in Armed Conflicts, Geneva, 1974-1977, XV, p. 269. CDDH/215/Rev. 1, paras. 30-31, quoted in *Commentary, supra* n. 78, para. 1469.

[84] *Commentary, supra* n. 78, para. 1475.

[85] The United Kingdom ratified the APs in 1995 and adopted legislation to implement them: Geneva Conventions (Amendment) Act, 1995, Chapter 27, 19 July 1995.

[86] Based on the author's correspondence with the UK Directorate of Army Legal Services (ALS 2).

tional law.[87] The United Kingdom has indicated that while it has not undertaken an Article 36 review of DU weapons, it has been monitoring the effects of DU, particularly on the environment.[88] The Ministry of Defence has stated that it 'continues to review the safety of all UK Armed Forces equipment, including Depleted Uranium ammunition.'[89] The US reviews of DU ammunition were based on the evidence available at the time. New evidence has since emerged which sheds greater light on the detrimental effects of exposure to DU. Although the United States is not bound by any continuing obligation that may exist in Article 36, as it is not a party to AP I, it would be desirable for it to undertake a new review into the legality of the use of DU based on current knowledge, and for all states that either possess or that are developing or considering acquiring DU weapons to do likewise.

Article 36 does not provide any indication as to what is required substantively of a legal review. But a review which is not based on all the available and most up-to-date scientific and other data, and which does not consider the legality of a weapon under all of the applicable law, could hardly be considered to conform with the obligation to ensure a weapon's lawfulness. One could argue that implicit in the duty to undertake a legal review is the proactive duty to acquire the necessary information to base it on. A state that fails to search out reliable and sufficient information upon which to base a legal review, or which negligently, recklessly or wilfully ignores evidence indicating that it causes superfluous injury or unnecessary suffering or poses serious dangers for civilians, could hardly comport with the legal duty to investigate the lawfulness of weapons.

What are the consequences where a state fails to undertake a legal review? The Commentary on AP I states that: 'If these measures are not taken, the State will be responsible in any case for any wrongful damage ensuing.'[90] In other words, there would seem to be no abstract responsibility for non-compliance with an Article 36 review absent any injury resulting from the prohibited use of a weapon.

---

[87] UK Military Manual, *supra* n. 14, p. 119, para. 6.20.1.

[88] Mr Spellar, Secretary of State for Defence, responding to a question by Ms Oona King, stated that: 'My Department has conducted extensive monitoring programmes of the environmental effects of depleted uranium munitions at the Defence Evaluation and Research Agency's firing ranges at Eskmeals and Kirkcudbright. The monitoring programme has confirmed that there are only low levels of DU contamination, which are well below anything that could be considered a health hazard. The results of this monitoring are published annually.' Parliamentary Question 96140, 8 November 1999. Interestingly, the monitoring of the effects of DU on health and the environment have been undertaken in the United Kingdom, where DU has only been used in testing, rather than in environments where it has actually been deployed, such as Iraq. The British Secretary of State for Defence, Mr. Doug Henderson stated that the MoD has not undertaken any specific review of the environmental and health effects of firing this ammunition in the Gulf. 'The Government do not have any information concerning the levels of DU currently present in Kuwait or southern Iraq and cannot, therefore, comment on this matter in any detail.' Nonetheless, Mr. Henderson was able to conclude with confidence: 'The use of DU rounds by UK forces was not in conflict with the United Nations Charter or any other International agreement or convention. In the use of these rounds and other weaponry, my Department complies strictly with international law.' Parliamentary Question 43051, 29 July 1998. This contrasts with the statement made by Blair a year later, and quoted above, denying that British forces deployed DU rounds against Iraq during the Gulf War.

[89] 'MOD publishes depleted uranium paper', Ministry of Defence (19 March 1999).

[90] *Commentary*, *supra* n. 78, para. 1466.

As to who should undertake the legal review, in the absence of international bodies empowered to make these determinations for States Parties, states should establish bodies at a national level charged with carrying out legal reviews that are both independent and impartial. The Commentary on Article 36 states: 'Article 36 does imply the obligation to establish internal procedures with a view to elucidating the problem of illegality and therefore the other Contracting Parties can ask for information on this point.'[91] Ideally, legal review bodies should be composed of experts in both military requirements and LOAC.

Not all States Parties to AP I undertake legal reviews of new weapons, despite being obliged to. So far only six states, including the US, have established formal mechanisms for weapons' reviews.[92] However, this lack of any serious resolution by states to comply with Article 36 does not relieve DU using, possessing, developing or acquiring states that are party to AP I from the obligation to assess the lawfulness of its use according to the applicable rules of LOAC.[93]

In conclusion, the following is advised for states with regard to conducting legal reviews of DU weapons:

- All states using, possessing, developing and acquiring DU that are party to AP I have a legal obligation to carry out a review of the weapon to ensure its compatibility with international law
- States that have already carried out a legal review (the United States) are advised to carry out a second one given that new facts regarding the effects of DU have come to light

2.2.2   Precautions in targeting: restricting the deployment of depleted uranium weapons in civilian areas

As noted by Carnahan in Chapter 4, DU ammunition was designed for use against tanks on the plains of Europe, not buildings in highly populated urban areas. Yet it seems that in Operation Iraqi Freedom, DU weapons were used in downtown Baghdad against both armoured and unarmoured targets.[94] According to the BBC,

---

[91] Ibid., para. 1482.

[92] Belgium, the Netherlands, Norway, Sweden, the United States and Australia. *A Guide to the Legal Review of New Weapons, Means and Methods of Warfare: Measures to Implement Article 36 of Additional Protocol I of 1977* (Geneva, ICRC January 2006), and conversation with the Guide's author, Kathleen Lawand (re. Australia's recent establishment of a review mechanism and the intention to update the Guide in 2007 to this effect). See also 'ICRC Expert Meetings on Legal Reviews of Weapons and the SIrUS Project – Jongny sur Vevey, 29-21 January 2001', 83 *IRRC* (2001) pp. 539 at 540-541; Daoust, et al., *supra* n. 79, p. 352.

[93] For a recent discussion of the obligations of states pursuant to Article 36 of AP I concerning the legal review of new weapons see *A Guide to the Legal Review of New Weapons,* ibid.

[94] According to journalist Scott Peterson: 'In Iraq [in 2003], DU was not just fired at armored targets. Video footage from the last days of the war shows an A-10 aircraft – a plane purpose-built around a 30-mm Gatling gun – strafing the Iraqi Ministry of Planning in downtown Baghdad. A visit to site yields dozens of spent radioactive DU rounds, and distinctive aluminium casings with two white bands, that drilled into the tile and concrete rear of the building'. S. Peterson, 'Remains of toxic bullets

'[r]eports from Baghdad speak of repeated attacks by US aircraft carrying DU weapons on high-rise buildings in the city centre'.[95] Fahey stated that: 'Based on available information, it appears that US and UK forces may have released approximately 100-200 metric tons of DU during combat in Iraq. It should be noted, however, that much of this expenditure appears to have been in or near urban areas, where the Iraqi people live, work, draw water, and grow and sell food. Therefore, the potential for DU exposures appears higher than in other conflicts...'[96] A reporter for the *Christian Science Monitor* measured radiation levels 1,000 to 1,900 times higher than normal in residential areas in Baghdad.[97] A team working for the *Seattle Post-Intelligencer* conducted tests at six sites from Basra to Baghdad. It 'found elevated levels of radiation at all of them. One destroyed tank near Baghdad was 1,500 times more radioactive than normal background radiation. Another was 1,400 times more radioactive than background.'[98]

One can question the military necessity of using DU weapons during aerial bombardment against objects such as buildings, unless an alternative weapon is not available for the reason of not being able to switch weapons mid-flight. As Fahey noted in Chapter 2, 90 percent of the DU dust settles within 50 metres of the target site in the case of a tank round and 100 metres in the case of DU dust created by aerial attacks;[99] in an urban environment this could have a detrimental impact on civilians located in the immediate or close vicinity of the impact site. Such high levels of radiation in urban areas cannot be considered as negligible and the risks to civilians from such levels of exposure cannot be considered as unforeseeable or insignificant. Even if weapons incorporating DU ammunition are precision weapons in the sense of hitting their targets accurately, it is not possible to limit their secondary effects, either spatially or temporally.

Given that the risk to human health is related to the extent of the exposure, in order to stay within the law and minimise health risks, it would be wise not to use DU weapons against large clusters of armoured vehicles in urban areas, risking very significant releases of DU dust, and preferable to desist from the use of DU weapons in such locales altogether. One must concede that commanders might resist adopting a policy limiting DU use to certain environments and restricting it in urban centres. If enemy tanks are to be engaged in towns, commanders will want to

---

litter Iraq', *The Christian Science Monitor* (15 May 2003), <http://www.csmonitor.com/2003/0515/p01s02-woiq.html>. Another journalist wrote: 'Reports from Baghdad speak of repeated attacks by US aircraft carrying DU weapons on high-rise buildings in the city centre.' A. Kirby, 'US rejects Iraq DU clean-up', BBC News Online (14 April 2003). <http://news.bbc.co.uk/1/hi/sci/tech/2946715.stm>. Fahey noted in Chapter 1 that press reports indicate that DU rounds were shot at a wide range of non-armoured targets, including Iraqi combatants, trucks and buildings'. Chapter 1, at p. 21.

[95] Kirby, Ibid.

[96] D. Fahey, The Use of Depleted Uranium in the 2003 Iraq War: An Initial Assessment of Information and Policies (24 June 2003), p. 5, <www.wise-uranium.org/pdf/duiq03.pdf>.

[97] Peterson, *supra* n. 94.

[98] L. Johnson, 'War's unintended effects: Use of depleted uranium weapons lingers as health concern', *Seattle Post-Intelligencer* (4 August 2003).

[99] Ibid.

use the most effective munitions to defeat those tanks. There is nothing in the law *per se* to prevent them from doing so, unless it can be shown that the humanitarian considerations arising out of the particular use of a particular weapon are such as to outweigh the military necessity of its use. Commanders may also question the feasibility of such a policy, and argue that it is not practical to stock different rounds, tungsten and DU, depending on whether the battle was to be fought in an urban or rural environment. These are valid considerations, but arguably the military necessity of the use of DU may not be such as to justify a potential violation of the principle of proportionality through the use of DU in densely populated areas. The risk to civilians from DU's known effects seems excessive in relation to the military necessity of the use of DU weapons in such cases.

Aside from the prohibitions against indiscriminate attacks in Article 51(4)((c) and disproportionate attacks in Article 51(5)(b), a legal basis for restricting DU use to situations where civilians are at risk of high levels of exposure lies in the principle of precaution in attack in Article 57(2)(a)(ii) of AP I, which requires parties to 'take all feasible precautions in the choice of means and methods of attack with a view to avoiding, and in any event to minimizing, incidental loss, injury to civilians and damage to civilian objects.'

In conclusion, the following precautions are advised when using DU weapons during the conduct of hostilities to avoid breaching a positive rule of LOAC:

- DU use, especially in large quantities in urban centres or in areas or circumstances particularly favourable to its dispersal, should be avoided in order to avoid a possible breach of the rule prohibiting indiscriminate /disproportionate attacks.
- Parties should avoid shooting DU near water sources, agricultural land, and other resources essential for the survival of the civilian population.

2.2.3   *Precaution in the aftermath of depleted uranium use*

A central plank of any risk reduction and minimisation strategy relates to the taking of precautionary measures in the aftermath of the use of DU weapons. Let us now examine what is needed in terms of remedial and risk reduction measures and whether there is any legal basis for requiring states to undertake them. Even if there is not, it would be in states' own interests to take whatever precautions are possible in the aftermath of using DU to minimise foreseeable and possible damage.

2.2.3.1   Remedial and risk reduction measures

As far as responsibility for clean-up of DU remnants and other risk reduction measures are concerned, *de lege ferenda* runs far ahead of the *lex lata*.

The official view of the United States and the United Kingdom is that no remedial responsibility relating to DU use arises. The United States said that it had no plans to remove debris from the Iraqi battlefield because research showed that it has no long-term effects.[100] According to the UK Ministry of Defence: 'A nation which has fired DU in conflict is under no legal obligation per se to return to the region post-conflict to clear up any DU that remains. The civil administration for the area concerned assumes responsibility for clean-up. The legality of this issue has developed through custom: there are no special policies or conventions which address clearance of DU residue.'[101] On the other hand, state practice indicates that while user states may not accept responsibility as a matter of legal obligation, they may take some steps towards remediation, although they are of course limited by the extent to which an affected state will allow them access to contaminated sites, in cases where they are not Occupying Powers. Although evidently not acting from a sense of legal obligation, the UK MoD has undertaken some collection of remnants of DU penetrators in Iraq in the aftermath of the 2003 war and it made some attempts to warn local populations to stay away from hit sites.[102]

There is no general provision in the law of armed conflict or any obligation in treaty or customary international law to oblige states that use a particular weapon to ensure that the target sites are cleaned up afterwards, although some remedial measures are now required in relation to one particular type of remnant, explosive remnants of war, under Protocol V to the CCW Convention.[103] However, even if the impact of an armoured piercing incendiary weapon with a hard target may produce remnants of sorts, remnants of DU weapons, in the form of unexploded shells, are not unexploded remnants of war within the meaning of Protocol V of the CCW Convention, nor can DU dust or residue be considered as remnants. Where Protocol V in Article 2(1) speaks of explosive ordnance, it means (1) 'explosives, with the exception of mines, booby traps and other devices as defined in Protocol II of this Convention as amended on 3 May 1996.' According to paragraph (2) '*Unexploded ordnance* means explosive ordnance that has been primed, fused, armed, or otherwise prepared for use and used in an armed conflict. It may have been fired, dropped, launched or projected and should have exploded but failed to do so.' DU is a

---

[100] Lieutenant-Colonel David Lapan, a Pentagon spokesman told the BBC that the 1990 study prepared for the US Army had been overtaken by more recent studies, for example, by the WHO and the Royal Society, 'into the health risks of DU, or the lack of them ... One thing we've found in these various studies is that there are no long-term effects from DU. And given that, I don't believe we have any plans for a DU clean-up in Iraq', Kirby, *supra* n. 94.

[101] Ministry of Defence, Depleted Uranium, Middle East 2003, <http://www.mod.uk/isssues/depleted_uranium/middle_east_2003.htm>.

[102] See the answers of UK Secretary for Defence Bernard Ingram to Michael Hancock during Parliamentary Questions, in which he stated that '(1) DU fragments on the surface are being removed from the battlefield as soon as they are discovered. (2) Local people have been warned through signs and leaflets that they should not go near, or touch, any debris they find on the battlefields. Military vehicles known to have been hit by DU munitions within the southern section of Iraq under British military control have been clearly marked.' Question 148055, 21 January 2004. See also D. Fahey, Chapter 1, at p. 22.

[103] Protocol on Explosive Remnants of War (Protocol V) to the Convention on Prohibitions or Restrictions on the Use of Certain Conventional Weapons Which May Be Deemed to Be Excessively Injurious or to Have Indiscriminate Effects, adopted on 28 November 2003, entered into force 12 November 2006, available at <http://www.icrc.org/ihl.nsf/FULL/610?OpenDocument>.

penetrator rather than an explosive, as DU missiles contain no explosive charge. Thus, the remedial obligations on states arising from the use of explosive remnants of war do not extend to DU weapons and cannot provide a legal basis for requiring states to clean-up after their use.

LOAC requires parties to armed conflicts to take precautionary measures to protect civilian from the effects of attacks. Article 58(c) of AP I, concerning the protection of civilians from the effects of attacks, enjoins parties to 'take the other necessary precautions to protect the civilian population, individual civilians and civilian objects under their control against the dangers resulting from military operations.' While generally this is considered to extend to precautionary measures taken during armed conflict, such as digging shelters, distribution of information and warnings and withdrawal of the civilian population to a safe place,[104] the language of the provision is in principle broad enough to encompass precautions from the immediate effects of hostilities but also long-term effects, where they are foreseeable. However, neither the ICRC's Commentary on AP I[105] nor its Customary International Humanitarian Law study address the question whether, *vis-à-vis* certain weapons, the obligation to take precautions against the effects of attacks could encompass post-deployment precautionary measures, such as remedial activities, indicating that states did not have such types of precautions in mind, and that there is no practice on this question.

While there is no specific provision in occupation law requiring an Occupying Power to clean-up the battlefield and, in particular, to remove the remnants of its own weapons, an Occupying Power is nonetheless responsible for restoring and ensuring, as far as possible, public order and the safety of the occupied population.[106] This might provide a legal basis for arguing that an Occupying Power which has deployed weapons whose remnants might endanger the health of the civilian population has a duty to take action to ameliorate the situation, to the extent that it is able to. Arguably, the longer the occupation continues, the greater this duty.

It was suggested by Koppe in Chapter 8 that the presumed customary obligation to observe a duty of care for the environment during armed conflict could provide a legal basis for measures aimed at remediation of the environment post DU-use. However, this remains somewhat tentative.

It is therefore almost true, as the user states assert, that in international law the primary responsibility for clean-up lies with the affected state. It is anyway usually the affected state that is best situated to carry out the remediation activities and engage in long-term monitoring of hit and no hit sites and its environment in general. However, an affected state will be a state that has suffered the effects of war, and may not be in a position – financially, technologically and logistically – to carry out remedial actions. It will therefore require the assistance of user states and/ or donor states and organisations. Given that the number of states that have used

---

[104] CIHL, *supra* n. 12, p. 70.

[105] *Commentary*, *supra* n. 78, pp. 694-695.

[106] Art. 43 Regulations Respecting the Laws and Customs of War on Land Annexed to the Hague Convention (IV) Respecting the Laws and Customs of War on Land.

DU is small, and that the biggest user is the United States, in this case it is clear that the user states are far better able to afford remedial action than the affected states. Their financial assistance in effecting clean-up and in undertaking long-term site monitoring, and their technical and other support, is therefore, essential if a serious attempt at clean-up of existing DU impact sites is to be undertaken. This is particularly the case in relation to the United States and the United Kingdom in Iraq and Afghanistan. Although they are no longer Occupying Powers, they shared effective control over the territory and were therefore ideally placed to undertake all of the post-deployment activities that are necessary in relation to DU use and had a moral if not a firm legal obligation to do so. In its resolution on the harmful effects of unexploded ordnance (landmines and cluster submunitions), adopted on 13 February 2003, the European Parliament expressed the willingness of states, including EU Member States to provide 'assistance, in the form of economic assistance, land clearance, social assistance and medical support, to those affected by such weapons.'[107] Such assistance could also be offered to states affected by DU.

The level of clean-up required will vary according to the amount of DU that was used, climatic, topographical and geographic conditions, and the proximity of the target site to civilian centres, *inter alia*. The priority should be urban and other densely populated areas where civilians are likely come into contact with DU remnants. While it may never be possible to remove all DU from the natural environment into which it has escaped, efforts should be made to ensure that the target site is returned to a level of contamination that is considered as safe. A precautionary approach would advise mapping and informing the local authorities of the sites targeted, and the isolation of all the target sites and their clean-up as soon as possible but, in any event, following the conclusion of the conflict.

Sites need to be cleaned-up and DU remnants and dust removed but civilians and troops also need to be warned to stay clear of them, or take the necessary precautions if they do come into contact with them, and children, in particular, need to be prevented from accessing them. The Pentagon's reluctance in the past to provide timely and accurate information regarding the amount of DU used in battle and maps of the target sites has not only made risk assessment difficult, and impeded clean-up and site monitoring, but it has exposed local populations and troops to unnecessary risks; these persons were unaware that they were running any. According to Fahey: 'Each time U.S. forces shot DU rounds, the U.S. Department of Defense neglected to warn ground troops and civilians about DU contamination.'[108] In Kosovo, the local population, as well as peacekeepers and humanitarian workers, were surprised to learn in 2001, two years after the NATO bombardment, that there was DU in their midst: 'Children had long been playing on destroyed equipment and in contaminated areas. In addition, adults had scavenged destroyed equipment for usable parts and scrap metal.'[109] No one had thought to secure DU

---

[107] Available at <http://www.idust.net/Docs/EuroParl03.htm>.

[108] Fahey, *supra* n. 4, pp. 189 at 191.

[109] Ibid., pp. 189 at 194.

contaminated sites or remove even the surface DU. There was no attempt to warn Bosnian civilians of the possible risks of contact with DU dust or contaminated material until 2001, six years after the end of the war there.[110]

In January 1998, the Pentagon's Office of the Special Assistant for Gulf War Illnesses admitted in relation to Gulf War veterans: 'Combat troops or those carrying out support functions generally did not know that DU contaminated equipment such as enemy vehicles struck by DU rounds required special handling ... The failure to properly disseminate such information to troops at all levels may have resulted in thousands of unnecessary exposures.'[111]

The report of the Royal Society into DU stated: 'It should be incumbent upon nations using DU munitions in future conflicts to advise the local population of the potential dangers of handling fragments of penetrators.'[112] The Royal Society and UNEP also recommend periodic testing of groundwater used for drinking adjacent to places where DU has been used.[113] The Royal Society issued a statement on 16 April 2003 saying that the hundreds of tonnes of DU which had been shot in Iraq should be removed to protect the civilian population. Both soldiers and civilians were in danger, in the short- and long-term, and children playing at contaminated sites were particularly at risk.[114]

In conclusion, the following precautions are advised following the use of DU weapons in order to minimise the risk of contamination of the natural environment and any consequential effects on humans (civilian and military):

- Mapping impact and no hit sites;
- Isolating impact and no hit sites;
- Warning local populations of the risks and to stay away;
- Ongoing monitoring of the environment close to impact sites;
- Clean-up of impact areas and removal and disposal of contaminated *matériel* and remnants;
- Warning clean-up personnel of the risks and advising the necessary precautions to minimise their levels of exposure.

2.2.3.2 Testing individuals and health monitoring

The cornerstone of a precautionary approach to DU weapons must be systematic and comprehensive programs of scientific and medical research into the effects of DU on human health, involving both military personnel and civilians. Such research can only be meaningful if widespread clinical testing of a substantial num-

---

[110] 'S. Castle, 'UN will send investigators to Bosnian DU target sites', *The Independent* (16 January 2001).

[111] Office of the Special Assistant to the Deputy Secretary of Defense for Gulf War Illnesses (OSAGWI), Annual Report, November 1996-November 1997 (Washington, DC January 1998) p. 30.

[112] Royal Society, *supra* n. 62, p. 24.

[113] See D. Fahey, Chapter 2, pp. 29 et seq.

[114] P. Brown, 'Scientists urge shell clear-up to protect civilians – Royal Society spells out dangers of depleted uranium', *The Guardian* (17 April 2003).

ber of those persons who have already been exposed to DU and those who are exposed to it in the future is undertaken, as well as ongoing monitoring of their health, using appropriate controls. As Toebes noted in Chapter 9, human rights law, especially the rights to health and information, may provide an international legal basis for exposed individuals to demand precautions in the aftermath of DU use as regards testing and monitoring, although this would admittedly be difficult to enforce at the national level, and in some cases it may not be necessary where such procedures are already available. However, there is no obvious international legal basis to insist that states undertake testing of persons. This is a step that both user and affected states are advised to take as a matter of good judgment, in order to be able to found considered legal and policy choices relating to the use of DU weapons.

The current incomplete state of knowledge about the effects of DU is not really due to the fact that insufficient time has passed for all the effects of this weapon to manifest themselves, even if some effects may still remain unknown. After all, DU was first fielded in 1991, 16 years ago at the time of writing. That is enough time for many problems to have manifested themselves, and they have. The difficultly has been conclusively linking them with DU exposure. While in the case of the Gulf War, many other causal factors cannot be excluded, it would be remiss to discount the deleterious effects of DU as a possible, or at least partial explanation for some pathologies, in some cases of actual exposure to DU. It seems that where DU may cause pathologies, it may do so relatively quickly. Thus, to the extent that DU's effects remain unpredictable it is because the work that should have been done to test all, most or even a statistically meaningful number of those who were exposed to DU in both the military and civilian populations, as soon as possible after exposure, and to monitor their health on an ongoing, long-term basis, has not been carried out.[115] One of the biggest problems with the research into DU that has been conducted so far is that it has been based on what is statistically almost meaningless baseline information. Obviously, no full picture of DU's effects can emerge so long as one does not look for it.

The paucity of proper testing and research into the effects of DU by the user states' military establishments, particularly the United States and its DU program,[116] is revealing, given the concerns raised by the little research that has been done. The more research into the effects of DU that is undertaken, the more pre-existing assumptions are being challenged. It had been widely believed that the kidneys are the main target organ for DU's effects but laboratory research is beginning to show that this may not be the case.[117] Such research has important implica-

---

[115] The Royal Society noted the 'lack of good quality data on some of the parameters that determine the extent of the exposure or the subsequent risk of disease. Most notable are the limited data on the levels of DU that might be inhaled in the impact aerosol within a tank pierced by a DU penetrator or that might be inhaled by soldiers entering the tank soon after the impact. There is also an almost complete absence of measurements of levels of uranium in samples of urine taken soon after exposure to DU.' Royal Society, *supra* n. 62, p. 21.

[116] See D. Fahey, Chapter 2, p. 29.

[117] See D. Fahey, Chapter 2, p. 29.

tions for testing. It may be that the wrong sort of testing is being done: mainly urinalysis for the presence of uranium, which may not be the most effective way of measuring the effects of exposure to DU. It is also clear that, in many cases, testing is taking place far too long following exposure. Another problem is the inadequacy of current testing methodology; there is a pressing need for more sensitive and accurate test methods, especially in the United States, and better selection of cohorts. Different tests may be needed for different types of exposure: the same test probably will not suffice for persons exposed to DU through ingestion or implantation and those exposed through inhalation, for example. Testing also needs to be sensitive to the fact that DU may affect men and women, and particularly children, in different ways.

In conclusion, the following precautionary measures are advised in relation to the testing of persons exposed to DU:

- Undertaking comprehensive testing of friendly personnel who are exposed to DU as soon as possible following exposure;
- Supporting efforts in (former) enemy states aimed at testing exposed combatants and ongoing health monitoring;
- Widespread testing and ongoing health monitoring of the civilian population in areas that have been particularly exposed, particularly, Baghdad and Basra.

2.3   A moratorium on depleted uranium use

The ultimate in precaution would be the adoption of a moratorium on the use of DU ammunition and armour until either its safety can be assured or its legality/illegality conclusively proven. As shown above, calls for the adoption of a moratorium on DU use have been getting louder in recent years. While NGOs and anti-DU activists have called for a temporary halt on the use of DU ammunition, calls for a moratorium have also been voiced in political fora such as the European Parliament. These initiatives are evidence of growing concern about the continued use of DU weapons while their harmlessness cannot be assumed.

A moratorium on the use of DU until further information about it is available would be the greatest step in risk minimisation, while leaving open the possible future use of DU in weapons if it turned out to be relatively safe. It could come about either through a formal or informal process. A formal process could see user and possessor states sign a declaration that they will commit to forego DU use for a specified or unspecified period of time. Penalties might or might not be specified in case a signatory state breaches the moratorium, although user states are more likely to sign up for the least onerous regime. Thus, the moratorium could consist of a mere commitment, without any enforcement regime. Given that there are only two or three user states, another approach might be for those states to make a unilateral commitment to a moratorium on DU use.

Realistically speaking, however, the prospects for any general moratorium on grounds of the possible health risks associated with DU use seem slim. In 2001 both the United States and the United Kingdom opposed a moratorium on DU use. According to news reports: 'The two NATO allies shot down a request from Italy during a meeting of alliance officials in Brussels for a halt on DU arms until they had been deemed safe.'[118] Their view has not changed in the meantime. It seems more likely that, unless states agree to a ban on the use of DU weapons – a prospect which, however, also seems remote – their use will cease only when states consider them no longer necessary because other, more effective, and hopefully safer, weapons have replaced them. As we have seen, this is already happening in the case of some types of DU ammunition.

Even if a general moratorium on DU use is not accepted by states, of great effect would be a voluntary operational moratorium on using DU weapons proximate to civilian habitations.

3.   CONCLUSION

The use of weapons in war is about all balancing military necessity and humanitarian considerations. It boils down to this: is the military necessity of the use of the weapon such as to outweigh the possible risks, including the risk of superfluous injury to combatants and harm to civilians?

In the case of DU, the military utility of its use is clear and proven while the humanitarian concerns it gives rise to are less concrete. This has worked to the advantage of those states wishing to use DU, because legally there is nothing to stop them unless its use can clearly be shown to violate LOAC in a particular case.

But, as we have seen, the military utility of DU has been somewhat overstated, given that its use has in practice been very different from that for which it was envisaged. The United States has not used DU against superior Soviet tanks but sometimes fairly decrepit armour. In many if not all cases an alternative to DU would suffice to get the job done. The military utility of DU is also diminishing as alternatives become increasingly effective and as the problems associated with its use become clearer. As more light is shed on the effects of DU, the extent of these humanitarian problems may become such as to outweigh any remaining military necessity for its use.

There are grounds for taking a more cautious approach to DU use. It is not true that problems associated with DU use are not foreseeable and cannot or should not be taken into account during military operations. While the available evidence is not conclusive, it gives rise to serious concern. Given what is already known and agreed upon about the effects of DU, states that avoid adherence to a precautionary approach risk possible liability. It is therefore in their interests that they should act prudently in minimising the risks that the use of DU poses.

---

[118] I. Geoghegan, 'NATO ducks uranium ban amid clamor for research', *The Daily News* (New York) (9 January 2001), <http://dailynews.yahoo.com/h/nm/20010109/sc/health_balkans_dc_10.html>.

POST SCRIPTUM

After the completion of this chapter, mounting concerns regards the effects of depleted uranium weapons galvanised 136 member states of the United Nations General Assembly to adopt the GA's first resolution dealing with DU in December 2007.[119] 'Taking into consideration the potential harmful effects of the use of armaments and ammunitions containing depleted uranium on human health and the environment', the resolution requested 'the Secretary-General to seek the views of Member States and relevant international organizations on the effects of the use of armaments and ammunitions containing depleted uranium, and to submit a report on this subject to the General Assembly at its sixty-third session'. It penciled in an agenda item entitled 'Effects of the use of armaments and ammunitions containing depleted uranium' for its next session.

The political battle over depleted uranium had only just begun.

---

[119] 'Effects of the use of armaments and ammunitions containing depleted uranium', A/RES/62/30, 5 December 2007.

LIST OF CONTRIBUTORS

Mr Théo BOUTRUCHE, Ph.D. candidate at the Graduate Institute of International Studies, Geneva, and Research and Teaching Assistant, International Law and International Organization Department, University of Geneva

Mr Burrus CARNAHAN, Professorial Lecturer in Law, The George Washington University, Washington, D.C.

Dr Guido DEN DEKKER, Attorney at Law, De Brauw Blackstone Westbroek, The Hague

Mr Dan FAHEY, Ph.D. candidate at the University of Berkley, California

Mr Tobias GRIES LL.M., Associate in the Berlin office of Kirkpatrick & Lockhart Preston Gates Ellis LLP

Dr Jann K. KLEFFNER, Assistant Professor of International Law, University of Amsterdam, and Managing Editor of the *Yearbook of International Humanitarian Law*

Dr Erik KOPPE, Managing Editor of the Netherlands International Law Review

Dr Avril MCDONALD, Research Associate at the TMC Asser Instituut and a Lecturer in International Humanitarian Law at the University of Groningen

Dr Manfred MOHR, European Affairs Officer at the German Red Cross

Dr Brigit TOEBES, Lecturer in Human Rights Law at the University of Aberdeen

Dr Marten ZWANENBERG, Legal Advisor at the Netherlands Ministry of Defence.

INDEX